Marine Navigation 2: Celestial and Electronic

Titles in the
Fundamentals of Naval Science Series

Seamanship

Marine Navigation 1: Piloting

Marine Navigation 2: Celestial and Electronic

Introduction to Shipboard Weapons

Introduction to Naval Engineering

Marine Navigation 2: Celestial and Electronic

Second Edition

Richard R. Hobbs
Commander
U.S. Naval Reserve

Naval Institute Press
Annapolis, Maryland

Fundamentals of Naval Science Series

Library of Congress Cataloging in Publication Data

Hobbs, Richard R.
Marine navigation.

(Fundamental of naval science series)
Includes index.
Contents: —2. Celestial and electronic.
1. Navigation. 2. Pilots and pilotage.
3. Nautical astronomy. 4. Electronics in naviga-
tion. I. Title. II. Series.
VK555.H67 1981 623.89 81-9538
ISBN 0-87021-363-6 (v. 2) AACR2

Printed in the United States of America

To Jean and Katherine

Who have kept me in serene waters
through life's stormy times

Contents

Foreword

Throughout the history of warfare at sea, navigation has been an important basic determinant of victory. Occasionally, new members of the fraternity of the sea will look upon navigation as a chore to be tolerated only as long as it takes to find someone else to assume the responsibility. In my experience, such individuals never make good naval officers. Commander Hobbs has succeeded in bringing together the information and practical skills required for that individual who would take the first step down the road toward becoming a competent marine navigator.

At the outset of this book, the author stresses the necessity for safe navigation, but there is another basic tenet of sea warfare that this book serves. The best weapons system man has ever devised cannot function effectively unless it knows where it is in relation to the real world, where it is in relation to the enemy, and where the enemy is in relation to the real world. Not all defeats can be attributed to this lack of information, but no victories have been won by those who did not know where they were.

William P Mack

W. P. MACK
VICE ADMIRAL, U.S. NAVY

Although these words were written by Admiral Mack for the first edition of this book almost ten years ago, they are certainly no less valid today than they were then. The modern student of navigation could well take the last sentence to heart not only as it applies to the practice of navigation, but also to life in general.

THE AUTHOR
ANNAPOLIS, 1981

Preface to the Second Edition

Marine Navigation 1: Piloting, the first of the two volumes comprising this work on marine navigation, deals with the safe navigation of a surface vessel in and near coastal and intracoastal waters contiguous to U.S. and foreign shores. Navigation in this environment is normally done primarily with reference to land and sea marks and is called *piloting*. After proceeding to sea beyond visual and radar range to land, the navigator can no longer use this means of position-fixing. He must then shift to alternative methods of determining position, which can all be broadly grouped into two main categories: *celestial* and *electronic navigation*. It is the purpose of this volume to set forth the basic principles and methods of celestial and electronic navigation in most common use by seagoing navigators to ascertain their position at sea.

To this end, the first eleven chapters of this volume are designed to acquaint the inexperienced navigator with the basic knowledge required for the successful practice of celestial navigation at sea. The next four chapters deal with the present state of the art of position-finding by electronic means, and the final chapter presents an example of a typical day's work in navigation at sea. After the student navigator has developed some mastery of the material presented herein, he should be well able to proceed on to other more advanced texts and procedures for both surface and air navigation, if that is desired. He may rest assured, however, that if he is proficient in the basic techniques presented in this volume, he need never fear becoming lost in the course of routine surface navigation at sea.

As was the case with Volume 1, much of the material covered is based on that contained in Bowditch's *American Practical Navigator* and *Dutton's Navigation and Piloting*. Excerpts from many of the publications dealing with celestial and electronic navigation distributed by the Defense Mapping Agency Hydrographic/Topographic Center and other government sources appear throughout. In order to desensitize the many excerpts and sample computations presented with respect to time, the year dates have been deleted in most examples, except where essential to the understanding of the material presented. The format of all navigational table and publication excerpts has been carefully checked in all cases, however, to ensure consistency with the existing format of the various publications as of the publication date of this volume. The student is cautioned to remember always to obtain and use the correct edition of the various publications covering the specific dates or periods in which the actual problem or situation to be dealt with takes place. The various almanac pages and propagation correction table excerpts reproduced in this volume are valid only for the specific years for which they were published; they cannot be used for obtaining data on which to base the solution of problems set in any other period of time.

Because this volume, like Volume 1, was originally written as a text for use by midshipmen at the U.S. Naval Academy and the various NROTC colleges and universities, the student will find a definite Navy orientation in much of the subject matter presented. Nevertheless, the bulk of the information and techniques related are just as equally applicable for the yachtsman or fisherman in a small boat at sea as they are to the navigator of a U.S. Navy surface combatant or the first mate of a merchant freighter. It is the author's hope that all categories of inexperienced mariners contemplating going to sea will continue to find this second edition at least as useful a bridge to the numerous other more advanced books and textbooks on navigation as their predecessors did the first edition.

The author is greatly indebted to Commander F. E. Bassett, USN, who originally conceived this text and under whose guidance and assistance the original edition was prepared, and to Commanders R. A. Smith, RN; J. L. Roberts, USN; F. A. Olds, USN; and A. J. Tuttle, USN; and Mr. E. B. Brown of the Defense Mapping Agency for their many suggestions and contributions. Special thanks are also due Mr. W. J. Clipson of the Educational Resources Graphic Arts Department, U.S. Naval Academy, who prepared many of the line drawings for this volume. Finally, the author wishes to express gratitude to Col. E. S. Maloney, USMC (Retired), for his expert review of the initial draft of this edition and his many suggestions for its improvement; and to the following for their thoughtful reviews and suggestions on the topics within their areas of expertise in the electronics chapters: Mr. C. F. Brown of the Raytheon Company, Mr. J. Podorsek of the Navy Astronautics Group, Mr. C. Powell of Racal-Decca Limited, and Mr. D. C. Scull and Commanders R. Vence and R. McFarland of the U.S. Coast Guard Omega and Loran-C sections.

Marine Navigation 2: Celestial and Electronic

1

Coordinate Systems of Celestial Navigation

In this first chapter of *Marine Navigation 2: Celestial and Electronic*, the basic facts and theories that constitute the foundations upon which the practice of celestial navigation depends will be set forth. First, the relationship of the earth with the surrounding universe and the other bodies of the solar system will be explored. Next, the terrestrial coordinate system already introduced in *Marine Navigation 1* will be reviewed, and two additional coordinate systems necessary for the practice of celestial navigation will be developed: the *celestial system* and the *horizon system*. Finally, the three systems will be combined to form the celestial and navigational triangles; the solution of this latter triangle is the basis of celestial navigation.

The Earth and the Universe

In order to simplify the explanation of the relationship of the earth with the surrounding universe, the rotating earth can be likened to a spinning gyroscope suspended in space. As the science of physics tells us, the spin axis of a gyroscope will remain forever oriented toward a particular point in space unless it is subjected to an outside force, in which case it will tend to precess in a direction at a right angle to the spin axis. The rotating earth is, of course, subjected to the gravitational pulls of the sun and the moon, as well as to the minute gravitational pulls of all the remaining bodies of the universe. Hence, the earth is subject to *precession*, which, together with its *rotation* about its axis and *revolution* in its orbit about the sun, constitute the three *major motions of the earth*. The period of the earth's precession is so long—about 25,800 years—that for relatively short periods of time the effects of precession can be disregarded, except when certain types of celestial tables designed for use over extended time periods are used.

By good fortune, it happens that in the present era of time the northern axis of the earth is aligned almost exactly with a star that was given the name Polaris—the pole star—by early Greek astronomers. Since the distance of this star is many thousands of times greater than the diameter of the earth's orbit about the sun, the orientation of the north pole of earth with respect to Polaris does not vary more than one or two degrees throughout the year. The beneficial effects of this situation will be fully developed in Chapters 9 and 10.

In addition to the aforementioned major motions, there are also three *minor motions of the earth* of which the navigator should be aware—*wandering of the terrestrial poles, variations in the rotational speed of the earth,* and *nutation.* The positions of the north and south terrestrial poles, like the magnetic poles discussed in Chapter 9 of *Marine Navigation 1,* are not stationary, but rather they move in a circular path approximately 100 feet in diameter. A complete cycle of movement is so slow, however, as to be almost immeasurable, and the effect can be disregarded for most navigational purposes. The rotational speed of the earth is presently slowing gradually at the rate of about .001 revolution per century; the effect of this minor motion is also disregarded for most navigational purposes. Nutation is an irregularity in the earth's precession caused by the influence of the moon and to a lesser extent the other bodies of the solar system; this effect, like precession, must be reckoned with when certain types of celestial tables designed for use over extended time periods are used.

Like the star Polaris, the other stars of the universe are so far from the earth that their positions seem to remain constant with respect to one another and to the earth's rotational axis. This fact led to the early misconception among ancient astronomers that the earth was located at the center of a hollow sphere upon which the stars were fixed; this sphere seemed to make one complete rotation around the earth every 24 hours. The concept lingered until 1851, when the rotation of the earth was conclusively proven by Jean Foucault with his now-famous pendulum experiment. The idea of such a *celestial sphere* is still useful even today, however, in visualizing the celestial coordinate system described later in this chapter.

Besides being responsible for their apparently unchanging locations on the celestial sphere, the vast distances of the stars from earth also cause some to appear brighter than others. Usually, but not always, those stars nearer the earth will appear brighter than those more distant. The relative brightness of a celestial body is expressed in terms of *magnitude,* based on the Greek astronomer Ptolemy's division of the visible stars into six groups according to their brilliance and the order in which they appeared at night. The first group of stars to become visible to the naked eye at twilight, called the *first magnitude stars,* are considered to be 100 times brighter than the faintest stars visible to the naked eye in full night, the *sixth magnitude stars.* Hence, a magnitude ratio of the fifth root of 100 or 2.512 exists between each magnitude group. The magnitude of a body with an intensity between two groups is denoted by a decimal fraction, as for example, a 2.6-magnitude star. The two brightest stars in the sky, Sirius and Canopus, are actually more than 100 times brighter than a sixth magnitude body. Consequently, they are assigned *negative* magnitudes of −1.6 and −0.9 respectively. In practice, all stars and planets of magnitude 1.5 or greater are collectively referred to as first magnitude bodies. As a point of interest, the moon varies in magnitude from −12.6 to −3.3, and the magnitude of the sun is about −26.7. The planets Venus, Mars, and Mercury also appear at negative magnitudes at certain times of the year.

The patterns in which many of the brighter stars were arranged reminded ancient man of various common terrestrial forms, and it was natural that these patterns or *constellations* became known by the names of the creatures they seemed to resemble, such as Ursa Major (the Big Bear), Scorpio (the Scorpion), and Aries (the Ram). The prominence of the early Greek and Roman mathematicians and astronomers in establishing astronomy as a science is indicated by the fact that most of the constellations as well as most of the individual celestial bodies visible to the naked eye bear names of Greek or Latin derivation.

In celestial navigation, the earth is usually considered to be a perfect sphere suspended motionless at the center of the universe. All heavenly bodies are assumed to be located on a second celestial sphere of infinite radius centered on the center of the earth. This *celestial sphere* is considered to rotate from east to west, with the rotational axis of the sphere concurrent with the axis of the earth; the north pole of the celestial sphere is designated by the abbreviation P_n, while the south pole is denoted by P_s. The sphere completes one rotation with respect to the earth every 24 hours.

To an observer located on the surface of the earth, it appears that all celestial bodies on the celestial sphere rise in the east, follow a circular path across the heavens, and set in the west. The circular path of each body across the heavens is referred to as its *diurnal circle;* if the heavens were photographed at night by a time-lapse camera, each celestial body would subtend a diurnal circle as it rotated across the heavens. Figure 1-1 depicts the celestial sphere and several diurnal circles.

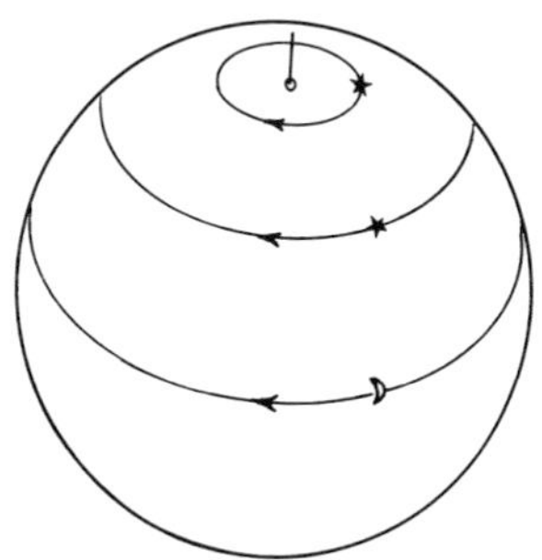

Figure 1-1. The celestial sphere and several diurnal circles.

The Solar System

The earth, together with eight other known planets, their moons, and the sun, constitute the *solar system.* Like the other planets of the solar system, the earth follows an elliptical path of revolution about the sun called an *orbit;* it takes about 365 days, or one year, for the earth to complete one circuit around the sun. The earth's axis of rotation is inclined at an angle of $23\frac{1}{2}°$ from the perpendicular with respect to the plane of the earth's orbit, called the *plane of the ecliptic.* As will be explained more fully later, this inclination is of special interest as it is primarily responsible for the different seasons experienced on the earth's surface, as well as the ever-changing length of the solar day. Figure 1-2 depicts the relationships between the earth, its orbit, and the sun.

Because of the tilt of the earth's axis with respect to the ecliptic, it seems to an observer on earth as though the sun is continually shifting its position with respect to the earth's equator as the earth moves in its orbit around the sun. When the earth reaches position A in Figure 1-2, the sun appears to be exactly over the $23\frac{1}{2}°$ north parallel of latitude. It is at its northernmost point with respect to the earth. This time, called the *summer solstice* in the northern hemisphere, occurs about June 21 of each year, about 12 days before *aphelion,* the orbital point farthest from the sun. Because the angle of incidence of the sun's rays is relatively large at this time in the northern hemisphere and relatively small in the southern hemisphere, the northern hemisphere experiences summer and the southern hemisphere winter at this point in the earth's orbit. As the earth continues

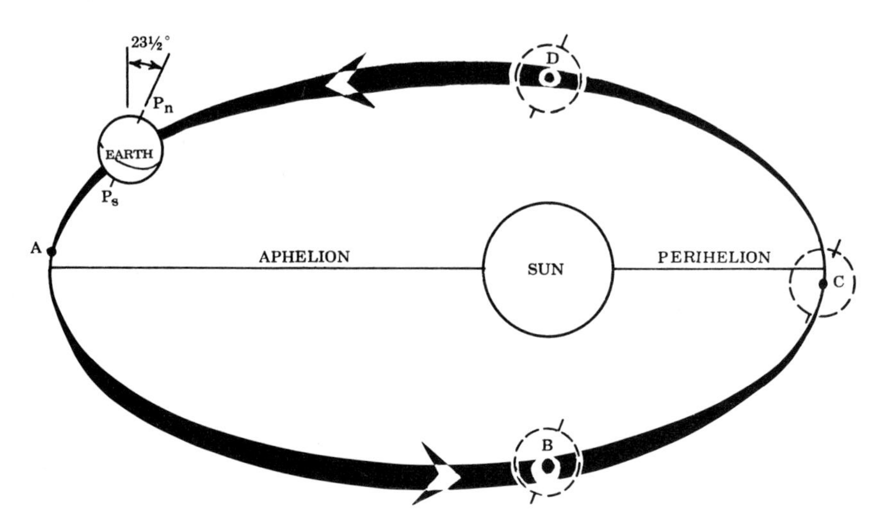

Figure 1-2. The earth-sun system.

in its orbit toward point B, the sun seems to move ever lower, until at position B in the figure it appears to cross over the equator from north to south latitudes. This time is referred to as the *autumnal equinox* in the northern hemisphere, and occurs about September 22. When the earth has proceeded to point C, the sun is at its extreme southern limit, $23\frac{1}{2}°$ south latitude; this time, which occurs about December 21, is termed the *winter solstice,* and precedes *perihelion,* the earth's closest approach to the sun, by about 12 days. The northern hemisphere is now experiencing winter, while in the southern hemisphere it is summer. Finally, the sun again crosses the equator at point D, this time from south to north latitudes, about March 21. This event is called the *vernal equinox,* and as noted below, it is of special significance in celestial navigation.

As the early astronomers observed the sun in the course of the earth's revolution about it, it seemed to them as though the sun were an independent ball moving across the heavenly sphere, transiting 12 different constellations in the course of a year. These 12 constellations comprise the *Zodiac.* Although with one important exception the zodiac has no navigational significance in modern times, it continues to be of importance even today in the practice of astrology.

When they observed the sun at the time of the vernal equinox, it seemed to the ancients as though the sun were located at a point within the constellation Aries. They chose this point, the *First Point of Aries,* abbreviated by the symbol (♈), as a reference point by which all bodies on the celestial sphere could be located. Although the sun today no longer lies in Aries at the vernal equinox as a result of the effects of the earth's precession, the First Point of Aries continues to be used as the fixed reference point for the location of bodies on the celestial sphere. The sun now appears to be in the constellation Pisces at the time of the vernal equinox.

With the exception of the outermost planet, Pluto, the other known planets of the solar system also revolve about the sun in about the same plane as the plane of the earth's orbit. Although they move at different speeds, all the planets revolve about the sun in the same direction. Mercury, the planet nearest the sun, and Venus, the second planet, are often referred to in celestial navigation as the *inferior planets,* as their orbits lie inside that of the earth, while Mars, Jupiter, and Saturn, the fourth, fifth, and sixth planets, are called the *superior planets* because they are the visible planets whose orbits are outside that of the earth. These three planets, together with Venus, are sometimes referred to as the *navigational planets.* Uranus, Neptune, and Pluto, the seventh, eighth, and ninth planets, are so far distant from the sun as to be of insufficient magnitude for navigational purposes most of the year. In addition to the nine known planets of the solar system, several astronomers and mathematicians currently theorize the existence of a possible tenth planet beyond the orbit of Pluto, based on certain aberrations in the orbits of Uranus and Neptune.

Because of the relative closeness of the other planets of the solar system, when they are viewed from earth they appear to move across the background of the unchanging stars and constellations. As will be discussed later, this motion necessitates the determination of the position of the visible planets and the moon in relation to the stars for each occasion on which they are used for celestial navigation purposes. Because the planets are illuminated by the reflected light of the sun, the inferior planets go through phases similar to the earth's moon, being "full" when on the opposite side of the sun from the earth, and "new" when on the same side. As the superior planets never pass between the earth and the sun, they never appear in the "new" phase.

The Moon

Just as the earth revolves about the sun, the moon revolves once about the earth in a period of about 30 days. As was mentioned in Chapter 11 of *Marine Navigation 1* in connection with tide, the moon revolves about the earth in the same direction as the earth rotates, thereby resulting in a time period of about 24 hours and 50 minutes for the earth to complete one rotation with respect to the moon. Since the moon itself completes one rotation every 30 days, the same side of the moon is always facing toward earth. Because the moon, like

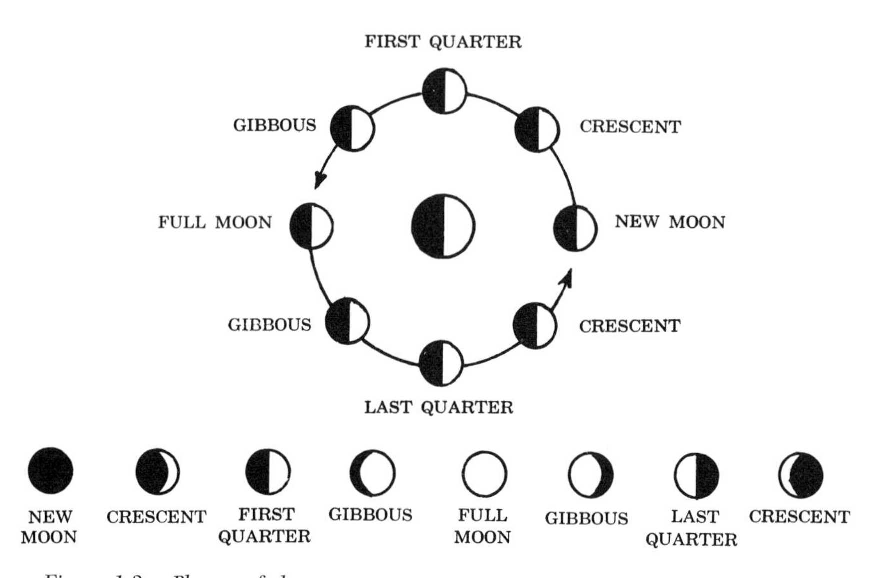

Figure 1-3. Phases of the moon.

the planets, is illuminated by the sun, the size and shape of the visible portion of the moon as viewed from the earth is constantly changing, as shown in Figure 1-3. The names given to the various visible portions or *phases* of the moon appear in the figure.

Occasionally the moon passes directly between the sun and the earth, thereby casting a shadow across the surface of the earth, as illustrated in Figure 1-4A. When viewed from within the shadow zone, it seems to the observer as though the sun disappears behind the superimposed moon. This result, termed a *solar eclipse,* is of great interest to solar scientists, as it provides an opportunity to study the sun's atmosphere or corona while the moon blocks the sun's surface from view. Locations on the borderline of the shadow zone on the earth's surface experience only a partial blockage of the sun's disk, a *partial solar eclipse.*

At other rather infrequent times, the moon's orbit takes it through the conical shadow zone cast by the earth, thereby creating a *lunar eclipse* visible to observers on the dark side of the earth. This phenomenon has little scientific value, but is interesting to observe. The position of the earth-sun-moon system during a lunar eclipse is illustrated in Figure 1-4B.

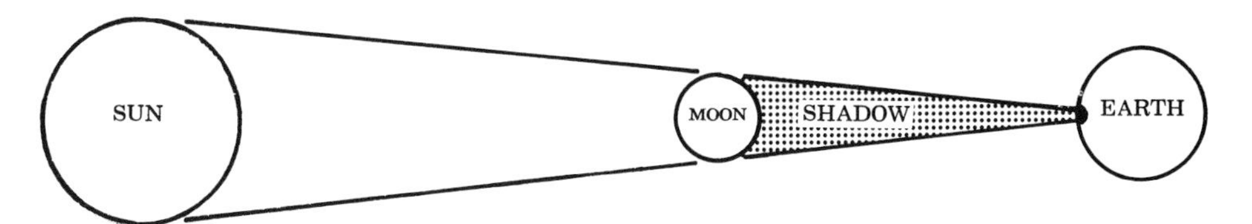

Figure 1-4A. The earth-sun-moon system during a solar eclipse.

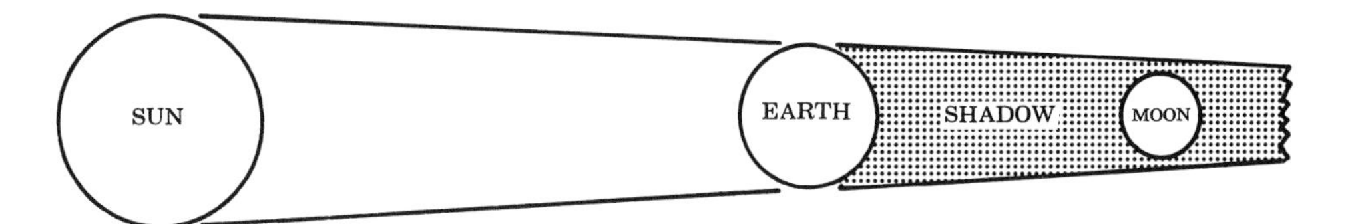

Figure 1-4B. The earth-sun-moon system during a lunar eclipse.

Review of the Terrestrial Coordinate System

As the celestial coordinate system is based in large measure upon the terrestrial coordinate system, it is necessary to review briefly the terrestrial system introduced in Chapter 4 of *Marine Navigation 1* before proceeding with a discussion of the celestial system. The terrestrial system, it may be recalled, is based on the equator and the prime meridian. The *equator* is defined as that great circle on the earth's surface formed by passing a plane perpendicular to the earth's axis midway between the poles. A *meridian* is any great circle on the earth's surface formed by passing a plane containing the earth's axis through the earth perpendicular to the equator; the *prime meridian* is the upper branch of the meridian which passes from pole to pole through the site of the Royal Observatory at Greenwich, England. The location of any position on the earth is specified by stating its location relative to the equator and the prime meridian. The *latitude* of a position is defined as the angular distance measured from the equator northward or southward through 90°; the direction of measurement is indicated by placing a suffix N (north) or S (south) after the angular measure. The *longitude* of a position is the angular distance measured from the prime meridian eastward or westward through 180°; the direction of measurement is indicated by placing a suffix E (east) or W (west) after the angular measure. Each degree of latitude or longitude is subdivided into 60 minutes, and each minute is further subdivided into either 60 seconds or tenths of a minute. Figure 1-5 depicts the equator, prime meridian, and several parallels of latitude and meridians.

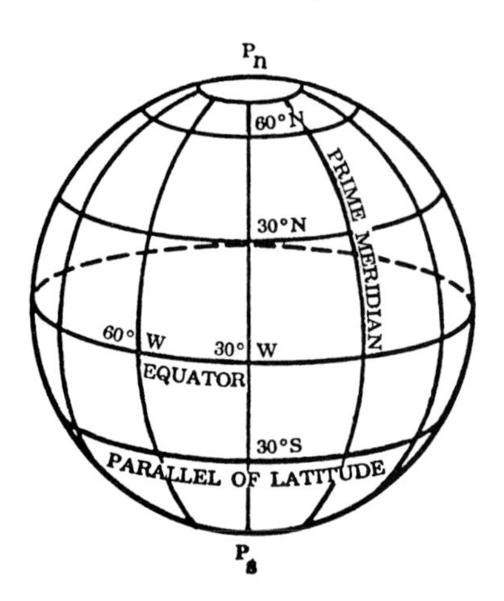

Figure 1-5. The terrestrial coordinate system.

As a final note, it was shown in Chapter 4, *Marine Navigation 1*, that every degree of arc of a great circle drawn upon the earth's surface is assumed for navigational purposes to subtend 60 nautical miles of distance, and every minute of such a great circle arc therefore subtends one nautical mile. This assumption is of great importance in celestial navigation, for reasons to be presented shortly.

The Celestial Coordinate System

Just as any position on the earth can be located by specifying its terrestrial coordinates, any heavenly body on the celestial sphere can be located by specifying its *celestial coordinates.* To form the *celestial coordinate system,* the terrestrial equator is projected outward onto the celestial sphere to form the *celestial equator,* sometimes called the *equinoctial.* The celestial equator is the reference for north-south angular measurements on the celestial sphere.

In similar fashion, terrestrial meridians can be projected outward to the celestial sphere to form *celestial meridians.* Because of the apparent rotation of the celestial sphere with respect to the earth, these projected celestial meridians appear to sweep continuously across the inner surface of the sphere, making them inconvenient to use as a basis for lateral measurements of position on the celestial sphere. Hence, a separate set of great circles are "inscribed" on the surface of the celestial sphere perpendicular to the celestial equator for use in describing the position of one point on the sphere relative to another. These great circles, called *hour circles,* are defined as follows:

> An *hour circle* is a great circle on the celestial sphere perpendicular to the celestial equator and passing through both celestial poles.

Every point on the celestial sphere has an hour circle passing through it. Just as the meridian passing through the observatory at Greenwich, England, was chosen as the reference for the lateral coordinate of a point on the terrestrial sphere, the hour circle passing through the First Point of Aries (♈) forms the reference for the lateral coordinate of a point on the celestial sphere. This hour circle is usually referred to simply as the "hour circle of Aries."

The celestial equivalent of terrestrial latitude is *declination,* abbreviated *dec.;* its definition follows:

> *Declination* is the angular distance of a point on the celestial sphere north or south of the celestial equator, measured through 90°.

Declination is labeled with the prefix N (north) or S (south) to indicate the direction of measurement; prefixes are used to differentiate declination from latitude. Figure 1-6A (next page) depicts the declination of a star located 30° north of the celestial equator; the terrestrial sphere is printed in black, while the celestial sphere is shown in blue.

The celestial equivalent of longitude is the *hour angle;* it is defined below:

> The *hour angle* is an angular distance measured laterally along the celestial equator in a westerly direction through 360°.

If it is desired to locate a body on the celestial sphere relative to the location of Aries, the hour circle of Aries is used as a reference for the measurement of the hour angle. Such hour angles, measured in a westerly direction from the hour circle of Aries to the hour circle of the body, are called *sidereal hour angles,* abbreviated *SHA.*

For the purposes of celestial navigation, it is not only desirable to locate a body on the celestial sphere relative to Aries, but also it is desirable to locate a body relative to a given position on earth at a

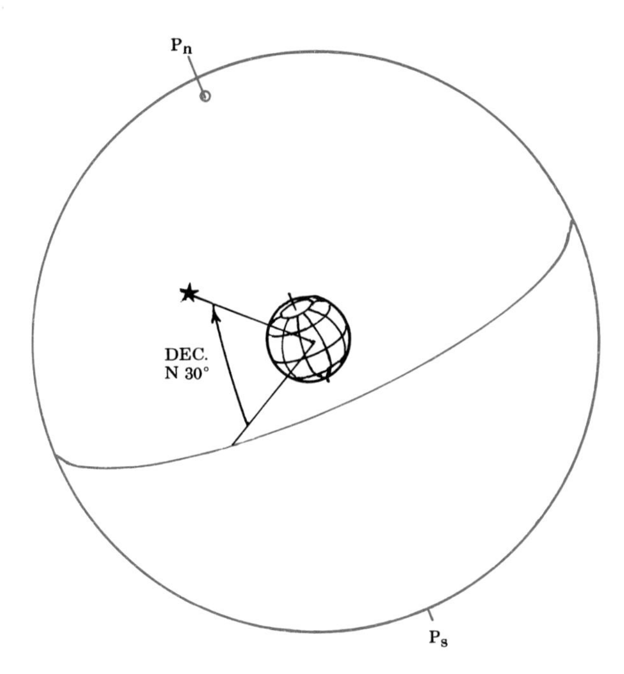

Figure 1-6A. Declination on the celestial sphere.

given time. To do this, two terrestrial meridians are projected onto the surface of the celestial sphere for use as references for hour angle measurements—the Greenwich meridian and the observer's meridian. The celestial meridians thus projected are termed the *Greenwich celestial meridian* and the *local celestial meridian,* respectively. Hour angles measured relative to the Greenwich celestial meridian are called *Greenwich hour angles,* abbreviated *GHA,* while those measured relative to the local celestial meridian are termed *local hour angles,* abbreviated *LHA.* Both Greenwich hour angles and local hour angles are measured westward from a projected terrestrial meridian to a celestial hour circle moving ever westerly with the rotating celestial sphere. Consequently, both GHA and LHA values grow increasingly larger with time, increasing from 0° to 360° once each 24 hours; they relate the rotating celestial sphere to the meridians of the earth. Sidereal hour angles, on the other hand, are measured between two hour circles on the celestial sphere; although the value of the SHA of the stars changes with time as the stars move through space relative to one another, the rate of change is exceedingly slow. Hence for purposes of celestial navigation, sidereal hour angles are considered to remain constant over a significant period of time.

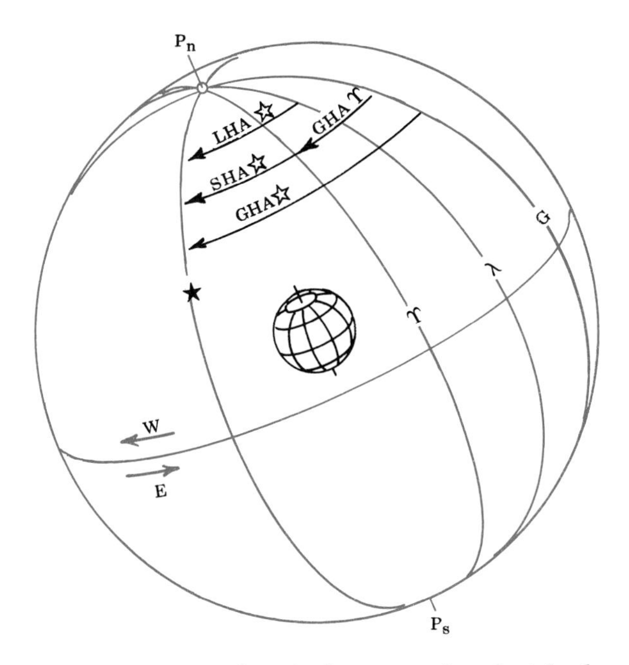

Figure 1-6B. SHA, GHA, and LHA of a star on the celestial sphere.

In Figure 1-6B, the hour circle of Aries and the projected Greenwich and observer's celestial meridians are shown superimposed in blue upon the celestial sphere of Figure 1-6A. The resulting sidereal, Greenwich, and local hour angles (SHA, GHA, and LHA) of the star at this given time are indicated.

As a point of interest, it can be seen from the figure that the GHA of the star (GHA☆) is equal to the sum of the GHA of Aries (GHA♈) plus the SHA of the star (SHA☆):

GHA☆ = GHA♈ + SHA☆

The importance of this relationship will become apparent later.

Astronomers use a different coordinate called *right ascension* (RA) to locate bodies on the celestial sphere. Right ascension is similar to SHA, but it is measured easterly from the hour circle of Aries, and is expressed in time units (hours) rather than arc. Right ascension finds only limited application in the practice of celestial navigation.

The Horizon Coordinate System

In order to obtain a celestial line of position by observation of a celestial body, a third set of coordinates, called the *horizon system*, is required. It differs from the celestial coordinate system described above in that it is based on the position of the observer, rather than on the projected terrestrial equator and poles. The reference plane of the horizon system corresponding with the plane of the equator in the terrestrial and celestial systems is the observer's celestial horizon, defined as follows:

> The *celestial horizon* is a plane passing through the center of the earth, perpendicular to a line passing through the observer's position and the earth's center.

The line mentioned in the definition, extended outward from the observer to the celestial sphere, defines a point on the sphere directly over the observer called his *zenith.* The observer's zenith is always exactly 90° of arc above the celestial horizon. The extension of the line through the center of the earth to the opposite side of the celestial sphere defines a second point directly beneath the observer called his *nadir.* The observer's zenith and nadir correspond to the terrestrial and celestial poles, while the *zenith-nadir line* connecting the observer's zenith and nadir corresponds to the axis of the terrestrial and celestial spheres.

The equivalent of a meridian in the terrestrial system and an hour circle in the celestial system in the horizon system is the *vertical circle:*

> A *vertical circle* is a great circle on the celestial sphere passing through the observer's zenith and nadir, perpendicular to the plane of the celestial horizon.

The vertical circle passing through the east and west points of the observer's horizon is termed the *prime vertical,* and the vertical circle passing through the north and south points is the *principal vertical.* The observer's principal vertical is always coincident with the projected terrestrial meridian (i.e., the local celestial meridian) passing through his position.

The equivalent of latitude in the horizon system is *altitude,* defined as follows:

> *Altitude* is the angular distance of a point on the celestial sphere above a designated reference horizon, measured along the vertical circle passing through the point.

The reference horizon for the horizon coordinate system is the celestial horizon of the observer, defined above as the plane passing through the center of the earth perpendicular to the zenith-nadir line of the observer. Altitude measured relative to the celestial horizon is termed *observed altitude,* abbreviated *Ho.* The observed altitude of a celestial body can be defined as the angle formed at the center of the earth between the line of sight to the body and the plane of the observer's celestial horizon. The other horizon used as a reference for altitude measurements is the observer's *visible* or *sea horizon,* the line along which sea and sky appear to meet. This is the reference horizon from which altitudes are measured by the sextant; such altitudes are called *sextant altitudes,* abbreviated hs. In practice, sextant altitudes must be converted to observed altitudes (*Ho*) to obtain an accurate celestial LOP; this conversion procedure will be explained in Chapter 4.

The equivalent of longitude in the horizon system is *true azimuth,* abbreviated *Zn*:

> *True azimuth* is the horizontal angle measured along the celestial horizon in a clockwise direction from 000°T to 360°T from the principal vertical circle to the vertical circle passing through a given point or body on the celestial sphere.

True azimuth can be thought of simply as the true bearing of a celestial body from the observer's position.

Figure 1-7 illustrates the horizon system of coordinates in red, and depicts the observed altitude and true azimuth of a star.

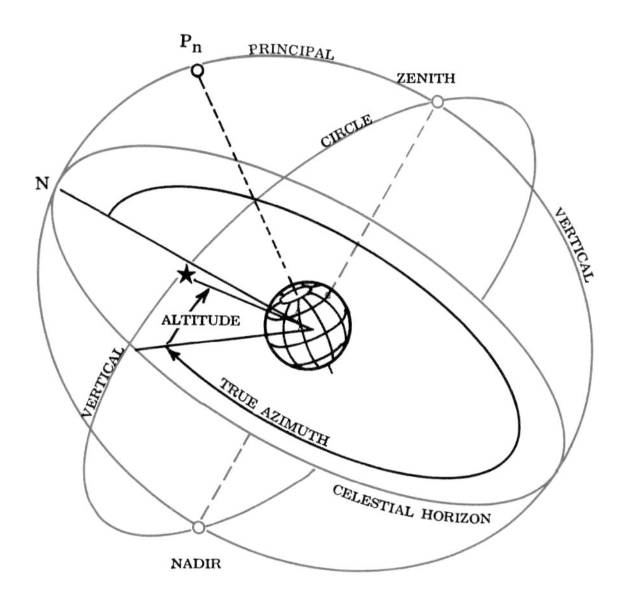

Figure 1-7. Observed altitude and true azimuth of a star in the horizon coordinate system.

The Celestial Triangle

For the purposes of celestial navigation, the terrestrial, celestial, and horizon coordinate systems are combined on the celestial sphere to form the *astronomical* or *celestial triangle.* When this triangle is related to the earth it becomes the *navigational triangle,* the solution of which is the basis of celestial navigation.

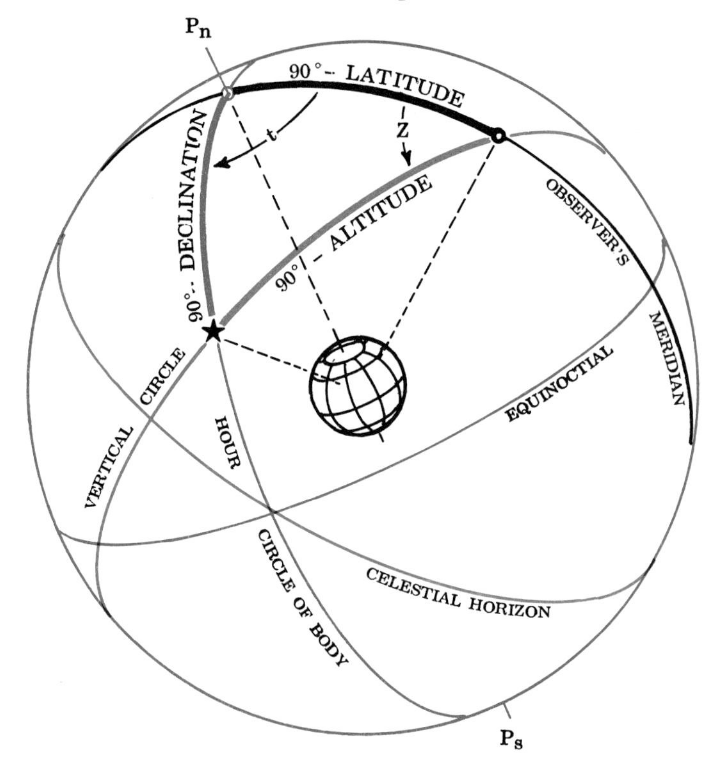

Figure 1-8. Astronomical or celestial triangle.

Each of the three coordinate systems examined in the foregoing sections of this chapter is used to form one of the sides of the celestial triangle. The three vertices of the triangle are the celestial pole nearest the observer, the observer's zenith, and the position of the celestial body. To illustrate, consider the celestial triangle depicted in Figure 1-8; in the figure, each of the coordinate systems discussed previously has been superimposed on the celestial sphere to form the celestial triangle, with the terrestrial system shown in black, the celestial system in blue, and the horizon system in red. The side of the triangle connecting the celestial pole with the observer's zenith is a segment of a projected terrestrial meridian; the side between the pole and the celestial body is a segment of the celestial hour circle of the body; and the side of the triangle between the observer's zenith and the position of the body is a segment of a vertical circle of the horizon system. In Figure 1-8, both the observer's zenith and the star being observed are located on the northern half of the celestial sphere. Other possible relationships will be discussed later.

In celestial navigation, the lengths of the sides of the celestial (later the navigational) triangle are of paramount importance. In Figure 1-8, the length of the side formed by the projected terrestrial meridian between the north celestial pole P_n and the observer's zenith expressed as an angle is 90° minus the observer's latitude. The length of the side concurrent with the hour circle of the body is in this case 90° minus the declination, but if the body were south of the celestial equator, this length would be 90° plus the declination. The length of the third side, which is measured along the vertical circle from the observer's zenith to the body, is 90° (the altitude of the zenith) minus the altitude of the body. Figure 1-8 should be studied carefully in order to fix these relationships in mind.

Only two of the angles within the celestial triangle are of concern in celestial navigation. The angle marked t in the figure at the north celestial pole is called the *meridian angle,* and is defined as the angle measured east or west from the observer's celestial meridian to the hour circle of the body. The meridian angle, like longitude, is measured from 0° to 180°, and is labeled with the suffix E (east) or W (west) to indicate the direction of measurement. The meridian angle bears a close relationship to the local hour angle (LHA), previously defined as the hour angle measured westwards from the observer's celestial meridian to the hour circle of a celestial body. If the LHA is less than 180°, the meridian angle t is equal to the LHA, and is west. If the LHA is greater than 180°, the meridian angle is equal to 360° minus the LHA, and is east.

The other angle of importance within the celestial triangle is the *azimuth angle,* which is located at the observer's zenith. It is abbreviated Z. The azimuth angle is defined as the angle at the zenith between the projected celestial meridian of the observer and the vertical circle passing through the body; by convention, it is measured from 0° to 180° either east or west of the observer's meridian. It is important to distinguish this azimuth angle of the celestial triangle from the true azimuth of the observed body, which, as mentioned earlier, can be likened to the true bearing of the body from the observer. Azimuth in this latter sense is always abbreviated *Zn,* and with altitude forms the two horizon coordinates by which a celestial body is located with respect to an observer on the earth's surface.

The third interior angle of the celestial triangle is called the *parallactic angle;* it is not used in the ordinary practice of celestial navigation, and will not be referred to henceforth.

The Navigational Triangle

In the determination of position by celestial navigation, the celestial triangle just described must be solved to find a *celestial line of position* passing through the observer's position beneath his zenith. While it would be possible to solve the celestial triangle by trigonometric methods (see Appendix A of this volume), in the usual practice of celestial navigation the solution is simplified by the construction of a second closely related *navigational triangle.* To form this triangle, the observer is imagined to be located at the center of the earth, and the earth's surface is expanded outward (or the celestial sphere compressed inward) until the surface of the earth and the surface of the celestial sphere are coincident.

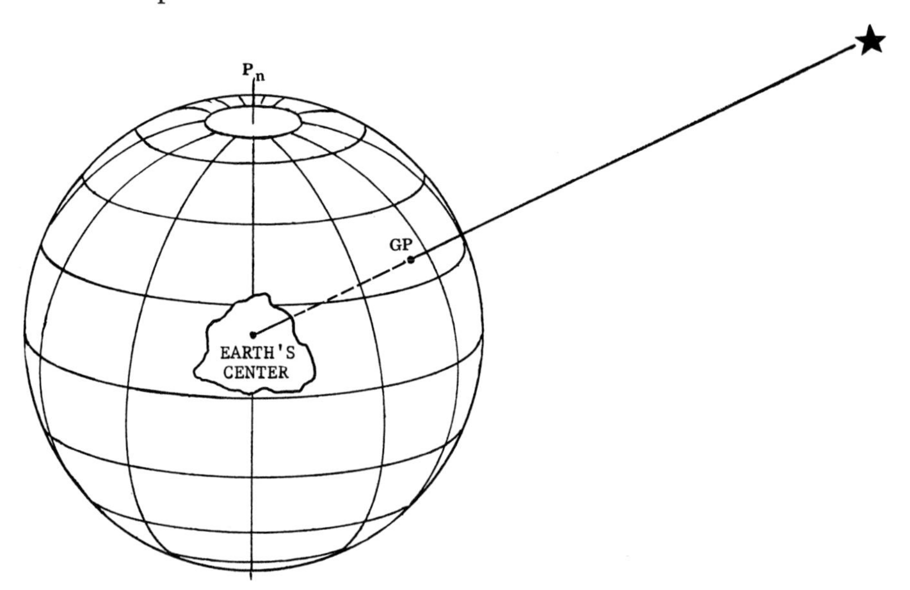

Figure 1-9. Geographic position (GP) of a celestial body.

After the earth's surface has been thus expanded (or the celestial sphere thus compressed), the position of the celestial body being observed becomes the *geographic position* (*GP*) of the body on the earth's surface. The concept of this geographic position or GP of a celestial body on the earth's surface is important and should be completely understood. Every celestial body has a GP located on the earth's surface directly beneath it, as shown in Figure 1-9. As the celestial sphere rotates about the earth, all geographic positions of celestial bodies move from east to west across the earth's surface. The GP of the sun is sometimes called the *subsolar point,* the GP of the moon the *sublunar point,* and the GP of a star the *substellar point.*

In every case, the diameter of the body is considered compressed to a point on the celestial sphere, located at the center of the body. The GP of the observed celestial body forms one vertex of the navigational triangle.

Since the coordinates of the observer's position on earth are not known, but rather are to be determined, the zenith of the observer in the celestial triangle becomes a hypothesized *assumed position* (*AP*) of the observer in the navigational triangle. This assumed position or AP forms a second vertex of the navigational triangle; the procedure for selection of the coordinates of the assumed position will be presented later.

The remaining vertex of the navigational triangle, the celestial pole, is termed the *elevated pole,* abbreviated P_n or P_s. The elevated pole is always the pole nearest the observer's assumed position; it is so named because it is the celestial pole above the observer's celestial horizon. Thus, the assumed position of the observer and the elevated pole are always on the same side of the celestial equator, while the geographic position of the body observed may be on either side. A sample navigational triangle appears in Figure 1-10.

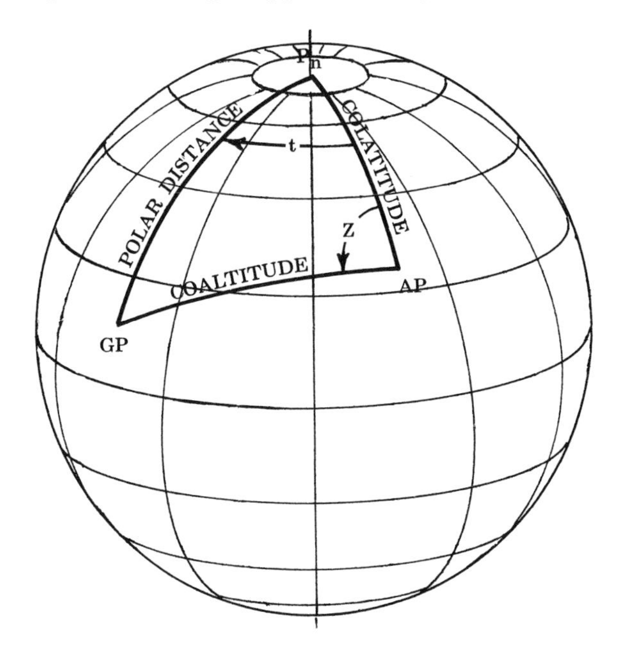

Figure 1-10. The navigational triangle.

Inasmuch as the positions of the AP of the observer and GP of the celestial body are unlimited, it follows that an infinite number of shapes are possible for the navigational triangle.

The three sides of the navigational triangle are called the *colatitude*, the *coaltitude*, and the *polar distance*. They are defined as follows:

The *colatitude* is the side of the navigational triangle joining the AP of the observer and the elevated pole. Since the AP is always in the same hemisphere as the elevated pole, the length of the colatitude is always 90° (the latitude of the pole) minus the latitude of the AP.

The *coaltitude*, sometimes called the *zenith distance*, is the side of the navigational triangle joining the AP of the observer and the GP of the body. Inasmuch as the maximum possible altitude of any celestial body relative to the observer's celestial horizon is 90°, the altitude of his zenith, the length of the coaltitude is always 90° minus the altitude of the body.

The *polar distance* is the side of the navigational triangle joining the elevated pole and the GP of the body. For a body in the same hemisphere, the length of the polar distance is 90° minus the declination of the GP; for a body in the opposite hemisphere, its length is 90° plus the declination of the GP.

The interior angles of the navigational triangle bear the same names as the corresponding angles of the celestial triangle, with only the *meridian angle t* at the elevated pole and the *azimuth angle* Z at the AP of the observer being of any consequence in the solution of the triangle. The measurement and labeling of the meridian angle was discussed in the preceding section. The azimuth angle is always measured from the observer's meridian toward the vertical circle joining the observer's AP and the GP of the body. Since the angle between the observer's meridian and the vertical circle of the body can never exceed 180°, the azimuth angle must always have a value between 0° to 180°. It is labeled with the prefix N (north) or S (south) to agree with the elevated pole, and with the suffix E (east) or W (west) to indicate on which side of the observer's meridian the GP lies. Labeling the azimuth angle in this manner is important, as it may be measured with reference to either the north or south poles, and in either an easterly or westerly direction. It may be recalled here that the meridian angle *t* is also measured either eastward or westward from the observer's meridian to the hour circle of the body, from 0° to 180°. Thus, the suffix of the meridian angle *t* and the suffix of the azimuth angle Z will always be identical.

In the ordinary practice of celestial navigation it is necessary to convert the azimuth angle of the navigational triangle to the true azimuth or bearing of the GP of the body from the AP of the observer. As an example of the conversion process, consider the navigational triangle in Figure 1-11A in which the south pole is the elevated pole.

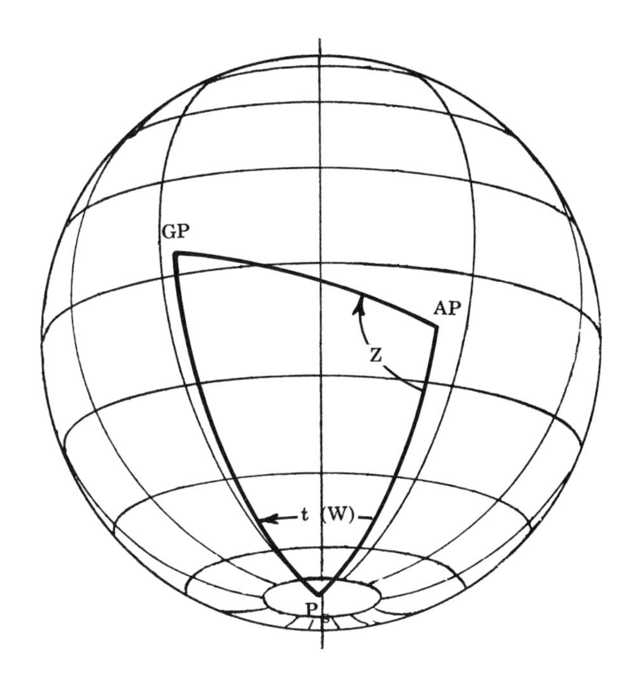

Figure 1-11A. A navigational triangle with the south pole the elevated pole.

In this figure, the south pole is the elevated pole because the AP of the observer is located in the southern hemisphere. At the time of the observation of the body, its GP has been determined to be north of the equator and to the west of the observer. Thus, the prefix for the azimuth angle Z in this case is S (south), to agree with the elevated pole, and the suffix is W (west), identical with the suffix of the meridian angle. Hence, if the size of the azimuth angle were 110°, the angle would be written S 110° W. To convert this azimuth angle to a true azimuth, it is helpful to draw a sketch of the directional relationships involved similar to that in Figure 1-11B. From the figure, it should be obvious that to convert the azimuth angle S 110° W to true azimuth, it is necessary only to add 180°. Thus, the true azimuth or bearing of the GP from the AP in this case is 180° + 110° = 290°T.

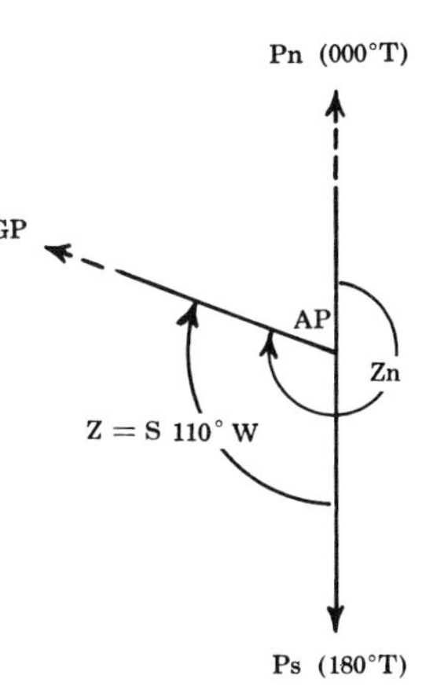

Figure 1-11B. Directional relationships for an azimuth angle.

In practice, an alternative method of conversion is frequently used that makes use of a set of rules based on the relationship of the azimuth angle with the latitude of the observer and the local hour angle of the body. This method will be discussed in a later chapter.

A complete understanding of the celestial

and navigational triangles is essential in celestial navigation. By solution of the appropriate triangle, the navigator can determine his position at sea, check his compass accuracy, predict the rising and setting of any celestial body, locate and identify any bodies of interest, and much more. It can well be stated that the solution of the celestial and navigational triangle *is* celestial navigation.

The Circle of Equal Altitude

To illustrate the basic concepts involved in obtaining a celestial line of position, suppose that a steel pole perpendicular to a level surface were raised, and a wire stretched from its top to the surface, such that the angle formed by the wire and the surface was 60°. If the end of the wire were rotated around the base of the pole, a circle would be described, as shown in Figure 1-12A. At any point on this circle, the angle between the wire and the surface would be the same, 60°. Such a circle is termed a *circle of equal altitude.*

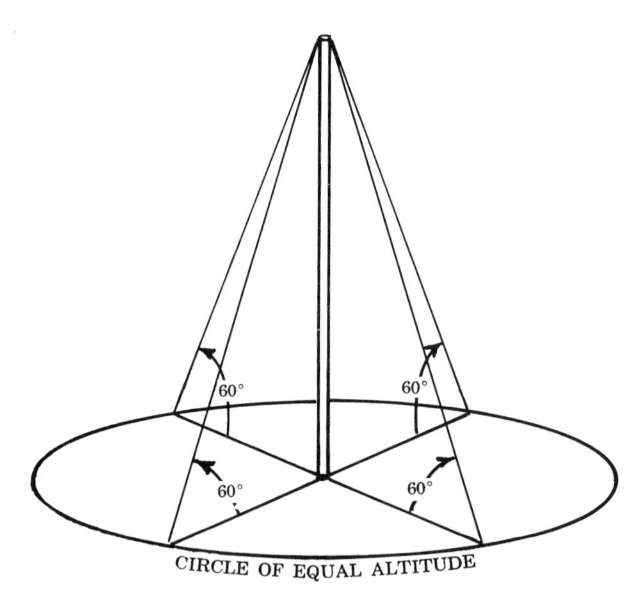

Figure 1-12A. A circle of equal altitude inscribed about a pole.

Now, suppose that the end of the pole were extended to an infinite distance. The angle formed by the wire anywhere on the flat surface would approach 90°, since the wire would be nearly parallel with the pole. If the surface were spherical, however, and the measurement made relative to a tangent plane, the angle would vary from 90° at the base of the pole to 0° at all points on the spherical surface 90° away from the location of the base. Figure 1-12B depicts two concentric circles of equal altitude inscribed on such a spherical surface, centered on the location of the base—the "GP"—of the pole. At all points on the circumference of the smaller circle, the angle formed by the wire and the tangent plane is 60°, while the angles measured along the larger circle are all 30°.

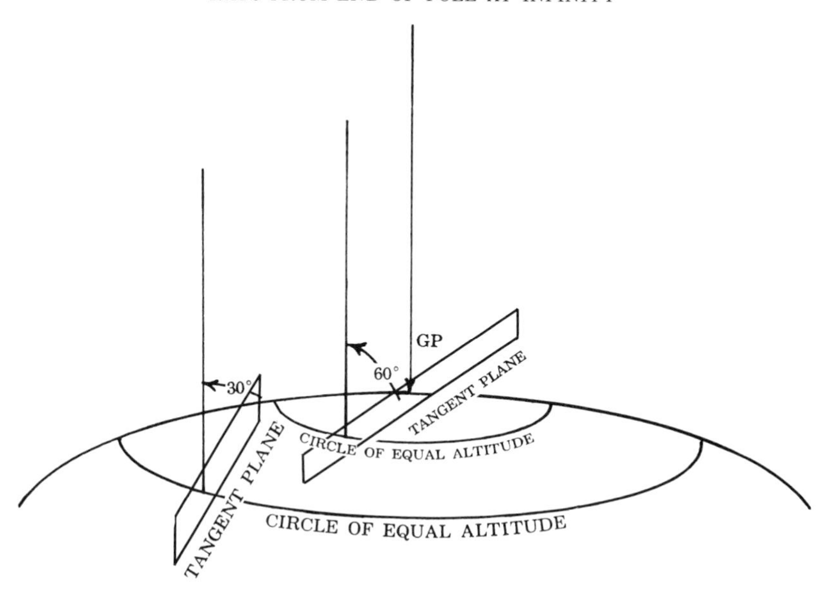

Figure 1-12B. Two concentric circles of equal altitude.

In celestial navigation, the situation is analagous to that depicted in Figure 1-12B. Suppose that a celestial body was observed and found to have an altitude of 60° above the observer's celestial horizon, and its GP at the moment of the observation was determined to have been located at 10° south latitude, 30° west longitude. Suppose further that the AP of the observer determined perhaps from his DR plot was 10° north latitude, 10° west longitude. Assuming a globe of sufficient scale were available, a navigational triangle similar to the one in Figure 1-13A on the following page could be constructed by plotting the coordinates of the GP and AP.

Now, let us plot a circle of equal altitude about the GP of the body, from which circle an altitude of 60° could be observed. To do this, let us assume for the moment that the observer's assumed position AP coincided with his actual position. If this were the case, the radius of the circle of equal altitude would have the same length as the coaltitude of the observer's navigational triangle; expressed as an angle, this is 90° (observer's zenith) − 60° (alti-

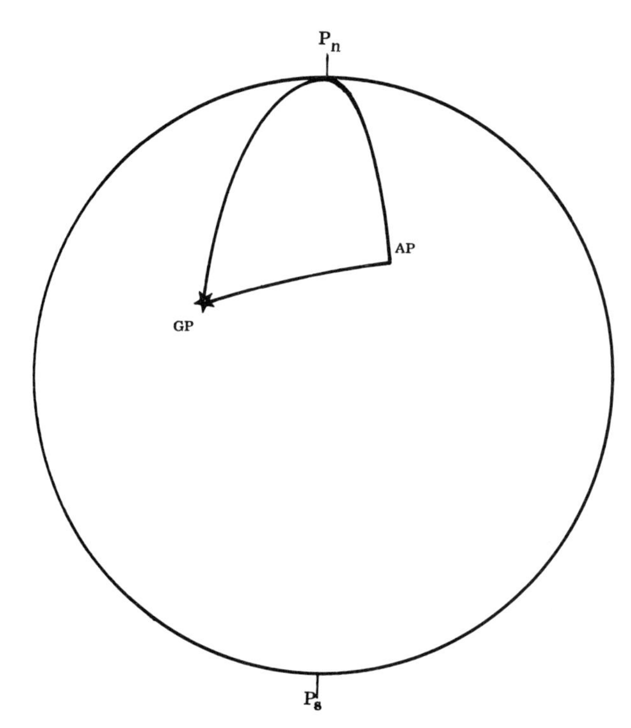

Figure 1-13A. A navigational triangle.

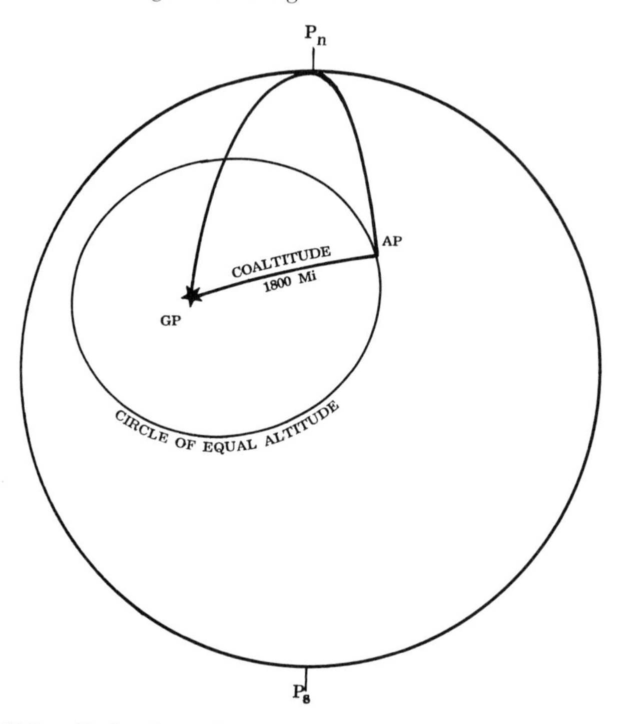

Figure 1-13B. Circle of equal altitude for Ho 60°.

tude of the body), or 30°. Since the coaltitude is a segment of a vertical circle that is itself a great circle, the important assumption that one degree of a great circle subtends 60 nautical miles on the earth's surface can be used to find the linear length of the coaltitude; it is 30° × 60 miles, or 1,800 miles. Thus, the circle of equal altitude for this observation could be formed by swinging an arc of radius 1,800 miles about the GP of the body. This circle, depicted in Figure 1-13B, represents a locus of all points, including the observer's actual position, from which it is possible to observe an altitude of 60° for this body at the time of observation.

Now suppose that the AP of the observer was in fact a small distance from his actual position, as is usually the case in practice. In this situation, the AP will probably lie off the circle as shown in Figure 1-13C, either closer to or farther away from the GP of the body.

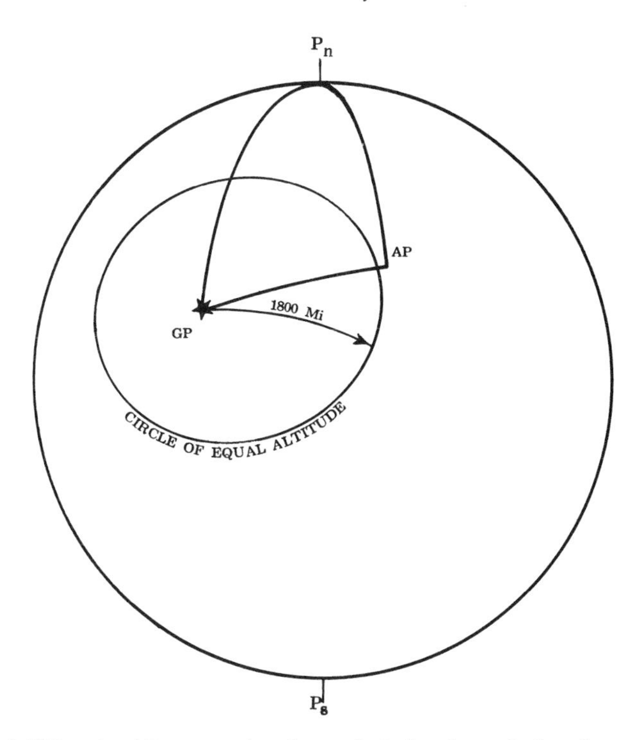

Figure 1-13C. An AP not on the observed circle of equal altitude.

Thus, if the observation of the body had in fact been made from the AP, a different altitude than that observed and therefore a different coaltitude (radius of the circle of equal altitude) would have been obtained. The significance of this difference in the lengths of the two coaltitudes corresponding with the observer's actual observation and his assumed position will become apparent in Chapter 2.

To find his exact position, i.e., the *celestial fix*, the observer could observe a second celestial body and plot a second circle of equal altitude around its GP. Normally the two circles would intersect in two places; the observer's position must be located at one of the intersections, probably the one nearest the AP of the observer, as depicted in Figure 1-13D.

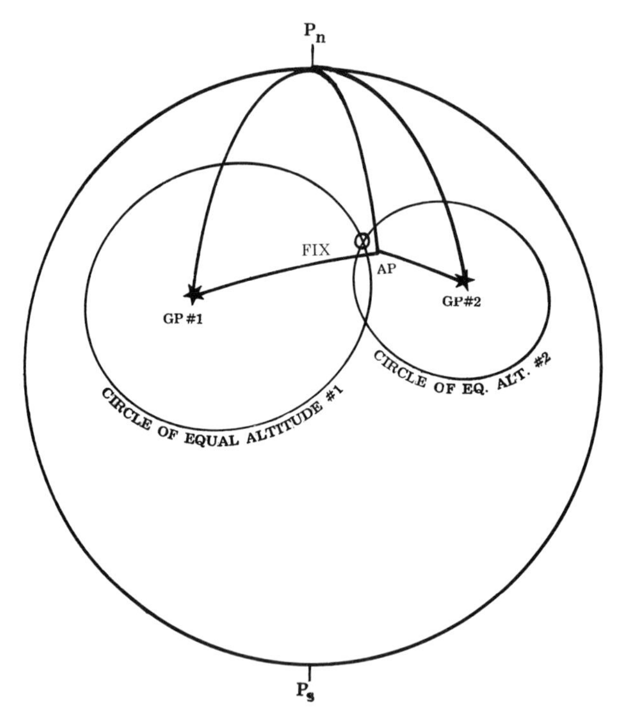

Figure 1-13D. Determining a celestial fix using two circles of equal altitude.

If the two intersection points were so close together as to cause confusion, the doubt could be resolved by observing a third body and plotting a third circle of equal altitude.

Because of the large-scale globe or chart that would have to be used to plot a celestial fix of meaningful accuracy by finding the intersection of two or more complete circles of equal altitude, this method is not normally used in the practice of celestial navigation. For bodies of low altitude, the length of the coaltitude is prohibitive. If the altitude of a body were 20°, for example, the coaltitude would be $(90° - 20°) \times 60$, or 4,200 miles long. To obtain any meaningful degree of accuracy, a chart hundreds of feet in length and width would have to be used to plot the resulting circle. Bodies of extremely high altitude are difficult to observe with accurate results using the marine sextant. An alternative method of plotting the celestial LOP, therefore, has been devised wherein only a small portion of the coaltitude and circle of equal altitude for each body observed is plotted. This method, called the *altitude-intercept method,* is the subject of the following chapter.

Summary

This chapter has set forth the basic theory upon which the remaining chapters of this book on the practice of celestial navigation are based. The earth's relationship to the universe and the solar system of which it is a part were examined in some detail, and the celestial and horizon systems of coordinates were developed. Finally, the terrestrial, celestial, and horizon systems were combined to form the celestial triangle, from which the navigational triangle is derived. The solution of the navigational triangle is basic to the practice of celestial navigation. Succeeding chapters will deal in detail with various aspects of the use of the navigational triangle in solving problems of particular interest to the navigator.

2

The Altitude-Intercept Method

As discussed in the last chapter, a celestial line of position can be produced from an observation of a celestial body by plotting a circle on the surface of the earth with radius equal to the coaltitude of the body, centered on the geographical position (GP) of the body at the time of the observation. A celestial LOP of this type, termed a *circle of equal altitude,* is considered impractical for bodies of altitude less than about 87°, because of the unwieldy length of the resulting coaltitude radius.

The *altitude-intercept method* is an alternative method of plotting the celestial LOP that eliminates the disadvantages of the circle of equal altitude. In the usual practice of celestial navigation, this method utilizes daily data tabulated in either one of two almanacs, the *Nautical Almanac* or the *Air Almanac,* in conjunction with one of two different *sight reduction tables,* to produce a computed altitude Hc to a body being observed from an assumed position (AP) of the observer. The computed altitude Hc is then compared with the observed altitude Ho to determine the position of the celestial LOP.

The Altitude-Intercept Method

The altitude-intercept method of plotting a celestial line of position was developed in 1875 by the Frenchman Marc St. Hilaire as an alternative to more cumbersome methods then in vogue. Originally the method employed two trigonometric equations called the "Cosine-Haversine" formulas, but because they were rather difficult to solve by manual methods, about 1930 Ogura, a Japanese, developed a more convenient solution based on the use of certain sight reduction tables. The use of similar but improved tables predominates today, although the advent of the miniature electronic calculator equipped with a capability of handling trigonometric functions has sparked renewed interest in the use of the original Cosine-Haversine formulas and others derived from them. (See Appendix A.) In its present form, the altitude-intercept method requires little other than the ability to plot on a chart, to look up required data in tables, and to add, subtract, and interpolate between numbers in a column.

In essence, in the altitude-intercept method only small segments of the radius and the circumference of the circle of equal altitude are plotted on a chart or plotting sheet. To do this, an assumed position (AP) for the time of the observation of a celestial body is chosen, and the true azimuth and computed altitude of the body observed are determined for this position by use of one of the sight reduction tables mentioned earlier. A line drawn from the AP of the observer toward the GP of the body in the direction of the true azimuth then represents a segment of the radius of the circle of equal altitude; a segment of the circumference of the circle is positioned along the true azimuth (radius) line by comparing the computed altitude with the observed altitude actually obtained.

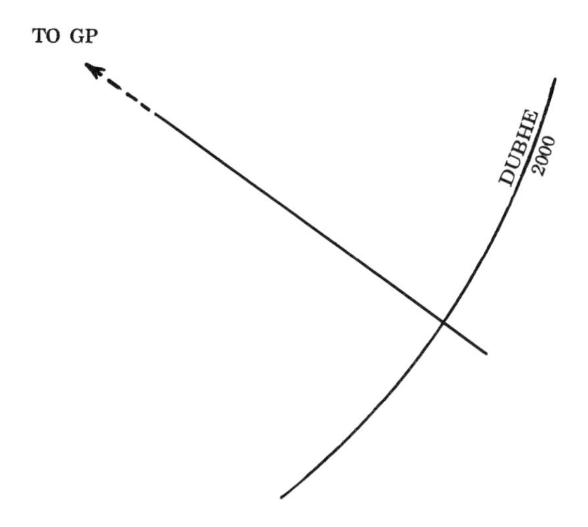

Figure 2-1A. Segment of a circle of equal altitude, GP nearby.

Consider for a moment a small segment of the radius and circumference of a circle of equal altitude in a situation in which the AP of an observer is assumed to be fairly close to the GP of the body, as in Figure 2-1A. Since the GP of the body is nearby, the portion of the circumference of the circle of equal altitude in the figure appears curved. If the distance to the GP were gradually increased, the portion of the circumference shown would appear increasingly less curved, finally approaching a straight line as the distance to the GP increased to a few hundred miles or more. This situation is depicted in Figure 2-1B on the following page. As can be seen from the figure, the circumference of the circle near the radius line has approached a straight line perpendicular to the radius at the point of intersection.

In the altitude-intercept method of plotting the celestial line of position, the LOP is always represented by a short segment of the

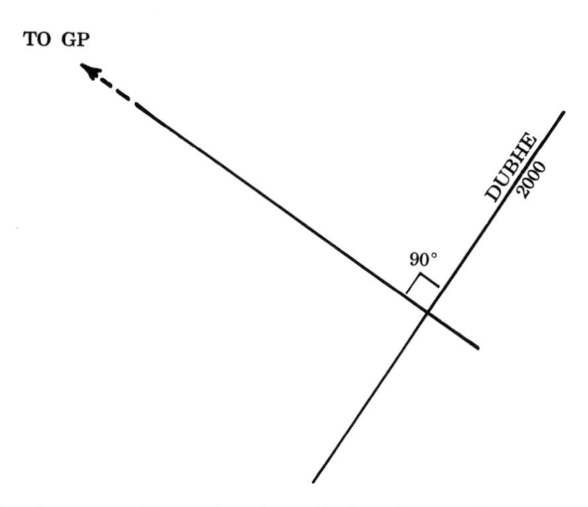

Figure 2-1B. Segment of a circle of equal altitude, GP distant.

circle of equal altitude such as that appearing in Figure 2-1B, drawn at right angles to the radius line. Since the radius of the circle of equal altitude always lies in the direction of the GP, the LOP can be thought of as being drawn perpendicular to the bearing or true azimuth (Zn) of the GP of the body. It is the positioning of the LOP along the true azimuth line that constitutes the basic problem solved by the altitude-intercept method.

As was mentioned earlier, the altitude of a given celestial body can be computed for the assumed position (AP) of the observer by use of an almanac in conjunction with a sight reduction table; this computed altitude is abbreviated Hc. If the observed altitude Ho were identical to the calculated altitude Hc, the circles of equal altitude corresponding to the computed and observed altitudes would be coincident and would both pass through the AP of the observer. If the observed altitude Ho were greater than the computed altitude Hc, the radius of the circle of equal altitude corresponding to the observed altitude would be smaller than the radius of the circle for the computed altitude. In this case, the observer must in reality be located closer to the GP of the body than his assumed position. Conversely, if the observed altitude were less than the computed altitude, the observer must be located farther away. Figure 2-2 illustrates the reasoning upon which the foregoing conclusions are based.

In the figure, three segments of circles of equal altitude are depicted, corresponding to an observed altitude Ho greater than, equal to, or less than a computed altitude Hc for a given assumed position AP. It should be noted that as the size of the observed altitude increases,

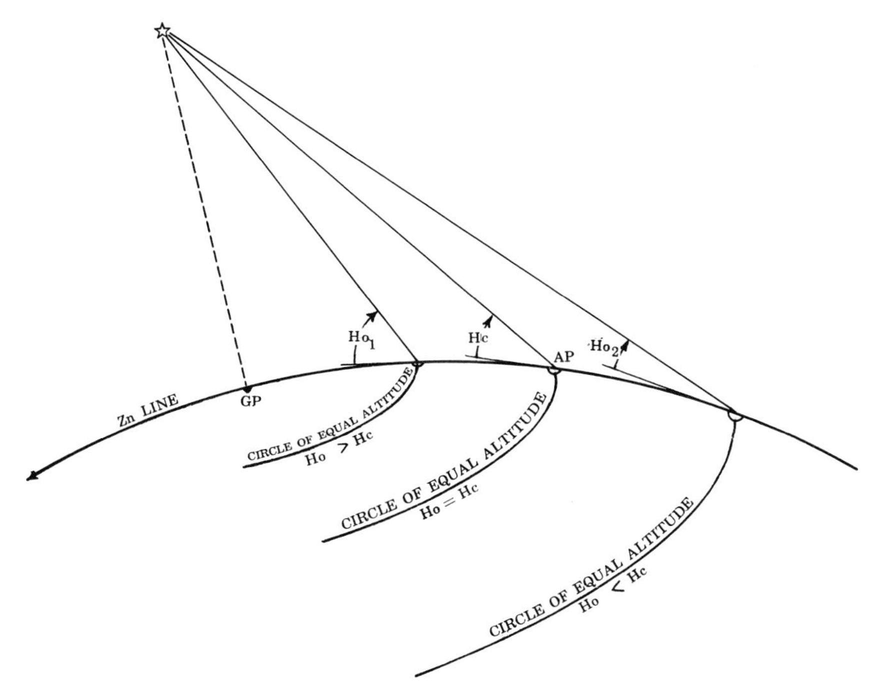

Figure 2-2. Effect of Ho *larger than, equal to, or smaller than* Hc. (*Note: Celestial body shown unrealistically close to earth to emphasize difference in size of angles.*)

the distance from the GP of the body decreases. This is reasonable, since the length of the coaltitude (90° − altitude), which is the radius of the circle of equal altitude, decreases as the altitude increases.

To calculate the distance between the observer and his AP along the true azimuth line, it is only necessary to find the difference between the observed and computed coaltitude distances from the GP of the body. The following algebraic expression represents the distance expressed in degrees of arc between the two coaltitudes:

$$(90^\circ - \text{Ho}) - (90^\circ - \text{Hc})$$

where 90° − Ho is the coaltitude of the observed altitude, and 90° − Hc is the coaltitude of the computed altitude. In practice, however, it is not necessary to compute the coaltitudes, since

$$(90^\circ - \text{Ho}) - (90^\circ - \text{Hc}) =$$
$$90^\circ - \text{Ho} - 90^\circ + \text{Hc} =$$
$$\text{Hc} - \text{Ho}$$

Thus, to find the distance from the AP along the true azimuth line at which the LOP corresponding with Ho should be drawn, it is only necessary to find the difference between the computed and observed altitudes. For every minute of arc difference, the intercept point of the LOP with the true azimuth line is moved one nautical mile from the AP, either toward the GP of the body if Ho is greater than Hc, or away from the GP if Ho is less than Hc. The distance thus determined between the intercept point and the AP is called the *intercept distance,* symbolized by a lower-case *a.*

Figure 2-3 shows an example of the computation of an intercept distance, in which an observed altitude Ho is greater than the computed altitude Hc.

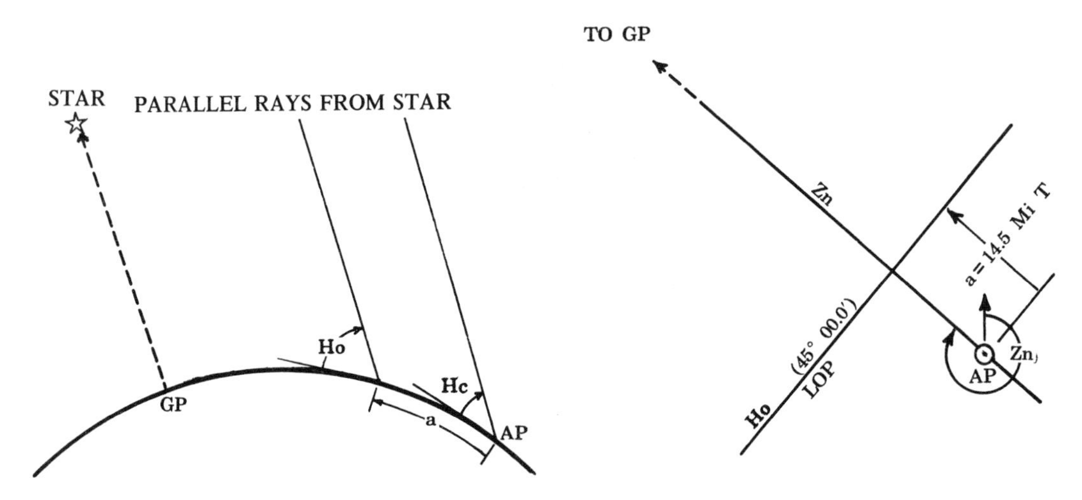

Figure 2-3. Altitude-intercept method, Ho *greater than* Hc.

In this example a star was observed having an Ho of 45° 00.0′. For an assumed position AP chosen for the time of the observation, an Hc of 44° 45.0′ was computed, as well as the true azimuth of the GP from the AP. To plot the celestial LOP corresponding to the observed altitude, the intercept point must be advanced along the true azimuth line toward the GP of the body (since Ho is greater than Hc) by an intercept distance equal to the arc-difference between Ho and Hc expressed in miles. This intercept distance is given by the following computation:

Ho	45° 00.0′
−Hc	44° 45.5′
a	14.5′ T

The computed intercept distance, 14.5 miles in this case, is labeled with the letter *T* to indicate that the intercept point must be moved *toward* the GP of the body. If the intercept distance is to be laid off *away from* the GP, in the case of Ho being less than Hc, the letter *A* is used to indicate this fact. Hence, in this example, the LOP corresponding with the observed Ho is drawn perpendicular to the true azimuth line through a point on the line 14.5 miles toward the GP of the star. Because the resulting LOP represents a portion of the circle of equal altitude along which the angle Ho could be observed, the observer's position must lie somewhere on this LOP.

To find the exact location of the observer on the LOP, the procedure described can be repeated for a second body observed simultaneously with the first. The intersection of the two simultaneous LOPs then defines the observer's position—his celestial fix—at the time of the observation. The plot of a celestial fix and running fix will be the subject of Chapter 7.

As a memory aid in determining whether the intercept distance should be laid off toward or away from the GP of the observed body when plotting a celestial LOP by the altitude-intercept method, two phrases have long been used. They are:

> "Coast Guard Academy" (Computed Greater Away), and
> "HoMoTo" (Ho More than Hc, Toward).

In order to achieve accuracy in the plot of the celestial LOP, and to a lesser extent to facilitate the plot by the altitude-intercept method, the assumed position is always chosen in such a way that the intercept distance is kept fairly short. The reason for this is that the true azimuth of the GP of an observed body from the AP of the observer actually lies along a great circle, while the intercept distance laid down on a Mercator chart follows a rhumb line. The longer the intercept distance is, the more the rhumb line representing it on the Mercator projection will diverge from the great circle true azimuth. Thus, an error is introduced into the plot of the celestial LOP, which increases in size in proportion to the length of the intercept distance. For relatively short intercept lengths of 30 miles or less, the error is insignificant, but for intercept distances longer than 60 miles, the effect of the error is considerable.

It may be apparent by now that the altitude-intercept method of plotting the celestial line of position takes its name from the fact that a computed *altitude* is first determined for an assumed position, then compared with the observed altitude to find the *intercept* distance a. The following section will briefly describe how the assumed position is chosen, and how the computed altitude and true azimuth are obtained for it.

The Determination of the Assumed Position, Hc, and Zn

The altitude-intercept method of solving the navigational triangle and plotting the resulting celestial line of position depends on the ability of the observer to determine the computed altitude Hc and the true azimuth Zn for an observed celestial body from a selected assumed position at the time of the observation. As was stated in the introduction to this chapter, the navigator usually employs one of two almanacs in conjunction with one of several different sight reduction tables to accomplish this determination. In Chapter 5, the use of the *Nautical Almanac* in conjunction with the *Sight Reduction Tables for Marine Navigation, No. 229,* is described, and in Chapter 6, the use of the *Air Almanac* with the *Sight Reduction Tables for Air Navigation, No. 249,* will be demonstrated.

The coordinates of the assumed position AP are chosen to provide the proper values of the quantities required for entering arguments in the sight reduction tables used, consistent with the restraint that the intercept distance from the AP to the resulting celestial LOP should be as short as possible. The rules for the selection of the assumed latitude and longitude will be presented in Chapters 5 and 6, as will the determination of the computed altitude and true azimuth using the two sight reduction tables.

Plotting the Celestial Line of Position

After the assumed position, intercept distance, and true azimuth have been determined by the solution of the navigational triangle, the celestial LOP can be plotted. Although the plot could be done on a fairly large-scale "working" chart of the area in which the ship is operating, it is the usual procedure to use a scaled *plotting sheet* for this purpose. The DMA Hydrographic/Topographic Center makes available a Large Area Plotting Sheet, or LAPS, covering several degrees of latitude and longitude, as well as a Small Area Plotting Sheet (SAPS), which covers two degrees of latitude and longitude. Only the latitude lines are labeled, with the labeling of the meridians left to the user. Each sheet can be used for either north or south latitudes by inverting the sheet. A typical SAPS centered on latitude 31° (SAPS-31) appears in Figure 2-4 opposite.

To plot a celestial line of position on a plotting sheet, a suitable SAPS for the latitudes involved is first selected, and the longitudes are labeled, increasing from right to left if in west longitudes, and from left to right if in east longitudes. Once the sheet has been properly prepared, the assumed position is plotted on it. From this AP a dashed line is drawn either in the direction of the true azimuth Zn for "Toward" intercepts, or 180° away from the Zn in the case of "Away" intercepts. The length of the intercept is then laid off along this line, and a point is inscribed on the line to indicate the length of the intercept distance. To complete the plot, the LOP is drawn through this point, perpendicular to the intercept line.

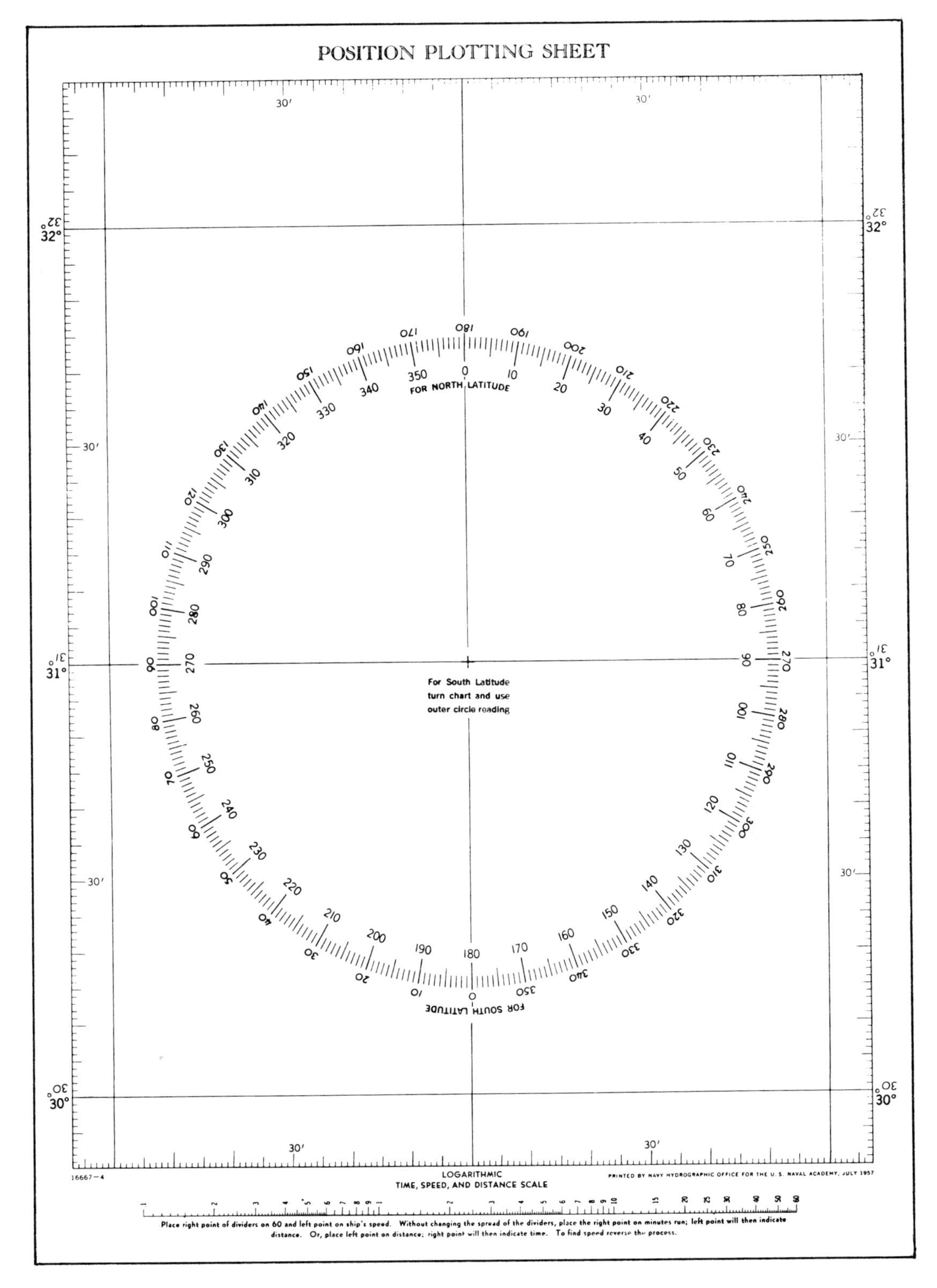

Figure 2-4. SAPS-31.

As an example of a plot of a typical celestial LOP, suppose that for an observation of the planet Venus an assumed position at L 34° S, λ 163° 08.4′ E had been selected, and the intercept distance and the true azimuth Zn were determined to be 14.8 miles "toward," and 095.1°T, respectively. The completed plot appears in Figure 2-5. Note that the LOP is labeled with the name of the body and the zone time of the observation to the nearest minute.

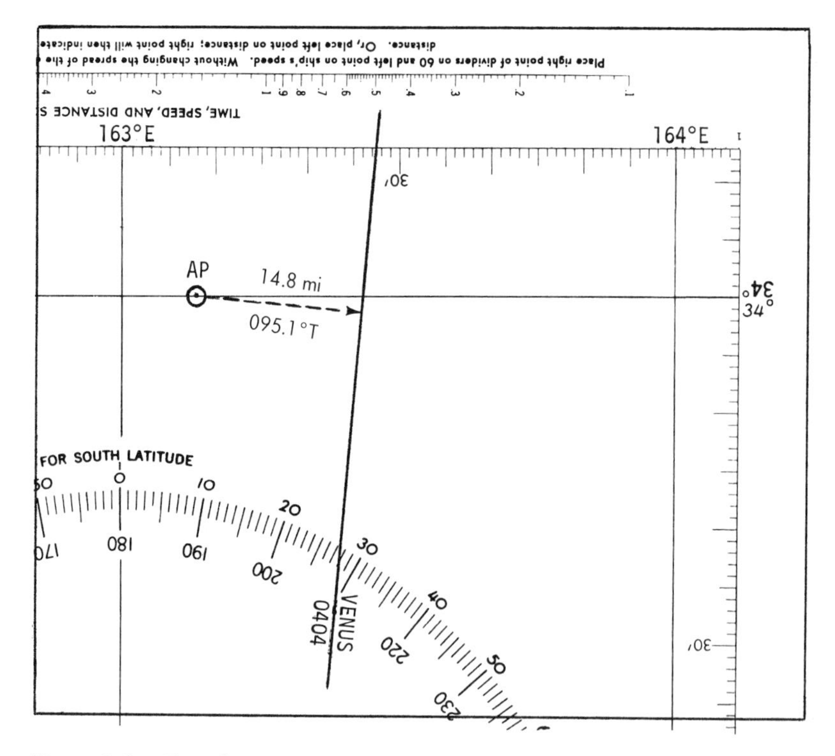

Figure 2-5. Plot of a Venus LOP.

The procedure for plotting several celestial LOPs on the same plotting sheet to obtain a celestial fix or running fix will be demonstrated in Chapter 7. In practice, after the fix has been plotted on the plotting sheet, its coordinates are then picked off by a pair of dividers or a drawing compass and the fix position is shifted onto the actual working chart.

Summary

The altitude-intercept method of solving the navigational triangle and plotting the resulting celestial LOP is of great use to the navigator engaged in the practice of celestial navigation at sea. After the assumed position has been selected and the computed altitude and true azimuth determined by the use of an almanac in conjunction with one of the several sight reduction tables, the celestial LOP is easily plotted either on an appropriate chart or plotting sheet. Once plotted, two or more celestial LOPs may be combined to form a fix or running fix. If a plotting sheet has been used for the plot, the coordinates of the fix are then transferred to the "working" chart of the area.

3

Time

The subject of time was introduced in *Marine Navigation 1* in connection with voyage planning. The fundamental concepts of time were set forth, including the reckoning of time according to two types of mean solar time: Greenwich mean time and local zone time. The procedures for time conversions from one time zone to another were also examined in some detail.

This chapter will complete the discussion of time, emphasizing its use and applications during the practice of celestial navigation. The first section will reexamine the various bases by which time is reckoned, including a hitherto unmentioned reference for the passage of time. Next, the relationship between time and longitude will be discussed, and an aid in visualizing these and other time relationships called the time diagram will be introduced. Finally, a discussion of the methods used for the timing of celestial observations will conclude the chapter.

The Bases of Time

From the beginning of recorded history, man has reckoned time according to the passage of two celestial bodies through the heavens—the sun and the moon. The science of archeology has produced much evidence of the fascination of ancient man with the passage of the sun across the sky by day, and with its migration in declination over the period of a year. It is theorized by many contemporary scientists that the ancient monument at Stonehenge, England, constructed about 1500 B.C., may well be the oldest known solar observatory. Many of the monuments of ancient Egypt were constructed and oriented in such a way that they marked significant points in the annual migration of the sun in declination, and the amazing accuracy of the calendars of the South and Central American Inca and Aztec Indian cultures that flourished in pre-Columbian times is well documented. In fact, almost every known culture since the dawn of history has made extensive use of the sun as the basis for dividing the passage of time into days and years. Use of the moon as the basis for grouping the days of the year into 30-day months was a natural corollary, given the approximately

30-day progression of its phases caused by its period of revolution about the earth.

Apparent Solar Time

Time reckoned according to the passage of the sun observable in the sky is termed *apparent solar time;* used in this context, the sun is referred to as the *apparent sun.* Until the latter half of the nineteenth century, when comparatively rapid means of transportation and communication in the form of the steam locomotive and the telegraph evolved, the passage of the apparent sun across the heavens was almost exclusively used as the basis by which all times were reckoned. When the sun's disk passed over the observer's meridian, his clocks were adjusted to read noon. Around 1800, it was of little consequence to a man living in Norfolk, Virginia, that his cousin located at the capital in Richmond kept a time that differed from his by about 4 minutes, since it took several days to make the journey between them on horseback.

Mean Solar Time

With the advent of the steam locomotive, however, which by 1890 could complete the Richmond to Norfolk transit in a few hours, maintaining schedules and computing arrival and departure dates and times by reference to the apparent sun became impossible for two reasons. First, if time is reckoned according to apparent solar time, a traveler's watch must be adjusted by one minute each time he traverses a distance of 15 minutes of longitude, as will be explained later. Second, as the earth proceeds along its elliptical path of revolution about the sun, the speed with which it moves in its orbit continually changes, as does the inclination of the earth's axis with respect to the sun. The net effect of these irregularities is that the length of the solar day varies at different times of the year.

To overcome these disadvantages of nonuniform apparent solar time, the railroads adopted a method of reckoning time, first proposed in the mid-eighteenth century, called *mean solar time.* Mean solar time employs as its reference a hypothesized *mean sun,* the hour circle of which is considered to move at a constant rate along the celestial equator. The rate at which the mean sun moves is the average rate of motion of the apparent sun during each mean solar day over a solar year. Since there are 24 hours in every mean solar day, and 360 degrees in the circumference of the earth, the mean sun moves at the constant rate of $360° \div 24$ hours, or 15° per hour. This fact is of fundamental importance in celestial navigation, since by this means degrees of arc as measured along a parallel of latitude are directly convertible into time units. As will be seen later, this convertibility of arc to time allows precomputation of the times of rising and setting of any celestial body, particularly the sun, as well as calculation of the times of twilight and the time at which apparent noon will occur. Spurred by railroad interests, mean solar time came into wide use throughout the United States, and from 1900 to the present it has been the predominant means by which the passage of time is reckoned throughout the world.

Equation of Time

Day	SUN Eqn. of Time 00h	12h	Mer. Pass.
	m s	m s	h m
25	03 13	03 10	11 57
26	03 07	03 03	11 57
27	03 00	02 57	11 57

Figure 3-1.
The equation of time, Nautical Almanac, *25, 26, 27 May.*

At certain times of the year, the apparent sun moves across the heavens at a slower rate than the mean sun, while at other times, the apparent sun travels faster. In the former case, the apparent sun may be as much as 15 minutes behind the mean sun, while in the latter case, the apparent sun may be as much as 15 minutes ahead. The difference between mean and apparent time at any instant is called the *equation of time.* It is tabulated for each day in the *Nautical Almanac,* in the format shown in Figure 3-1.

As can be seen in the figure, there are three entries in the *Almanac* for each day. The first column gives the number of minutes and seconds by which the apparent sun either lags or leads the mean sun as these bodies pass over the lower branch of the Greenwich meridian—the 180th meridian (00^h)—and the second column contains the time differences applying at the upper branch, or prime meridian (12^h). The figures in the third column are simply the local mean times rounded to the nearest whole minute at which the apparent (actual) sun will pass over or *transit* the Greenwich meridian. During the period 25-27 May, for example, the local mean time at which the apparent sun will transit the Greenwich meridian is approximately 1157. Thus, the apparent sun must be ahead of the mean sun on these dates, by the amount of minutes and seconds listed in the first two columns for each day in the 3-day period covered. Conversely, were the apparent sun lagging behind the mean sun, the times of meridian passage listed in the third column would be 1200 or later, with the figures in the other two columns indicating the exact amount of the lag to the nearest second. Because the equation of time will vary by only a few seconds as the sun continues around the earth in the course of each day, the local mean time of the sun's meridian passage at Greenwich tabulated in the *Almanac* for each day can also be used for all other meridians. On 25 May, for instance, the apparent sun will transit the upper branches of all meridians approximately 3 minutes ahead of the mean sun. Hence, the apparent sun will cross the standard meridians of all time zones at 1157 local mean time (equivalent to 1157 local zone time).

The equation of time is used primarily to determine the local mean times of meridian passage of the apparent sun at Greenwich for tabulation in the *Nautical Almanac.* In Chapter 9 the use of these tabulated local mean times of meridian passage to determine the zone time of apparent noon at any position on earth will be demonstrated. The zone time of apparent noon thus determined is referred to as *local apparent noon,* or *LAN.*

Sidereal Time

In addition to apparent and mean solar time, there is one other basis of celestial time which the navigator may occasionally see mentioned in some advanced texts. This time, called *sidereal* time, uses the earth's rotation relative to the stars for its basis, rather than its rotation relative to the sun. Because the earth rotates in the same direction as its direction of revolution about the sun—counterclockwise as viewed from above the plane of the ecliptic—each sidereal day is about 3 minutes and 56 seconds shorter than the mean solar day, as illustrated in Figure 3-2.

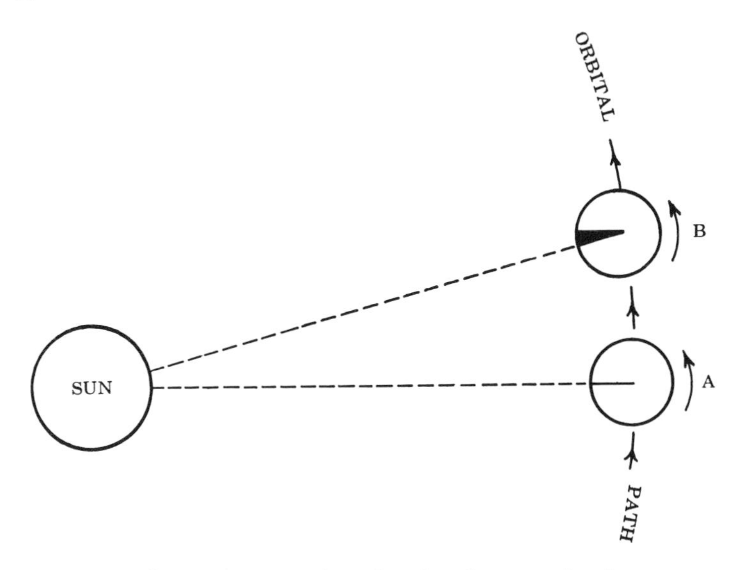

Figure 3-2. Difference between the sidereal and mean solar days.

In this figure, the earth begins one complete rotation at position A. At position B, the rotation has been completed with respect to the stars, but with respect to the sun, the earth must still rotate an amount equal to the shaded arc. Thus, the sidereal day is about 23 hours 56 minutes in length, while the mean solar day is exactly 24 hours long.

Sidereal time does not have extensive application in the usual practice of celestial navigation on earth. It is used mainly as the basis by which star charts and star finders are constructed.

Atomic Time Standards

In contrast to the aforementioned bases of time dependent upon the spatial relationship of the earth with the sun and the surrounding universe, there is another basis of time independent of these relationships that is coming into ever wider use today because of the extreme accuracy it can afford, particularly in scientific applications. This is *Universal Coordinated Time* (UTC), based on the frequency of vibrations of excited atoms of the radioactive element cesium. Because UTC is based on an unchanging atomic time standard, and mean time is affected by variations in the rate of rotation and revolution of the earth, there can be as much as a .9-second difference between times reckoned on the two bases. The exact difference at any time is called the *DUT1 correction;* its value is held between ±.9 seconds by the occasional addition or subtraction of a so-called *leap second* adjustment to UTC applied as required, normally at the end of either June or December.

For some years now, UTC has been exclusively used as the basis of time for the various radio time signals broadcast worldwide by designated radio stations of the various maritime nations. As will be discussed later in this chapter, most navigators rely on these signals to set their chronometers and to compare stopwatches when at sea. The data entries in the various navigational almanacs, however, are tabulated for *Greenwich mean time* (GMT), which as discussed in *Marine Navigation 1* and in the following section is mean time reckoned according to the relationship of the mean sun with the Greenwich meridian. Fortunately, though, for most practical navigation at sea, the differences between UTC and GMT may be safely ignored, since even at times of maximum difference between UTC and GMT, the greatest positional error that could result is only .2 minutes of the longitude determined from celestial observations made at that time (equivalent to .2 mile near the equator).

Time and Longitude

In celestial navigation, as in the voyage planning discussed in *Marine Navigation 1,* time is a major concern of the navigator. For the most part, the navigator uses mean solar time reckoned according to the travel of the mean sun with respect to one of three reference meridians—the Greenwich or prime meridian, the central meridian of the time zone in which he is located, and the meridian passing through his position. Time reckoned according to the position of the mean sun relative to the prime meridian is called *Greenwich mean time,* abbreviated *GMT.* Time using the central meridians of the various time zones as reference meridians is called *zone time,* abbreviated *ZT,* and time using the observer's meridian as a reference is referred to as *local mean time,* or *LMT.* The first two types of time, GMT and ZT, are extensively discussed in Chapter 15 of *Marine Navigation 1;* local mean time, LMT, will be examined below. Before proceeding to the following material, it would be helpful to review the discussion of Greenwich mean and zone time in *Marine Navigation 1.*

Local mean time, as stated above, is based on the relationship of the mean sun with the observer's meridian. When the mean sun is at *lower transit* of the observer's meridian, meaning that it is passing over the lower branch of his meridian, the LMT is midnight. At *upper transit* of the mean sun, when the upper branch of the observer's meridian is crossed, it is noon in LMT. The concept of local mean time by itself does not have much importance in celestial navigation, but the relationship of local mean time with local zone time and with Greenwich mean time is of great importance in determining times of sunrise and sunset, moonrise and moonset, and local apparent noon.

Local mean time differs from zone time by the amount of time required for the mean sun to traverse the difference in longitude between the observer's meridian and the standard meridian of his time zone. As was shown in the preceding section, the mean sun is considered to move around the earth at the constant rate of 15° of arc (longitude) per hour, or 1° every four minutes. Thus, differences in longitude, written $d\lambda$, can be directly converted into differences in time. To facilitate such conversions of arc to time, the table partially shown in Figure 3-3 (next page) is included in the *Nautical Almanac.*

As an example of the use of the table in Figure 3-3, suppose that an observer located at longitude 64° 13.3′ W desired to compute the time required for the mean sun to travel from the central meridian of his time zone, 60° W, to his meridian. The difference in longitude is $64° \ 13.3' - 60° = 4° \ 13.3'$. Entering the first column of the table, a value of time of 0 hours 16 minutes is first extracted for 4° of the $d\lambda$. To find the additional time increment for the remaining 13.3 minutes of arc, the right-hand portion of the table is used. First, the horizontal row containing entries for 13′ is located, then the value under the column heading closest to the decimal fraction desired is read. Since .3 is closest to .25, the value for 13.3′ of 0 minutes 53 seconds is extracted. This result is then added to the value of time previously obtained for 4° to yield a final figure of 0 hours, 16 minutes, 53 seconds for the conversion of 4° 13.3′ of arc to time. Rounding off to the nearest minute, it could be stated that it will require about

CONVERSION OF ARC TO TIME

0°–59°		60°–119°		120°–179°		180°–239°		240°–299°		300°–359°			0′·00	0′·25	0′·50	0′·75
°	h m	°	h m	°	h m	°	h m	°	h m	°	h m	′	m s	m s	m s	m s
0	0 00	60	4 00	120	8 00	180	12 00	240	16 00	300	20 00	0	0 00	0 01	0 02	0 03
1	0 04	61	4 04	121	8 04	181	12 04	241	16 04	301	20 04	1	0 04	0 05	0 06	0 07
2	0 08	62	4 08	122	8 08	182	12 08	242	16 08	302	20 08	2	0 08	0 09	0 10	0 11
3	0 12	63	4 12	123	8 12	183	12 12	243	16 12	303	20 12	3	0 12	0 13	0 14	0 15
4	0 16	64	4 16	124	8 16	184	12 16	244	16 16	304	20 16	4	0 16	0 17	0 18	0 19
5	0 20	65	4 20	125	8 20	185	12 20	245	16 20	305	20 20	5	0 20	0 21	0 22	0 23
6	0 24	66	4 24	126	8 24	186	12 24	246	16 24	306	20 24	6	0 24	0 25	0 26	0 27
7	0 28	67	4 28	127	8 28	187	12 28	247	16 28	307	20 28	7	0 28	0 29	0 30	0 31
8	0 32	68	4 32	128	8 32	188	12 32	248	16 32	308	20 32	8	0 32	0 33	0 34	0 35
9	0 36	69	4 36	129	8 36	189	12 36	249	16 36	309	20 36	9	0 36	0 37	0 38	0 39
10	0 40	70	4 40	130	8 40	190	12 40	250	16 40	310	20 40	10	0 40	0 41	0 42	0 43
11	0 44	71	4 44	131	8 44	191	12 44	251	16 44	311	20 44	11	0 44	0 45	0 46	0 47
12	0 48	72	4 48	132	8 48	192	12 48	252	16 48	312	20 48	12	0 48	0 49	0 50	0 51
13	0 52	73	4 52	133	8 52	193	12 52	253	16 52	313	20 52	13	0 52	0 53	0 54	0 55
14	0 56	74	4 56	134	8 56	194	12 56	254	16 56	314	20 56	14	0 56	0 57	0 58	0 59
15	1 00	75	5 00	135	9 00	195	13 00	255	17 00	315	21 00	15	1 00	1 01	1 02	1 03
16	1 04	76	5 04	136	9 04	196	13 04	256	17 04	316	21 04	16	1 04	1 05	1 06	1 07
17	1 08	77	5 08	137	9 08	197	13 08	257	17 08	317	21 08	17	1 08	1 09	1 10	1 11
18	1 12	78	5 12	138	9 12	198	13 12	258	17 12	318	21 12	18	1 12	1 13	1 14	1 15
19	1 16	79	5 16	139	9 16	199	13 16	259	17 16	319	21 16	19	1 16	1 17	1 18	1 19
20	1 20	80	5 20	140	9 20	200	13 20	260	17 20	320	21 20	20	1 20	1 21	1 22	1 23
21	1 24	81	5 24	141	9 24	201	13 24	261	17 24	321	21 24	21	1 24	1 25	1 26	1 27
22	1 28	82	5 28	142	9 28	202	13 28	262	17 28	322	21 28	22	1 28	1 29	1 30	1 31
23	1 32	83	5 32	143	9 32	203	13 32	263	17 32	323	21 32	23	1 32	1 33	1 34	1 35
24	1 36	84	5 36	144	9 36	204	13 36	264	17 36	324	21 36	24	1 36	1 37	1 38	1 39
25	1 40	[illegible]	[illegible]	145	9 40	[illegible]	[illegible]	[illegible]	[illegible]	[illegible]	[illegible]	25	1 40	[illegible]		
26	[illegible]															
[illegible]																

Figure 3-3. Extract from the "Conversion of Arc to Time" table, Nautical Almanac.

17 minutes for the mean sun to traverse the indicated arc distance.

It is the constant rate of travel of the mean sun that makes conversions of arc to time, and thence conversions between zone time and local mean time, possible. Differences between *any* two solar mean times based on the travel of the mean sun can be thought of as being equal to the difference of longitude between their reference meridians, converted to units of time. Thus, GMT differs from zone time by the longitude of the standard meridian of the zone converted to time; GMT differs from local mean time by the longitude of the place converted to time; and zone time differs from local mean time by the arc-time difference corresponding to the difference of longitude between the standard meridian of the zone and the meridian of the observer. The only rule that must be remembered when applying arc-time differences is that a location that is east of another has a later time than the more westerly place, and a location that is west of another has an earlier time.

Several examples of conversion of local zone time to Greenwich mean time and vice versa, based on the difference of longitude between the standard meridian of the time zone and the Greenwich meridian, were given in Chapter 15 of *Marine Navigation 1.* In Figure 3-4A, suppose that observers at three meridians, 57° 45′ W, 60° W, and 64° W, wished to convert the local mean time of the sun's meridian passage at their meridians to the local zone times of apparent noon (LAN) at these same meridians on 25 May. Earlier it was shown that on this date the local mean time of upper transit of the sun was 1157. Thus, at each of the three meridians, the sun will transit at 1157 local mean time, as depicted in the figure.

Since the 60° W meridian is the standard meridian of this time zone,

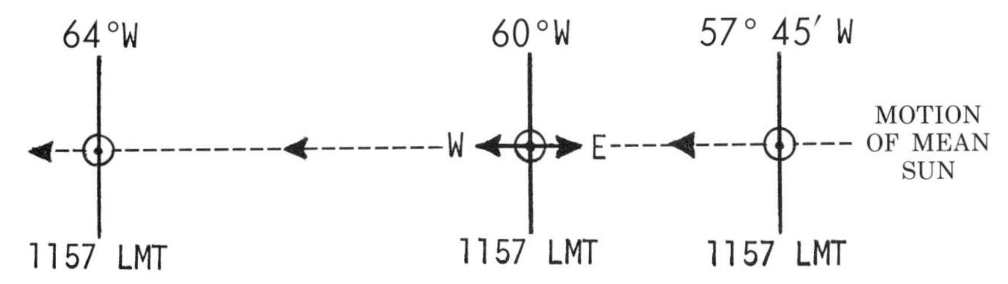

Figure 3-4A. Local mean time of noon is 1157 at each of three meridians in the +4 time zone on 15 May.

and local mean time and zone time are equivalent along the standard meridians of all zones, it follows that at 60° W, the LMT and ZT of meridian passage of the sun (LAN) are the same, 1157. All meridians within the zone east of 60° W will experience LAN at an earlier zone time, and all meridians to the west will experience LAN at a later zone time, with the differences being the amount of time it takes the mean sun to travel the arc between the local meridian and the standard meridian of the zone. From the arc-time conversion table in Figure 3-3, the time required for the mean sun to traverse the 2° 15′ of arc between 57° 45′ W and 60° W is 9 minutes. Hence, LAN will occur at 57° 45′ W at 1157 – 9 = 1148 zone time. In like manner, converting the 4° of arc between 60° and 64° W yields an arc-time difference of 16 minutes, which when applied to 1157 results in the zone time of LAN at 64° W of 1213. These results are indicated in Figure 3-4B.

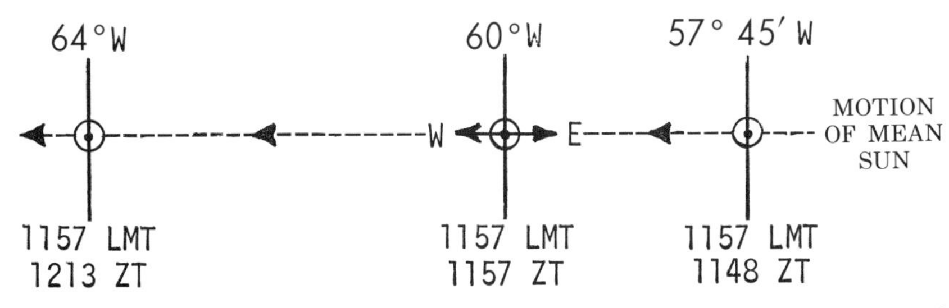

Figure 3-4B. LMT of apparent noon converted to ZTs at 57° 45′ W, 60° W, and 64° W.

The procedures for determination of the time of local apparent noon will be discussed in greater detail in Chapter 9.

The Format of a Written Time

In celestial navigation, when writing a complete time expressed in hours, minutes, and seconds, the format shown in Figure 3-5 is generally used. Hours, minutes, and seconds are expressed using two digits, and each quantity is separated from the others by a dash.

__ __ - __ __ - __ __

HOURS MINUTES SECONDS

Figure 3-5. Format of a 6-digit time.

As an example of a written time, if the apparent sun crossed the central 60° W meridian in the example in the preceding section at exactly 1157 zone time, it would cross an observer's meridian at 64° 13.3′ W at 12-13-53 zone time. If any quantities to be written are less than 10 in value, zeros are used to fill in the blank spaces, for example, 01-02-03 for 1 hour, 2 minutes, and 3 seconds. Whenever celestial observations are made, they are always recorded to the nearest second in the format given.

The Time Diagram

To assist the navigator in visualizing the time relationships existing among the hour circle of the sun, the hour circle of the celestial body being observed, the Greenwich meridian, and the meridian passing through his assumed position, an aid called the *time diagram* is frequently used. This diagram is especially helpful when converting from either local mean or local zone time to GMT, a conversion of basic importance because GMT is the time used in all almanacs as the basis for tabulation of the coordinates of all celestial bodies.

Essentially, a time diagram is nothing more than a sketch of the earth centered on the north or south pole, with the hour circles and meridians of interest depicted as radial lines. By convention, the south pole is generally selected as the center of the diagram, for ease in labeling the various hour angles and the meridian angle. As an example, consider the time diagram shown in Figure 3-6A. The circumference of the circle in Figure 3-6A represents the equator as seen from the south pole, P_s, located at the center. East is in a clockwise direction, and west counterclockwise; all celestial bodies, therefore, can be imagined as revolving in a counterclockwise direction about the circle. By convention the upper branch of the observer's meridian is always drawn as a solid vertical line extended up from the center; it is customarily labeled with a capital *M*, as shown. The dashed line extending down from the center is the lower branch of the observer's meridian, usually labeled with a lower-case *m*.

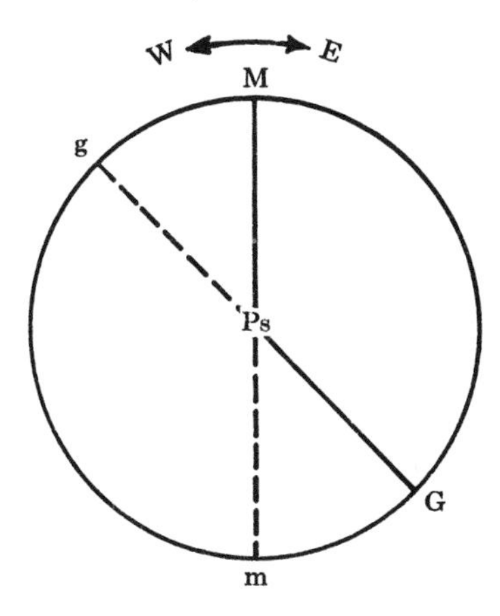

Figure 3-6A. A time diagram.

After the observer's meridian has been drawn, the Greenwich meridian is then located, based on the observer's assumed longitude. For this figure the observer's longitude was assumed to be 135° west, so the upper branch of the Greenwich meridian was drawn 135° clockwise from the observer's meridian M. The upper branch of the prime meridian is ordinarily labeled with a capital *G*, while the lower branch, represented by a second dashed line, is denoted by a lower-case g.

The hour circle of the sun is next located on the diagram by referring to the time of the observation being depicted. In positioning the sun's hour circle, the time diagram may be thought of as the face of a 24-hour clock, with m representing ZT 2400/0000, and M 1200 ZT. Thus, if it were 0600 zone time, the sun would be located 90° clockwise from M; if it were 1800, the sun would be 90° counterclockwise. For this example, suppose that it were 2100 local zone time. The hour circle of the sun would then be located as shown in Figure 3-6B, with the symbol ⊙ representing the sun itself. To complete the diagram, the hour circle of the celestial body being observed, if other than the sun, is plotted. To do this, its Greenwich hour angle must first be obtained from an almanac. As the first step of this process, the local zone time of the observation must be converted to GMT; it is here that the time diagram is of probably the greatest value.

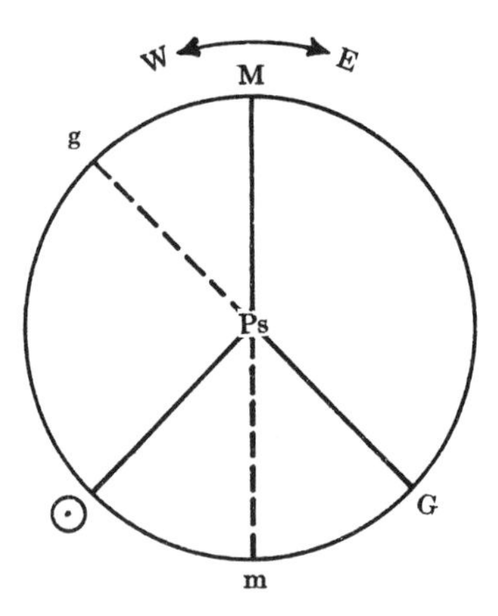

Figure 3-6B. Locating the sun on the time diagram.

Since the sun is the basis of Greenwich mean time as well as zone time, the approximate GMT and date can be determined by inspection from the time diagram. In the diagram of Figure 3-6B, for instance, the hour circle of the sun is located about 90° to the west of the lower branch of the Greenwich meridian. Hence, it must be about 6 hours after midnight at Greenwich, or 0600 GMT. Since the sun has not yet reached the lower branch of the observer's meridian, the date at Greenwich must be one day later than the date at the observer's position.

In general, a difference between the dates at Greenwich and at the location of the observer is always indicated by the time diagram if the hour circle of the sun falls between the angle formed by the lower branches of the Greenwich and the observer's meridians. The meridian whose lower branch is to the west of the sun will always have the earlier date.

After the GMT time and date of the observation have been calculated, the time diagram can then assist in determining the value of the local hour angle of the body observed; the importance of this determination will become apparent in later chapters.

To use the time diagram for this purpose, the hour circle of the body must first be plotted on the diagram. If a body other than a star is observed, the value of its Greenwich hour angle can be obtained from the daily pages of an almanac; its hour circle can then be located on the diagram relative to the Greenwich meridian. Figure 3-7A depicts the hour circles of the moon and the planet Venus (note the Venus symbol ♀) positioned on the time diagram of Figure 3-6B by use of their GHAs. The LHAs of the bodies are indicated.

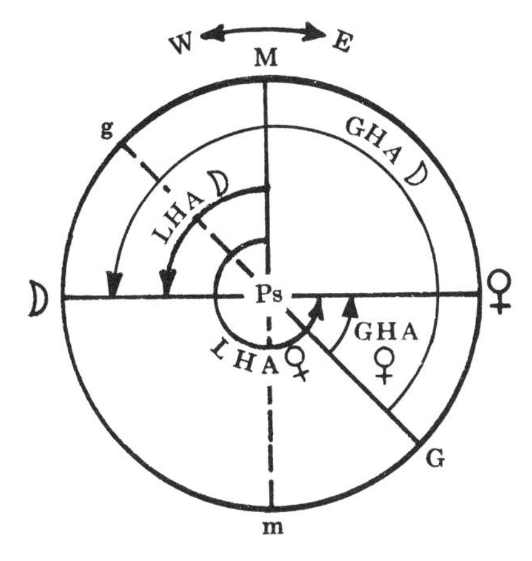

Figure 3-7A. Time diagram, showing the LHA of the moon and Venus.

If the body observed is a star, its GHA is not directly obtainable from an almanac, but rather it must be computed from the sum of the GHA of Aries plus the SHA of the star as explained in Chapter 1. Once the GHA has been thus determined, the hour circle of the star can be plotted on the diagram. In Figure 3-7B the hour circles of Aries and a star are shown on the time diagram, and the LHA of the star is identified.

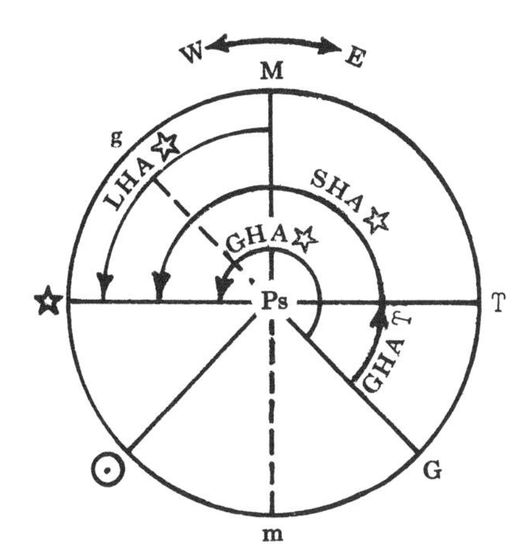

Figure 3-7B. Time diagram, showing the LHA of a star.

Additional examples of the use of the time diagram as an aid in converting local zone time to GMT and in determining the LHA of an observed body will be presented in Chapter 5.

Timing Celestial Observations

As mentioned earlier, the coordinates of celestial bodies are tabulated in almanacs with respect to the Greenwich mean time and date. It is necessary, therefore, that the GMT of each celestial observation be recorded to the nearest second, in order to be able to extract the required data from the almanacs with sufficient precision to make an accurate determination of position possible. There are two methods of determining the correct time to the required accuracy on board ship—the *chronometer* and *radio time signals*. The chronometer and an associated timepiece, the *comparing watch*, will be discussed in the remainder of this section, while radio time signals will be the subject of the balance of this chapter.

The chronometer is described in Chapter 7 of *Marine Navigation 1* on navigational instruments; its description will not be repeated here. Figure 7-14 (*Marine Navigation 1*) shows the Hamilton chronometer, the most widely used chronometer in the U.S. Navy at the present time. This and a number of other makes of chronometer are available commercially, with several models ranging in price from $150–300 featuring battery-powered quartz crystal movements accurate to within one minute per year.

Most Navy ships the size of a destroyer or larger have an allowance for three of these instruments. They are usually kept together in a specially constructed box in the chartroom. Several Navy publications, including *NavShips Technical Manual* and certain other NavShips directives contain instructions concerning the care, winding, transportation, and use of the chronometer. The models like the Hamilton chronometer that require periodic winding are wound once every 24 hours, to ensure that they will never intentionally run down. This responsibility is usually assigned to a quartermaster called the clock petty officer, who in addition to winding the chronometers, also periodically resets and winds all other standard wall clocks on board the ship.

Even a chronometer cannot keep absolutely exact time, but the characteristic that distinguishes this instrument from a normal clock or watch is that its rate of gain or loss of time is nearly constant. The difference between the time indicated on the chronometer or *chronometer time*, usually abbreviated *C*, and the correct GMT at any instant is referred to as the *chronometer error*, or *CE*. It is labeled either (F) or (S), depending on whether the chronometer time is ahead (faster than) or behind (slower than) GMT. Because chronometers are normally not reset on board ship, an accumulated error may become large in time. If, for example, the chronometer lost three seconds each day, in one 30-day month it would indicate a chronometer time, C, 1 minute and 30 seconds behind or "slower" than GMT. The CE in this case would be written (S) 01-30. Thus, if the chronometer read 01-02-03 at a particular time, the conversion to GMT would be as follows:

C	01-02-03
CE	+(S) 01-30
GMT	01-03-33

Slow chronometer errors are always *added* to chronometer time to obtain GMT, while *fast* errors are always *subtracted.*

OPNAVINST 3120.32 requires that every chronometer be checked for accuracy at least once each day, and the results entered into a log book called the *Navigational Timepiece Rate Book.* This procedure is normally accomplished on board ship by means of the radio time signal described in the next section. The navigator usually arranges for the time signal to be fed into the chartroom by the shipboard radio communications facility at a given time each day. The error of each chronometer relative to the broadcast time signal is observed and recorded into the *Rate Book.* Differences between successive daily chronometer errors are computed and entered for each day, and average daily rates are computed periodically by the following formula:

$$\text{ADR (Average daily rate)} = \frac{\text{Error on last day observed} - \text{error on first day observed}}{\text{difference in dates of observation}}$$

A sample page of the *Navigational Timepiece Rate Book* from a ship having three chronometers appears in Figure 3-8 on the next page.

It is humanly impossible to observe a chronometer error much more accurately than to the nearest half second, so the daily errors (successive daily rates) are entered in the *Rate Book* to this precision. This is the reason why the successive daily rates shown in Figure 3-8 are not all the same. For most chronometers, however, the average daily rate should not change much from one computation to another. If it does, it may indicate that the chronometer is in need of overhaul; it should be exchanged for a new chronometer from the chronometer pool when the ship returns to her home port.

Since the chronometer cannot be removed and brought topside when celestial observations are to be made, a small hand-held watch the size of a stopwatch called a *comparing watch* is often used to time actual observations. A typical comparing watch is shown in Figure 3-9 on the following page.

The comparing watch differs from a normal stopwatch in that it usually has a sweep hand that can be stopped at the moment of observation, while the remainder of the watch movement continues in mo-

DATE	A					B					C					OBSERVATION	
YEAR 19__	MAKE HAMILTON TYPE SIZE 85 SERIAL NO. 192 W					MAKE HAMILTON TYPE SIZE 85 SERIAL NO. 1135					MAKE HAMILTON TYPE SIZE 35 SERIAL NO. 156 B						
MONTH JULY	ERROR RELATIVE TO G.C.T. + = FAST − = SLOW			SUCCESSIVE DAILY RATES		ERROR RELATIVE TO G.C.T. + = FAST − = SLOW			SUCCESSIVE DAILY RATES		ERROR RELATIVE TO G.C.T. + = FAST − = SLOW			SUCCESSIVE DAILY RATES		LOCAL TIME TO NEAREST MINUTE	
DAY	±	MIN.	SECONDS	±	SECONDS	±	MIN.	SECONDS	±	SECONDS	±	MIN.	SECONDS	±	SECONDS	TIME	INITIALS
1	+	11	49.0	+	1.5	−	6	02.0	−	1.0	+	16	22.0	+	2.0	1130	RC
2	+	11	51.5	+	2.5	−	6	02.5	−	0.5	+	16	24.0	+	2.0	1130	RC
3	+	11	53.0	+	1.5	−	6	03.0	−	0.5	+	16	25.0	+	1.0	1125	RC
4	+	11	54.5	+	1.5	−	6	04.0	−	1.0	+	16	23.0	−	2.0	1130	GG
5	+	11	55.5	+	1.0	−	6	04.0		0.0	+	16	21.5	−	1.5	1130	GG
24	+	12	[illegible]	+	2.0	[illegible]	[illegible]	[illegible]	[illegible]	[illegible]	+	16	20.0	−	[illegible]	[illegible]	[illegible]
25	+	12	30.5	+	[illegible]	−	6	13.5		0.0	+	[illegible]	[illegible]	[illegible]	[illegible]	1130	RC
26	+	12	32.0	+	1.5	−	6	13.5		0.0	+	16	22.5	+	1.0	1130	RC
27	+	12	34.0	+	2.0	−	6	14.0	−	0.5	+	16	23.0	+	0.5	1135	RC
28	+	12	35.5	+	1.5	−	6	15.0	−	1.0	+	16	24.0	+	1.0	1130	RC
29	+	12	37.0	+	1.5	−	6	16.0	−	1.0	+	16	25.5	+	1.5	1140	GG
30	+	12	39.0	+	2.0	−	6	16.5	−	0.5	+	16	26.5	+	1.0	1130	GG
31	+	12	41.5	+	2.5	−	6	17.0	−	0.5	+	16	27.0	+	0.5	1130	DAH

+1.75 AVERAGE DAILY RATE −0.5 AVERAGE DAILY RATE +0.17 AVERAGE DAILY RATE

NOTE: FOR COMPUTATION OF AVERAGE DAILY RATE SEE PARAGRAPH 5 UNDER INSTRUCTIONS.

16—66873-1

Figure 3-8. Sample page from a Navigational Timepiece Rate Book.

tion. After the time of the observation has been recorded, the indicator sweep can then be returned to run with the second sweep of the main movement, until such time as it is again stopped to record the next observation. If a comparing watch is not available, an ordinary stop-watch may be used as the timing device. In either case, it is a good

Figure 3-9. A comparing watch.

practice to read and record first the seconds, then the minutes, and finally the hours for a celestial observation. This procedure is, of course, particularly desirable if the watch does not have the capability of being stopped to "freeze" the time of the observation.

When a round of celestial observations is to be made, the comparing or stopwatch is set as close as possible to either the local zone time (ZT) or to GMT, usually by reference to the chronometer. It is also possible to use a radio time signal to set the watch; many small-boatmen will do this, particularly if they use a comparing or a quartz wristwatch in lieu of a chronometer at sea. In either case, the difference between watch time (W) and ZT or GMT is then determined. This difference is the *watch error,* abbreviated *WE*; like the chronometer error discussed earlier, watch error is labeled either fast (F) or slow (S). After recording the watch time of each celestial observation, fast watch errors are subtracted, and slow watch errors added, in order to obtain the correct ZT or GMT for each observation. If the watch is set to local zone time, the zone time of each observation is converted to GMT by applying the zone difference figure.

In cases in which a comparing watch is set by reference to the chronometer, the usual procedure is to set the watch to the indicated chronometer time. The watch error is then identical to the algebraic sum of any difference between the watch time and chronometer time and the chronometer error.

As an example of the determination and application of watch error, suppose that for a given observation the navigator's comparing watch read 20-11-02. Earlier, the comparing watch had read 19-29-01 when the chronometer read 19-30-50. The chronometer error (CE) on GMT on the date in question is determined to be (F) 00-30-01. To obtain the GMT of the observation, it is first necessary to determine the watch error, which in this case is the difference between the time indicated by the watch and the correct GMT.

When the watch time (W) was compared to the chronometer time (C), the watch time was found to be 1 minute 49 seconds slower than the chronometer time:

C	19-30-50
W	19-29-01
C−W	(S) 01-49

To find the watch error (WE), this time difference (C − W) must be added algebraically to the chronometer error (CE):

C−W	(+S)	01-49
CE	(−F)	30-01
WE	(−F)	28-12

Having obtained the watch error, the GMT of the observation can be determined by applying the WE to the watch time of the observation:

W	20-11-02
WE	(−F)28-12
GMT	19-42-50

Again, a positive "plus" sign is associated with all slow errors, and a negative "minus" sign is used in conjunction with fast errors.

Radio Time Signals

Radio time signals are broadcast worldwide by radio stations of many foreign maritime nations and by two U.S. Bureau of Standards and several Navy stations. The two Bureau of Standards stations, WWV at Fort Collins, Colorado, and WWVH at Kauai, Hawaii, transmit continually, while Navy stations such as NAM at Norfolk, Virginia, and NPN at Guam, transmit periodically for only a few hours each day. As mentioned earlier in this chapter, the time signals transmitted by the U.S. and most foreign stations are based on Coordinated Universal Time (UTC), which can be considered as equivalent to GMT for most practical navigational purposes.

The DMAHTC *Publications 117A* and *117B, Radio Navigational Aids,* contain descriptions of the various time signals and the frequencies at which they are transmitted; the former covers U.S. and European stations receivable in the Atlantic and Mediterranean, and the latter those stations receivable in the Pacific and Indian Ocean regions. There are several standard time signal transmission formats used throughout the world. The WWV and WWVH format is depicted in Figure 3-10A. As indicated in the diagram, voice announcements of the UTC time are made each minute, with a female voice on WWVH and a male voice on WWV. During their hours of transmission, all Navy stations, as well as a number of foreign stations, transmit a tone pattern called the *United States System* shown in Figure 3-10B, beginning 5 minutes before each hour of UTC.

The value of DUT1 at the time of each broadcast by the U.S. stations is indicated by a pattern of either double ticks or emphasized second tones transmitted for each of the 15 seconds immediately following each whole minute tone. The number of second ticks or tones emphasized indicates the numerical value of DUT1. When DUT1 is positive, seconds 1 through 8 are emphasized, corresponding to a DUT1 value of from +.1 to +.8 seconds. When DUT1 is zero, no seconds are emphasized. When DUT1 is negative, seconds 9 through 15 are emphasized, corresponding to a DUT1 value of from −.1 to −.7 seconds. In

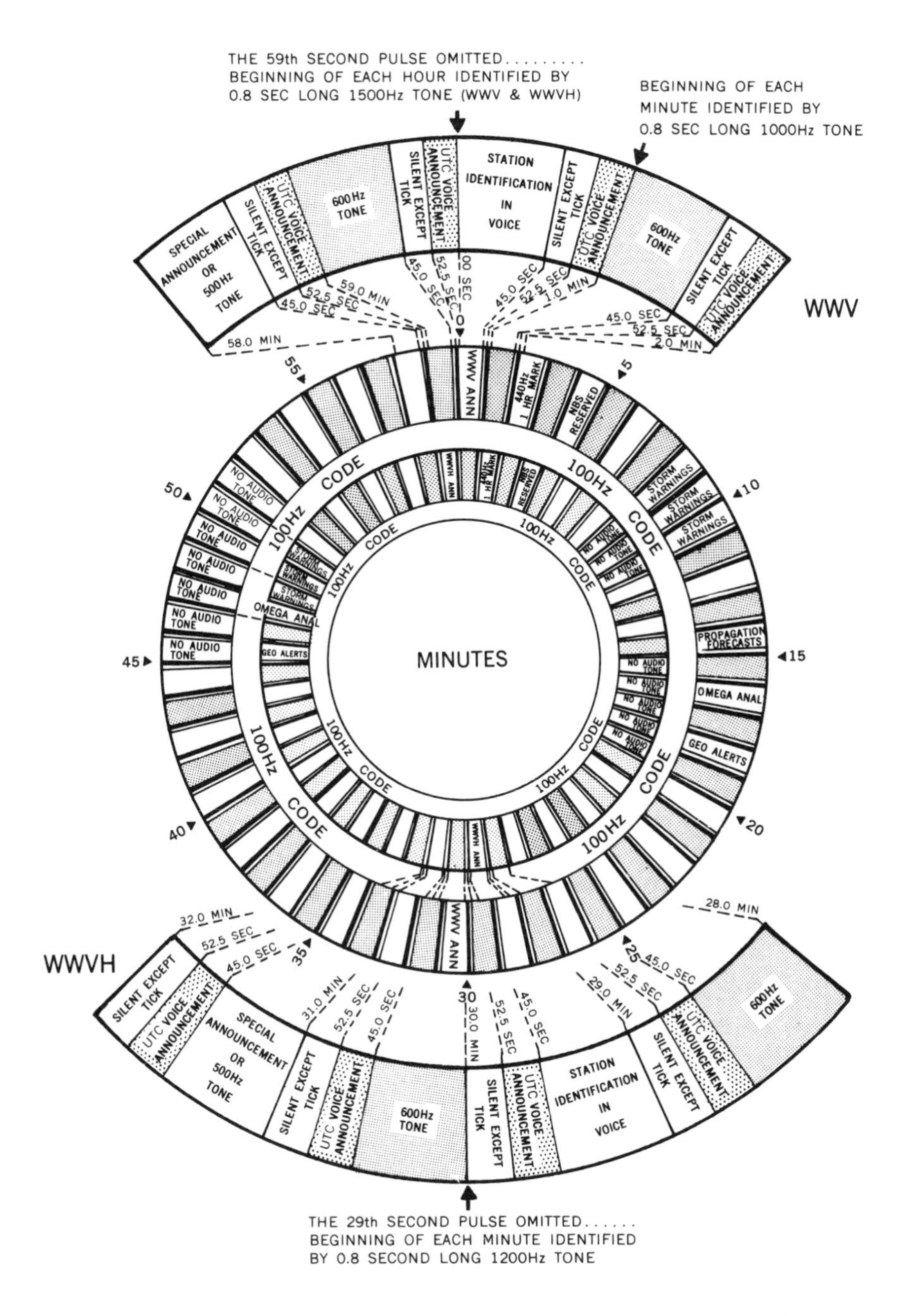

Figure 3-10A. The WWV and WWVH radio time broadcast format. Detailed information on the various segments of the format is contained in both volumes of the DMAHTC publication Radio Navigational Aids.

Minute	Second										
	50	51	52	53	54	55	56	57	58	59	60
55	-		-	-	-	-					-
56	-	-		-	-	-					-
57	-	-	-		-	-					-
58	-	-	-	-		-					-
59	-										—

Figure 3-10B. The United States System *of time signal tones.*

situations requiring extreme accuracy, the DUT1 correction thus obtained is added algebraically to UTC to yield GMT correct to within .1 second.

In practice, when one or more chronometers are carried on board a vessel at sea, a time signal is used to determine the error of the most accurate one once each day. The remaining chronometers, if any, are then compared with this reference chronometer to obtain their individual errors. In the case of the small-boatman who may not have a chronometer, he will generally use the time signal much more often to check the accuracy of his boat's timepieces, and to set his comparing or stopwatch prior to every round of celestial observations.

Summary

In this chapter, the fundamental aspects of time as it applies to the practice of celestial navigation have been discussed, including the bases by which time is reckoned, the time diagram, and the use of the chronometer, radio time signal, and comparing watch in recording the time of celestial observations. The material in succeeding chapters will relate the various applications of time and time theory to such diverse problems as the solution of the navigational triangle for a celestial LOP, the determination of times of rising and setting of the sun and moon, the calculation of the duration and commencement of twilight, and the computation of the time of local apparent noon.

4

The Marine Sextant

The instrument most commonly associated with the practice of celestial navigation is the *marine sextant.* The sextant can be simply described as a hand-held instrument designed to measure the angle between two objects with great precision. In practice, it is usually used to measure the altitudes of celestial bodies above the visible sea horizon, but, as stated in *Marine Navigation 1,* it can also be employed to measure the horizontal angle between two terrestrial objects in order to obtain a terrestrial line of position.

The name sextant is derived from the Latin *sextans,* meaning the sixth part; its arc is approximately one-sixth of a circle. Because of the optical principles incorporated in the sextant, however, modern instruments can measure angles up to about 145°. The sextant has been a symbol of the practice of navigation at sea for more than 200 years, and the professional navigator regards the quality of his sextant and the skill with which he uses it as sources of great pride.

Characteristics of the Marine Sextant

There are dozens of models of sextants of varying quality, precision, and cost in use by modern navigators; they range in price from about $25 for a plastic model useful for training and as a spare for emergency use, to over $1,200 for one of high-quality aluminum and brass construction. In the U.S. Navy, the Mark 2 Mod 0 model pictured in Figure 4-1 (next page) has been in common use for the last twenty years, and is still the model most often found on board most Navy ships.

The nomenclature of the principal parts of this sextant, which are identified by the letters in the figure, is representative of nearly all varieties of the micrometer drum sextant:

A The *frame* is usually constructed of either brass or aluminum in the form shown in the figure. It is the basic part of the sextant to which all others are attached.

B The *limb* is the bottom part of the frame, cut with teeth on which the micrometer drum rides.

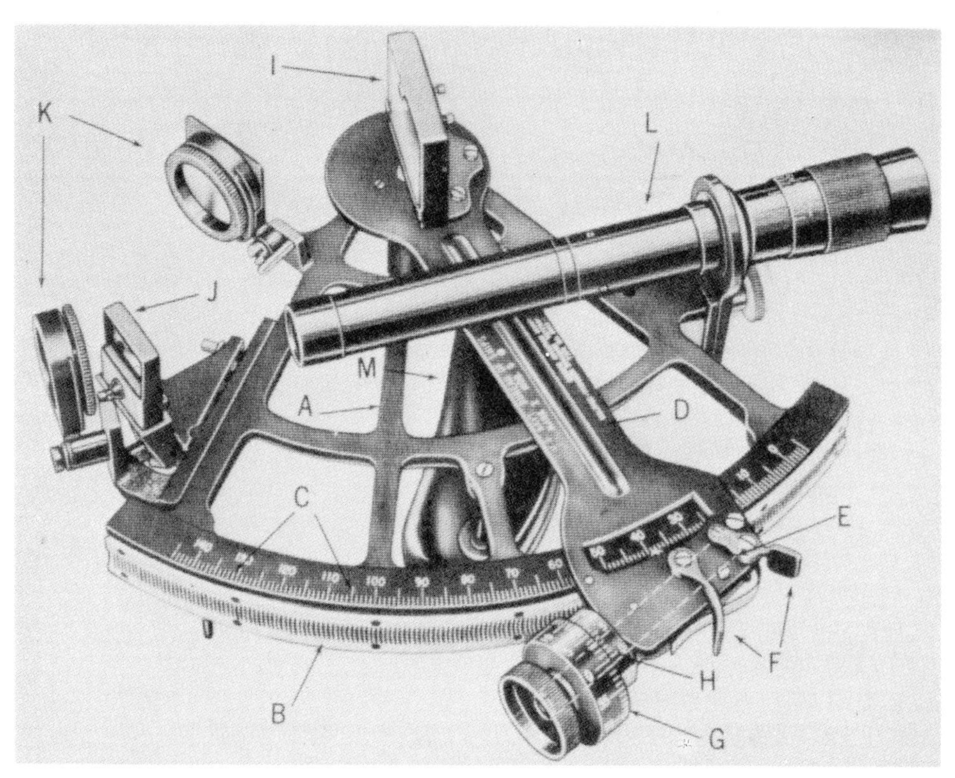

Figure 4-1. The U.S. Navy standard Mark 2 Mod O marine sextant.

C The *arc* refers to the altitude graduations of arc inscribed on the limb. On many sextants the arc is marked in a strip of brass, silver, or platinum inserted along the side of the limb.

D The *index arm* is a movable bar pivoted about the center of curvature of the limb, on which the index mirror and micrometer drum are fixed.

E The *tangent screw* (not visible in the figure) is a screw gear mounted on the end of the micrometer drum shaft; it engages the teeth of the limb. Turning the *micrometer drum* (G) turns the tangent screw, which in turn has the effect of moving the index arm along the arc of the sextant.

F The *release levers* are spring-actuated clamps that hold the tangent screw in place against the teeth of the limb. By compressing the release levers the tangent screw can be disengaged, allowing rapid movement of the index arm along the arc.

G The *micrometer drum* is graduated in 60 minutes of arc around its circumference; one complete turn of the drum moves the index arm one degree of altitude along the arc, thus allowing readings of minutes of arc between whole degrees to be made.

H The *vernier*, adjacent to the micrometer drum and fixed to the index arm, allows readings to be made to the nearest tenth of a minute of arc.

I The *index mirror* is a piece of silvered glass mounted on the index arm, perpendicular to the plane of the instrument, and centered directly adjacent to the pivot of the index arm.

J The *horizon glass*, similar in construction to the index mirror of a stadimeter, is constructed of silvered glass inserted in the right half of the horizon glass frame, perpendicular to the plane of the sextant.

K *Shade glasses* of variable darkness are mounted on the frame in front of the index mirror and horizon glass. They can be rotated into the line of sight between the mirrors and between the observer and the horizon to reduce the intensity of the light reaching the eye of the observer.

L The *telescope* screws into an adjustable collar in line with the horizon glass, and amplifies both the reflected and direct images observed.

M The *handle*, made of wood or plastic, is designed to be held in the right hand during sextant observations.

There has been an ongoing project underway for several years to develop an advanced sextant for shipboard use that would incorporate an electronic digital readout instead of a micrometer drum to indicate the angle measured on the sextant. It would have the capability of accepting a night vision telescope for observations using the visible horizon after dark, or a device that could provide an artificial horizon if the natural horizon is obscured by clouds or fog. Eventually, the electronic output from this sextant would be fed directly into a small computer for virtually instantaneous solution for the resulting celestial LOP; several such LOPs would then be compared by mathematic algorithms to produce a computer solution for the celestial fix. Unfortunately for practitioners of celestial navigation, the increased availability of and tendency for reliance upon modern long-range electronic position-fixing systems and equipment has caused this development program to be deemphasized somewhat, so it may be several more years before this space-age sextant becomes available for general use.

Optical Principle of the Sextant

The optical principle of the sextant is illustrated in Figure 4-2, with the solid line representing the path of an incoming light ray from a celestial body being observed. The instrument is constructed in such a way that the angle BDC between the body and the horizon is always equal in value to twice the angle between the index mirror and horizon glass, angle BGC, which is measured along the arc of the sextant. Thus, an arc encompassing one-sixth of a circle can be graduated so that angles up to 120° can be read. The arc of most modern sextants is extended slightly beyond one-sixth of a circle, so that angles up to 145° can be measured.

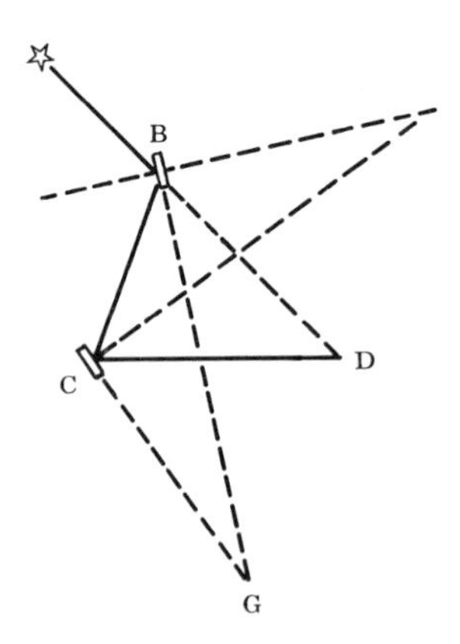

Figure 4-2. Optical principle of the sextant.

Reading the Sextant

When an angle representing either the altitude of a celestial body or the difference in bearing of two terrestrial objects is to be read with the sextant, a three-step procedure is always used. The sample sextant reading pictured in Figure 4-3 will illustrate the steps. First, the number of whole degrees is read by observing the position of the arrow on the index arm in relation to the arc. In this case, the arrow lies between 29° and 30°, so a value of 29° is obtained. Next, the minutes are read by noting the position of the zero mark on the vernier with respect to the graduations of the micrometer drum. In Figure 4-3 the zero falls between 42′ and 43′; hence, a value of 42′ results. Finally, the tenths of a minute are read by noting which of the 10 marks on the vernier is most nearly opposite one of the graduations on the micrometer drum. In this case, a value of .5 is indicated. Thus, the angle depicted in the figure is 29° 42.5′.

If the index arm arrow is very close to a whole degree mark on the arc, care must be taken to obtain the correct angle by referring

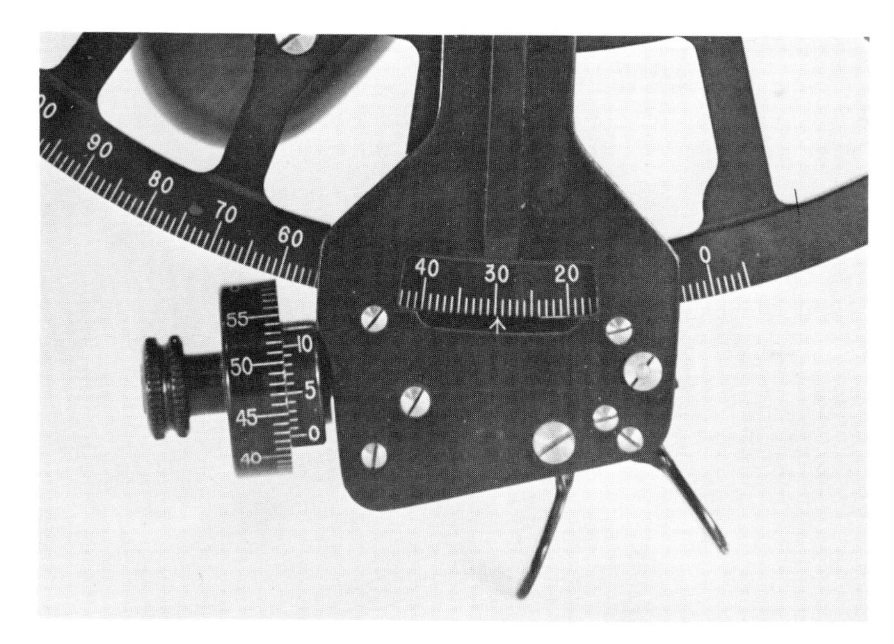

Figure 4-3. A sextant angle of 29° 42.5′.

to the micrometer drum. If the arrow were pointing directly at 45° on the arm, for instance, and 57′ on the micrometer drum were opposite the zero of the vernier, the correct reading would be 44° 57′, not 45° 57′. Similarly, any doubt as to the correct minute can be resolved by noting the fraction of a minute indicated on the vernier.

Preparation for Use of the Sextant

Prior to going to sea, the navigator should always check his sextant carefully to ensure that all correctable mechanical error that may be present in his instrument is eliminated. There are seven major sources of this so-called *instrument error* in the micrometer drum sextant. Of these seven, four are adjustable, and three are not. The adjustable instrument errors include the following:

Lack of perpendicularity of the frame and index mirror;

Lack of perpendicularity of the frame and horizon glass (side error);

Nonparallelism of the index mirror and horizon glass with each other at the zero setting; and

Nonparallelism of the telescope with the frame (collimation error).

The major nonadjustable instrument errors are:

Prismatic error, occurring because of nonparallelism in the faces of the shade glasses, index mirror, and horizon glass;

Graduation error resulting from improper calibration of the scales of the arc, micrometer drum, or vernier; and

Centering error as a result of the index arm not being pivoted at the exact center of curvature of the arc.

In the U.S. Navy it is the usual practice to leave the adjustment of the sextant to the optical repair facility of either a tender or a shipyard, but if necessity requires, an experienced navigator may attempt to adjust the instrument himself. The *American Practical Navigator* (Bowditch) and *Dutton's Navigation and Piloting* both contain an excellent description of the adjustment procedures, and can be used as a reference if the need arises.

After all adjustable errors have been reduced or eliminated insofar as possible, the sextant will still retain some residual variable adjustable error as well as a small fixed nonadjustable instrument error. Additionally, a small variable error called *personal error* may often be produced as a result of the eye of the observer acting in conjunction with the optical system of the sextant. In practice, the fixed component of instrument error and any personal error are usually considered to be insignificantly small for most practical purposes.

One component of variable instrument error, however, must be taken into account each time sextant observations are to be made. This is *index error,* which results from nonparallelism of the horizon glass and index mirror. Inasmuch as this error continually changes over time as external conditions vary, it can never be completely adjusted out, and it is usually of a significant size. Index error, therefore, must be separately determined on each occasion that the sextant is to be used for a round of observations.

Figure 4-4A. Sextant set at zero, no index error.

By day, the index error is best determined by an observation of the sea horizon. With the sextant set at 0° 0.0′ of arc, the horizon should appear as shown in Figure 4-4A. If the horizon appears as in Figure 4-4B, there is some index error present. To determine the magnitude and direction of the error, the micrometer drum is slowly rotated until the direct and reflected images of the horizon are adjacent. If after so doing the micrometer drum reads *more than* 0.0′, the error is *positive;* conversely, if the reading on the drum is *less than* 0.0′, the

Figure 4-4B. Sextant set at zero, index error present.

error is *negative.* In the former case, all angles read on the instrument would be too large, and in the latter case, all angles would be too small. Consequently, when a sextant has a *positive* index error, sometimes referred to as being "on the arc," all subsequent sextant altitudes must be corrected by *subtracting* the amount of the error. When a sextant has a *negative* ("off the arc") index error, all subsequent observations must be corrected by *adding* the amount of the error. This correction is referred to as the *index correction,* abbreviated IC; it is always equal in amount but opposite in sign to the index error. The following mnemonic aid has been coined to aid in determining the sign of the index correction:

> When it (the index error) is *on* (positive), it (the IC) is *off;* when it's *off,* it's *on.*

As an example, suppose that the navigator observed his visible horizon and found that when it was aligned as in Figure 4-4A, his sextant read as shown in Figure 4-5 on the following page. Here, the micrometer drum has been rotated on the negative side of the 0° mark to 57.5′. Hence, the index error is −(60.0′ − 57.5′) or −2.5′ "off the

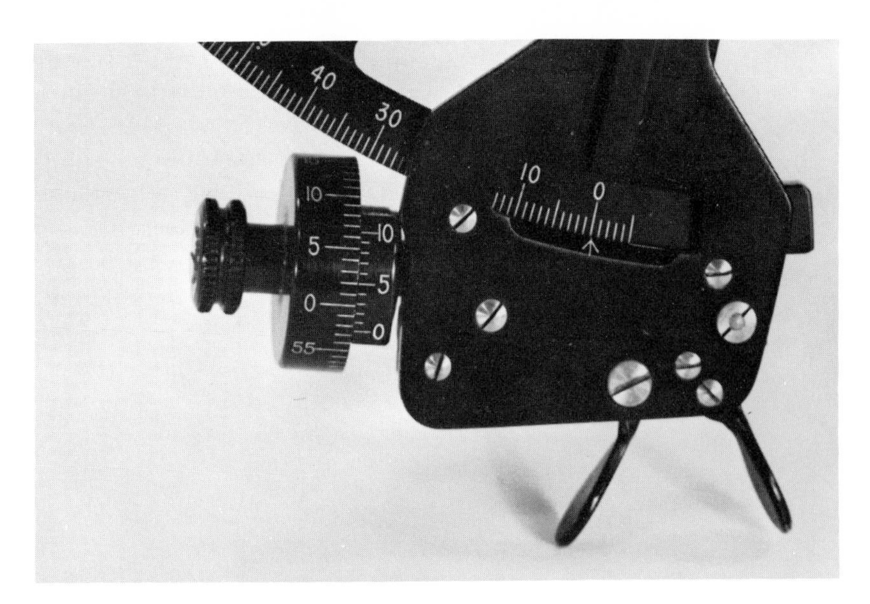

Figure 4-5. A sextant index error.

arc." All subsequent sextant altitudes, therefore, must be corrected by adding "on" an index correction, IC, of +2.5′.

At night, the index error may be determined by observing a star. The direct and indirect images are either brought into coincidence or aligned directly alongside one another as described above for the horizon, and the amount by which the micrometer drum was rotated away from the zero setting to achieve this result is the index error.

Other examples of the application of the index correction to the altitude of a celestial body as observed with the sextant will be given later in this chapter.

The Techniques of Sextant Altitude Observations

After having determined the index error of his instrument, the navigator is ready to proceed with the altitude observations of his selected celestial bodies. In Chapter 1, the observed altitude of a celestial body, Ho, is defined as the angle formed at the center of the earth between the line of sight to the body and the plane of the observer's celestial horizon. This angle is always measured along the vertical circle passing through the observer's zenith and nadir and the body under observation.

Altitude observations made with the hand-held marine sextant measure the vertical angle along the vertical circle between the observer's visible sea horizon and the body at the time of observation. The angle thus measured is termed the *sextant altitude* of the body;

it is abbreviated *hs.* Because the observer is located on the surface of the earth rather than at its center, the observer's visible sea horizon and his celestial horizon are not coincident. In fact, as will be discussed in more detail later, the line of sight to the observer's sea horizon in most cases is not even parallel to the plane of his celestial horizon, owing to his height of eye above the earth's surface. Hence, certain corrections to be discussed later in this chapter must always be applied to the sextant altitude hs to obtain the desired observed altitude Ho. In the remainder of this section, the technique of using the sextant to obtain the sextant altitude hs of the sun, moon, planets, and stars will be examined.

When the sun is to be observed, the sextant is initially set at 0°, and then it is held vertically in the right hand with the line of sight directed at the sea horizon below the position of the sun. Suitable shade glasses are moved into position, depending on the brightness of the horizon and of the sun, and the index arm is moved outward by means of the release levers until the reflected image of the sun is brought down and appears in the horizon glass roughly alongside the direct view of the horizon. Next, the micrometer drum is slowly rotated until the sun appears to be resting exactly on the horizon. To check the perpendicularity of the sextant, the instrument should then be tilted slightly to either side around the axis of the telescope, causing the sun to appear to swing in an arc across the horizon glass, as in Figure 4-6A; this process is called *swinging the arc.*

Figure 4-6A. Swinging the arc of a sextant.

After swinging the arc several times, the image of the sun should be adjusted to its final position at the bottom of the arc tangent to the horizon as shown in Figure 4-6B, and the altitude should be read.

Figure 4-6B. The sun at the instant of tangency.

The lower edge of the sun is referred to as its *lower limb,* and the upper edge, its *upper limb.* Although most navigators prefer to observe the lower limb of the sun when measuring the sun's altitude, it is also possible to obtain an altitude by observation of the upper limb. For an upper limb observation, the sun's image is brought below the horizon until the upper limb is tangent. Upper limb observations are particularly useful if the lower limb of the sun is obscured by clouds, making only the top portion of the sun's disk sharply defined.

When the moon is observed, the same procedure outlined above is employed, except that the shade glasses are not used. As is the case with the sun, either the upper or lower limb may be brought tangent to the visible horizon, with upper limb observations being the only technique available if the phase of the moon is such that only the upper limb is illuminated.

When a star or planet is observed, a different procedure than the one just described for the sun or moon must usually be employed, because of the reduced intensity and small apparent size of these bodies. There are three methods that can be used to observe a star or planet. In the first method, the altitude and true azimuth of the body at the approximate time of observation are first precomputed by means of a set of tables or a device called a *starfinder;* both of these methods of precomputation of the position of a star or planet will be examined in a later chapter. When the time of observation draws near, the sextant is set to the predetermined altitude of the body and aimed in the direction of the precomputed true azimuth. The body should then appear in the horizon glass approximately on the visible horizon. By swinging the arc and adjusting the micrometer drum, the center of the body is brought into exact coincidence with the horizon as shown in Figure 4-7, and its altitude is read.

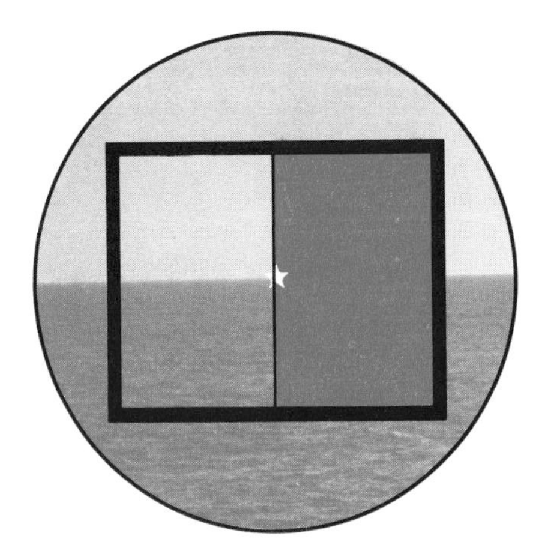

Figure 4-7. A star at the moment of altitude observation.

The second method of observation of a star or planet used if the altitude and true azimuth have not been precomputed is to set the sextant at zero and elevate it until the body appears in the field of view of the telescope. The release levers are then compressed and the index arm gradually moved forward, while at the same time the sextant is slowly depressed. In this manner, the body is kept always in the telescope field, until the horizon appears and is brought approximately tangent to the star. At this point, the release levers are expanded, and the micrometer drum is used to align the center of the body exactly on the visible horizon.

The third method that can be used to bring a star or planet into the field of view for observation consists of inverting the sextant and sighting the body through the telescope, as shown in Figure 4-8 on the following page. The release levers are then compressed with the fingers of the right hand, and the index arm is rotated forward until the horizon is brought "up" to approximate tangency with the body in the horizon glass. The sextant is then shifted to its normal vertical position in the right hand, with care being taken not to disturb the setting of the index arm. The desired star or planet should appear about on the horizon, and the micrometer drum is then used to adjust the final position of the body.

The choice of the method used for star or planet observations depends on the preference of the individual navigator and the circumstances under which the observations are being conducted.

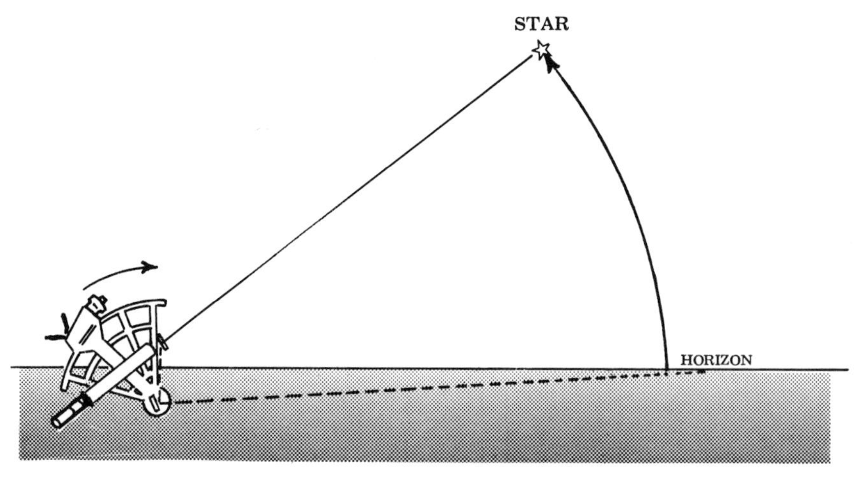

Figure 4-8. Bringing the horizon up to a star.

When a sextant observation is to be made, it is the usual practice for the navigator to make the actual observation and for an assistant to record the exact watch time at the moment of observation. The observer should give a "Standby," when his measurement is nearly complete, and a "Mark!" at the instant when he makes the reading. The navigator then lowers his sextant and reads off the altitude to the nearest tenth of a minute, while the time is recorded to the nearest second.

Much practice is usually required with the sextant before the observation technique of the inexperienced navigator is perfected. Precision in the observation of the sextant altitude of a celestial body is very important. In the altitude-intercept method of plotting the celestial LOP, every minute of error in the observed altitude of a celestial body causes the plotted line of position to be off by one mile.

Care of the Sextant

The modern marine sextant is a well built, very precise optical instrument capable of rendering years of service if it is properly maintained. Its usefulness can be greatly impaired, however, by careless handling or neglect. If the sextant is ever dropped, some error is almost certain to be introduced in all subsequent sightings.

When not in use, the sextant should always be kept secured in its case, and the case itself should be securely stowed in a location free from excessive heat, dampness, and vibration. In particular, the sextant should never be left unattended on the chartroom chart table or other furnishings.

Next to careless handling, moisture is the greatest enemy of the sextant. The sextant mirrors and telescope lens should always be wiped dry with a piece of lens paper after each use; cloth of any type tends to retain dust particles that may scratch the mirror or lens surface. Moisture has an extremely deleterious effect both on the silvering of the mirrors and the graduations of the arc. Should resilvering of the mirrors become necessary, this task, like instrument alignment, is best left to an optical instrument repair facility. Materials can be procured to perform resilvering of the mirrors on board ship, however; the *American Practical Navigator* contains a description of the resilvering procedure. The arc and teeth of the limb should always be kept lightly coated with a thin film of oil or petroleum jelly.

If the sextant is normally stowed in an air-conditioned space and the ship is operating in a humid climate, it is a good practice to bring the sextant in its case out into the open well before use to prevent condensation from forming on the mirror surfaces.

Sextant Altitude Corrections

For purposes of solving the navigational triangle in order to obtain a celestial line of position, the sextant altitude hs must be corrected to the value that would represent the observed altitude of the center of the body above the celestial horizon for an observer located at the center of the earth. This observed altitude is symbolized Ho. The corrections that must be applied to the sextant altitude hs to obtain the observed altitude Ho may be grouped into five categories, each of which will be described in a following section of this chapter:

Corrections for inaccuracy in reading the sextant;

Corrections for deviation from the horizontal reference plane;

Corrections for bending of light rays from the body;

Adjustment to the equivalent reading at the center of the body;

Adjustment to the equivalent reading at the center of the earth.

As will be explained below, not all of these corrections are applied to each sextant observation, but every sextant altitude must be adjusted by certain of them. To aid in the systematic application of the appropriate corrections for each body, a standardized *sight reduction form* is usually employed. Several examples of the application of the various corrections using a sight reduction form developed by the U.S. Naval Academy will be presented in the following pages, for each of the types of bodies observed in celestial navigation.

Corrections for the effects listed above that must be applied to the sextant altitude hs to obtain the desired observed altitude Ho are obtained from a publication known as an *almanac.* Two such almanacs are used in the normal practice of celestial navigation by U.S. Navy and most civilian navigators; each will be the subject of a later chapter of this text. In the *Nautical Almanac,* designed primarily for surface navigators, all corrections to the sextant altitude are tabulated precise to the nearest tenth of a minute of arc. In the *Air Almanac,* designed primarily for air navigators, corrections are precise only to the nearest minute. In marine navigation it is usually desirable to take advantage of the increased accuracy afforded by using the more precise tables of the *Nautical Almanac,* whereas in air navigation the higher speeds of travel render this degree of precision impractical. In the following sections, the use of the *Nautical Almanac* to obtain all necessary corrections for each type of celestial body observed will be featured, as this text is oriented primarily to the practice of marine surface navigation.

Corrections for Inaccuracy in Reading the Sextant

Corrections compensating for inaccuracy in reading the sextant must be applied to every sextant altitude observed. The sources of error leading to inaccuracy in reading the sextant have already been discussed earlier in this chapter: fixed instrument error, personal error, and variable index error. Of these three, only the last—index error—is of significance for most sextant observations.

As mentioned previously, the *index correction,* or *IC,* is always equal in amount to the index error, but opposite in sign. Hence, if a sextant micrometer drum setting were 01.5′ on the arc when the horizon was aligned in the horizon glass, the following expression would represent a sextant altitude of 34° 31.6′ corrected for index error:

hs	34° 31.6′
IC	−01.5′
	34° 30.1′

Once determined, the index correction is considered constant for all angles subsequently read with the sextant on that occasion.

Correction for Deviation from the Horizontal Reference Plane

As is the case with the index correction, corrections for the difference between the line of sight to the observer's sea horizon and a horizontal reference plane parallel to his celestial horizon must be made for every sextant altitude observation. The necessity for this correction is illustrated in Figure 4-9.

In Figure 4-9, the angle measured with the sextant between the incoming light ray from a celestial body and the observer's visible horizon is the sextant altitude hs. As can be seen, it is always larger than the *apparent altitude ha* between the incoming ray and a horizontal reference plane parallel to the observer's celestial horizon, by the amount of dip of the visible horizon beneath the horizontal plane. This *dip angle,* as it is called, results primarily from the height of eye of

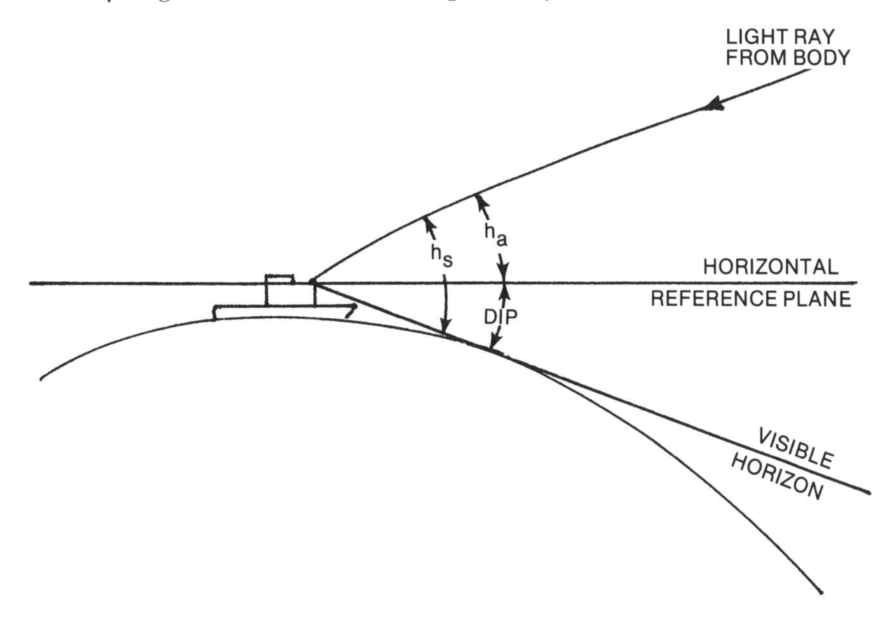

Figure 4-9. Necessity for dip correction illustrated. The sextant altitude hs *will always be greater than the apparent altitude* ha *by the amount of the dip angle.*

the observer and the curvature of the earth and is exaggerated somewhat by atmospheric refraction near the surface. As his height above the earth's surface increases, the distance to his visible sea horizon, and therefore the size of the dip angle, also increases. Because the sextant altitude ha is used as the entering argument in the *Air* and *Nautical Almanacs* for all additional corrections required to produce the desired observed altitude Ho, all altitudes observed with the sextant must be decreased by the amount of the dip angle. This correction, known as the *dip correction* or simply *dip,* is therefore always negative; its amplitude slowly increases with the increased height of eye of the observer.

Although an approximation to the dip correction could be determined mathematically using spherical trigonometry, the usual procedure is simply to look up its value in a table such as that which appears on the inside front cover of the *Nautical Almanac.* An excerpt from this table appears in Figure 4-10 on the following page.

The values of dip in the table are computed by the formulas

$$D = -0.97\sqrt{h} \quad \text{or} \quad D = -1.76\sqrt{m}$$

where h is the height of the observer in feet and m is the height of the observer in meters; normal conditions of atmospheric refraction are assumed. It has been found that under unusual atmospheric refraction conditions, such as those existing in polar regions or in mid-latitudes after the passage of a squall line, the actual values of dip may be up to 15′ in excess of the tabulated value computed by these formulas. Hence, the navigator should always be aware of the possibility that the actual dip may be greater than the tabulated value when unusual atmospheric conditions prevail.

To use the dip table shown in Figure 4-10, the observer's height of eye expressed either in meters or in feet is used as the entering argument. The basic table is a *critical value table,* meaning that

DIP				
Ht. of Eye	Corrⁿ	Ht. of Eye	Ht. of Eye	Corrⁿ
m	′	ft.	m	′
2·4		8·0	1·0	− 1·8
	−2·8		1·5	− 2·2
2·6		8·6	2·0	− 2·5
	−2·9		2·5	− 2·8
2·8		9·2	3·0	− 3·0
	−3·0		See table ←	
3·0		9·8	m	′
	−3·1		20	− 7·9
3·2		10·5	22	− 8·3
	−3·2		24	− 8·6
3·4		11·2	26	− 9·0
	−3·3		28	− 9·3
3·6		11·9	30	− 9·6
	−3·4		32	−10·0
3·8		12·6	34	−10·3
	−3·5		36	−10·6
4·0		13·3	38	−10·8
	−3·6		40	−11·1
4·3		14·1	42	−11·4
	−3·7		44	−11·7
4·5		14·9		11·9
	−3·8			
4·7		15·7		
	−3·9			
5·0		16·5		
	−4·0			
5·2		17·4		
	−4·1			
5·5		18·3		
	−4·2			
5·8		19·1		
	−4·3			
6·1		20·1		
	−4·4			
6·3		21·0		
	−4·5			
6·6		22·0		
	−4·6			
6·9		22·9		
	−4·7			
7·2		23·9		
	−4·8			
7·5		24·9		
	−4·9			
7·9		26·0		
	−5·0			
8·2		27·1		
8·5				

Figure 4-10. Excerpt from dip table, Nautical Almanac.

corrections to be extracted from the table are tabulated for ranges of values of the entering argument. The dip correction extracted is the one tabulated for the interval containing the actual height of eye of the observer. If the observer's height of eye falls exactly on a tabulated height of eye argument, the dip correction corresponding to the preceding interval is extracted.

If the observer's height of eye is less than 2.4 meters or 8.0 feet, or greater than 21.4 meters or 70.5 feet, the inserts to the right of the basic critical value table are used with interpolation to determine the dip correction. Thus, if the observer's height of eye were 2.3 meters, an interpolated dip correction of −2.7′ would be obtained from the table insert; if the height of eye were 4.8 meters, a dip correction of −3.9′ would be extracted from the basic table; and if the height of eye were 26.0 feet, an exact tabulated argument, the dip correction −4.9′ from the preceding interval would apply.

The inside back cover of the *Nautical Almanac* also contains a small dip table identical to the basic table inside the front cover. It is convenient to use with the adjacent moon "Altitude Correction Tables" used for observations of the moon.

If a clear view of the sea horizon is not obtainable because of close proximity of a ship or a shoreline, an altitude can be measured by using the waterline of the obstruction as a reference horizon if its distance from the observer is known. In such cases, Table 22 of the *American Practical Navigator* (Bowditch) may be used to obtain the value of the dip correction; this table is partially reproduced in Figure 4-11 on the next page. As an example of the use of this table, if the sea horizon of an observer with a height of eye of 30 feet were obscured by a column of ships two miles distant, the dip correction extracted from Table 22 in Figure 4-11 would be −9.3′. This contrasts with a value of −5.3′ determined for a similar height of eye from the dip table of the *Nautical Almanac* for the visible sea horizon.

After the value of the dip correction has been determined, it is usually added algebraically to the IC. The resulting aggregate correction is then applied to the sextant altitude hs to obtain the apparent altitude ha of the body. As mentioned above, all remaining corrections to be applied in order to obtain the observed altitude Ho are tabulated in the *Air* and *Nautical Almanacs*, using ha as the entering argument.

Corrections for Bending of Light Rays from the Body

Light is assumed to travel through a transparent medium of uniform properties in a straight line at a constant speed. When a light ray passes into a medium of different properties, particularly different density, the speed of light, it is theorized, changes slightly. Moreover, if this ray enters the second medium at an angle, the change of speed does not take place simultaneously across the ray. The effect of this sequential change of speed across the light ray is to cause it to change direction upon entering the second medium; this change in direction is termed *refraction.* A light ray entering a denser medium at an oblique angle is bent toward a line perpendicular to its surface, and a ray entering a medium of less density is bent away from the perpendicular. The greater this *angle of incidence* between the incoming light ray and the perpendicular to the surface of the new medium, the greater will be the angle of change of direction or *angle of refraction.*

TABLE 22

Dip of the Sea Short of the Horizon

Dis-tance	Height of eye above the sea, in feet										Dis-tance
	5	10	15	20	25	30	35	40	45	50	
Miles	′	′	′	′	′	′	′	′	′	′	*Miles*
0. 1	28. 3	56. 6	84. 9	113. 2	141. 5	169. 8	198. 0	226. 3	254. 6	282. 9	0. 1
0. 2	14. 2	28. 4	42. 5	56. 7	70. 8	84. 9	99. 1	113. 2	127. 4	141. 5	0. 2
0. 3	9. 6	19. 0	28. 4	37. 8	47. 3	56. 7	66. 1	75. 6	85. 0	94. 4	0. 3
0. 4	7. 2	14. 3	21. 4	28. 5	35. 5	42. 6	49. 7	56. 7	63. 8	70. 9	0. 4
0. 5	5. 9	11. 5	17. 2	22. 8	28. 5	34. 2	39. 8	45. 5	51. 1	56. 8	0. 5
0. 6	5. 0	9. 7	14. 4	19. 1	23. 8	28. 5	33. 3	38. 0	42. 7	47. 4	0. 6
0. 7	4. 3	8. 4	12. 4	16. 5	20. 5	24. 5	28. 6	32. 6	36. 7	40. 7	0. 7
0. 8	3. 9	7. 4	10. 9	14. 5	18. 0	21. 5	25. 1	28. 6	32. 2	35. 7	0. 8
0. 9	3. 5	6. 7	9. 8	12. 9	16. 1	19. 2	22. 4	25. 5	28. 7	31. 8	0. 9
1. 0	3. 2	6. 1	8. 9	11. 7	14. 6	17. 4	20. 2	23. 0	25. 9	28. 7	1. 0
1. 1	3. 0	5. 6	8. 2	10. 7	13. 3	15. 9	18. 5	21. 0	23. 6	26. 2	1. 1
1. 2	2. 9	5. 2	7. 6	9. 9	12. 3	14. 6	17. 0	19. 4	21. 7	24. 1	1. 2
1. 3	2. 7	4. 9	7. 1	9. 2	11. 4	13. 6	15. 8	17. 9	20. 1	22. 3	1. 3
1. 4	2. 6	4. 6	6. 6	8. 7	10. 7	12. 7	14. 7	16. 7	18. 8	20. 8	1. 4
1. 5	2. 5	4. 4	6. 3	8. 2	10. 0	11. 9	13. 8	15. 7	17. 6	19. 5	1. 5
1. 6	2. 4	4. 2	6. 0	7. 7	9. 5	11. 3	13. 0	14. 8	16. 6	18. 3	1. 6
1. 7	2. 4	4. 0	5. 7	7. 4	9. 0	10. 7	12. 4	14. 0	15. 7	17. 3	1. 7
1. 8	2. 3	3. 9	5. 5	7. 0	8. 6	10. 2	11. 7	13. 3	14. 9	16. 5	1. 8
1. 9	2. 3	3. 8	5. 3	6. 7	8. 2	9. 7	11. 2	12. 7	14. 2	15. 7	1. 9
2. 0	2. 2	3. 7	5. 1	6. 5	7. 9	9. 3	10. 7	12. 1	13. 6	15. 0	2. 0
2. 1	2. 2	3. 6	4. 9	6. 3	7. 6	9. 0	10. 3	11. 6	13. 0	14. 3	2. 1
2. 2	2. 2	3. 5	4. 8	6. 1	7. 3	8. 6	9. 9	11. 2	12. 5	13. 8	2. 2
2. 3	2. 2	3. 4	4. 6	5. 9	7. 1	8. 3	9. 6	10. 8	12. 0	13. 3	2. 3
2. 4	2. 2	3. 4	4. 5	5. 7	6. 9	8. 1	9. 2	10. 4	11. 6	12. 8	2. 4
2. 5	2. 2	3. 3	4. 4	5. 6	6. 7	7. 8	9. 0	10. 1	11. 2	12. 4	2. 5
2. 6	2. 2	3. 3	4. 3	5. 4	6. 5	7. 6	8. 7	9. 8	10. 9	12. 0	2. 6
2. 7	2. 2	3. 2	4. 3	5. 3	6. 4	7. 4	8. 4	9. 5	10. 6	11. 6	2. 7
2. 8	2. 2	3. 2	4. 2	5. 2	6. 2	7. 2	8. 2	9. 2	10. 3	11. 3	2. 8
2. 9	2. 2	3. 2	4. 1	5. 1	6. 1	7. 1	8. 0	9. 0	10. 0	11. 0	2. 9
3. 0	2. 2	3. 1	4. 1	5. 0	6. 0	6. 9	7. 8	8. 8	9. 7	10. 7	3. 0
3. 1	2. 2	3. 1	4. 0	4. 9	5. 9	6. 8	7. 7	8. 6	9. 5	10. 4	3. 1
3. 2	2. 2	3. 1	4. 0	4. 9	5. 7	6. 6	7. 5	8. 4	9. 3	10. 2	3. 2
3. 3	2. 2	3. 1	3. 9	4. 8	5. 7	6. 5	7. 4	8. 2	9. 1	9. 9	3. 3
3. 4	2. 2	3. 1	3. 9	4. 7	5. 6	6. 4	7. 2	8. 1	8. 9	9. 7	3. 4
3. 5	2. 2	3. 1	3. 9	4. 7	5. 5	6. 3	7. 1	7. 9	8. 7	9. 5	3. 5
3. 6	2. 2	3. 1	3. 8	4. 6	5. 4	6. 2	7. 0	7. 8	8. 6	9. 4	3. 6
3. 7	2. 2	3. 1	3. 8	4. 6	5. 4	6. 1	6. 9	7. 7	8. 4	9. 2	3. 7
3. 8	2. 2	3. 1	3. 8	4. 6	5. 3	6. 0	6. 8	7. 5	8. 3	9. 0	3. 8
3. 9	2. 2	3. 1	3. 8	4. 5	5. 2	6. 0	6. 7	7. 4	8. 1	8. 9	3. 9
4. 0	2. 2	3. 1	3. 8	4. 5	5. 2	5. 9	6. 6	7. 3	8. 0	8. 7	4. 0
4. 1	2. 2	3. 1	3. 8	4. 5	5. 1	5. 8	6. 5	7. 2	7. 9	8. 6	4. 1
4. 2	2. 2	3. 1	3. 8	4. 4	5. 1	5. 8	6. 5	7. 1	7. 8	8. 5	4. 2
4. 3	2. 2	3. 1	3. 8	4. 4	5. 1	5. 7	6. 4	7. 0	7. 7	8. 4	4. 3
4. 4	2. 2	3. 1	3. 8	4. 4	5. 0	5. 7	6. 3	7. 0	7. 6	8. 3	4. 4
4. 5	2. 2	3. 1	3. 8	4. 4	5. 0	5. 6	6. 3	6. 9	7. 5	8. 2	4. 5
4. 6	2. 2	3. 1	3. 8	4. 4	5. 0	5. 6	6. 2	6. 8	7. 4	8. 1	4. 6
4. 7	2. 2	3. 1	3. 8	4. 4	5. 0	5. 6	6. 2	6. 8	7. 4	8. 0	4. 7
4. 8	2. 2	3. 1	3. 8	4. 4	4. 9	5. 5	6. 1	6. 7	7. 3	7. 9	4. 8
4. 9	2. 2	3. 1	3. 8	4. 3	4. 9	5. 5	6. 1	6. 7	7. 2	7. 8	4. 9
5. 0	2. 2	3. 1	3. 8	4. 3	4. 9	5. 5	6. 0	6. 6	7. 2	7. 7	5. 0
5. 5	2. 2	3. 1	3. 8	4. 3	4. 9	5. 4	5. 9	6. 4	6. 9	7. 4	5. 5
6. 0	2. 2	3. 1	3. 8	4. 3	4. 9	5. 3	5. 8	6. 3	6. 7	7. 2	6. 0
6. 5	2. 2	3. 1	3. 8	4. 3	4. 9	5. 3	5. 7	6. 2	6. 6	7. 1	6. 5
7. 0	2. 2	3. 1	3. 8	4. 3	4. 9	5. 3	5. 7	6. 1	6. 5	6. 9	7. 0
7. 5	2. 2	3. 1	3. 8	4. 3	4. 9	5. 3	5. 7	6. 1	6. 5	6. 9	7. 5
8. 0	2. 2	3. 1	3. 8	4. 3	4. 9	5. 3	5. 7	6. 1	6. 5	6. 9	8. 0
8. 5	2. 2	3. 1	3. 8	4. 3	4. 9	5. 3	5. 7	6. 1	6. 5	6. 9	8. 5
9. 0	2. 2	3. 1	3. 8	4. 3	4. 9	5. 3	5. 7	6. 1	6. 5	6. 9	9. 0
9. 5	2. 2	3. 1	3. 8	4. 3	4. 9	5. 3	5. 7	6. 1	6. 5	6. 9	9. 5
10. 0	2. 2	3. 1	3. 8	4. 3	4. 9	5. 3	5. 7	6. 1	6. 5	6. 9	10. 0

Figure 4-11. "Table 22," American Practical Navigator (*Bowditch*).

Light from a celestial body travels through the vacuum of space in a relatively straight line until it encounters the earth's atmosphere. Being a denser medium, the atmosphere has the effect of bending incoming light rays toward the earth's surface. Since the atmosphere itself is not uniform, but increases in density as the earth's surface is approached, a light ray emanating from a celestial body striking the atmosphere at an oblique angle is bent in a gradual curving path, as shown in Figure 4-9 and Figure 4-12 below. This gradual bending of an incoming light ray in the earth's atmosphere is called *atmospheric refraction.* The greater the angle of incidence of the ray with the atmosphere, the greater will be the angle of refraction toward the surface. Hence, if a celestial body has an apparent altitude of near 90°, the effect of atmospheric refraction is negligible, but as its altitude decreases, the refraction effect increases to a maximum of about 34.5′ of arc for a body located on the visible horizon.

As is shown in Figure 4-12, the effect of atmospheric refraction is always to cause the celestial body observed to appear to have a greater altitude than is the case in reality. Thus, the refraction correction for all celestial observations would normally always be negative. In the *Nautical Almanac,* however, the refraction correction for the sun and the moon is combined with other predominantly positive corrections described below to form a so-called aggregate *altitude correction.* Except in the case of upper limb and extremely low altitude lower limb observations of the sun, the absolute value of the various positive corrections is always greater than the negative refraction correction; this results in the net altitude correction being positive for all lunar and most lower limb solar observations. The term altitude correction is also

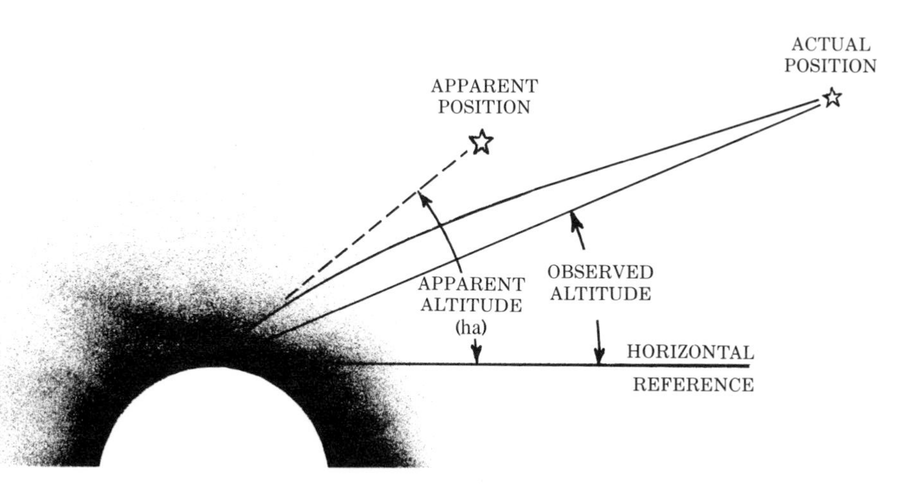

Figure 4-12. Effect of atmospheric refraction. It will always cause the apparent altitude ha to appear to be greater than the observed altitude Ho to the center of the body.

used for the refraction correction for the planets and stars in the *Nautical Almanac.* Being based entirely on refraction in the case of these bodies, it is always negative for them.

The altitude correction tables for the sun, planets, and stars are located inside the front cover of the *Nautical Almanac.* Correction tables for apparent altitudes for these bodies between 10° and 90° are printed on the inside front cover, while corrections for ha less than 10° are located on the facing page. These tables are reproduced in Figures 4-13A and 4-13B on pages 76 and 77. Altitude correction tables incorporating the refraction correction for the moon are located inside the back cover, with corrections for ha between 0° and 35° on the left-hand page, and those for ha between 35° and 90° on the right. Figures 4-14A and 4-14B, pages 78 and 79, depict these tables. The tables on the left side of the inside front cover, shown in Figure 4-13A, are, like the dip table, *critical value tables,* i.e., tables in which values of the correction to be found are tabulated for the ranges of values of the entering argument. The tabulated altitude correction is correct for any ha between those printed half a line above and half a line below. Should an apparent altitude fall exactly on a tabulated argument, the correction a half line above is used. Interpolation to the nearest tenth is required in the remaining tables of Figures 4-13B, 4-14A, and 4-14B, for entering arguments of apparent altitude between the tabulated values.

To illustrate the use of the altitude correction tables of the *Nautical Almanac* to obtain the correction for an observation of a star, suppose that the star Dubhe was observed with a sextant having an IC of +1.2′, and its hs was recorded as 24° 37.7′. The observer's height of eye is 37.6 feet. To find the apparent altitude and then the altitude correction, the foregoing information is first entered on a sight reduction form, the applicable portion of which is reproduced below. Note that on the form positive corrections are entered on the left side of the column, while negative corrections are written on the right. The

Body	DUBHE
IC	+ 1.2' -
Dip (Ht 37.6 ')	-5.9'
Sum	-4.7'
hs	24° 37.7'
ha	24° 33.0'
Alt. Corr	
Add'l.	
H.P. ( )	
Corr. to ha	
Ho (Obs Alt)	

apparent altitude ha is found by applying the sum of the IC and dip corrections, −4.7′, to the hs. Entering the Altitude Correction Table of the *Nautical Almanac* shown in Figure 4-13A with an ha of 24° 33.0′, a correction of −2.1 is obtained. Applying this correction to the ha, the form is completed as follows:

Body	DUBHE
IC	+ 1.2' −
Dip (Ht 37.6 ')	-5.9'
Sum	-4.7'
hs	24° 37.7'
ha	24° 33.0'
Alt. Corr	-2.1'
Add'l.	0
H.P. ( )	
Corr. to ha	-2.1'
Ho (Obs Alt)	24° 30.9'

If, as is usually the case for a star observed under normal atmospheric conditions, no "additional" correction applies, the observed altitude Ho is obtained by subtracting the altitude correction from ha as shown above.

The altitude correction tables of the *Nautical Almanac* are based on the assumption that near normal atmospheric conditions of temperature and barometric pressure prevail; the assumed temperature is 50° F (10° C) and the assumed pressure is 29.83 inches (1010 millibars) of mercury. If actual atmospheric conditions differ markedly from these values, the atmospheric density and therefore its refractive characteristics would be affected to some degree. Under these conditions, it is necessary to apply an additional refraction correction to the apparent altitude, especially when the ha is 10° or less. A combined correction table for nonstandard air temperature and nonstandard atmospheric pressure is given on page A4 of the *Nautical Almanac*, reproduced in Figure 4-15 on page 80. To use the table, the top half is first entered, using as a vertical argument the temperature and as a horizontal argument the pressure. The point at which imaginary lines from these arguments cross locates a zone letter. Using as arguments this zone letter and the apparent altitude, a correction is then found; interpolation to the nearest tenth is necessary for apparent altitudes between tabulated values. The resulting correction is entered in the "Add'l" space on the sight form, and applied to the ha along with the altitude correction for standard conditions to obtain the Ho.

In practice, corrections for unusual temperatures and barometric pressure conditions are generally not applied to apparent altitudes

ALTITUDE CORRECTION TABLES 10°-90°—SUN, STARS, PLANETS

SUN

OCT.—MAR. App. Alt.	Lower Limb	Upper Limb	APR.—SEPT. App. Alt.	Lower Limb	Upper Limb
° ′	′	′	° ′	′	′
9 34	+10·8	−21·5	9 39	+10·6	−21·2
9 45	+10·9	−21·4	9 51	+10·7	−21·1
9 56	+11·0	−21·3	10 03	+10·8	−21·0
10 08	+11·1	−21·2	10 15	+10·9	−20·9
10 21	+11·2	−21·1	10 27	+11·0	−20·8
10 34	+11·3	−21·0	10 40	+11·1	−20·7
10 47	+11·4	−20·9	10 54	+11·2	−20·6
11 01	+11·5	−20·8	11 08	+11·3	−20·5
11 15	+11·6	−20·7	11 23	+11·4	−20·4
11 30	+11·7	−20·6	11 38	+11·5	−20·3
11 46	+11·8	−20·5	11 54	+11·6	−20·2
12 02	+11·9	−20·4	12 10	+11·7	−20·1
12 19	+12·0	−20·3	12 28	+11·8	−20·0
12 37	+12·1	−20·2	12 46	+11·9	−19·9
12 55	+12·2	−20·1	13 05	+12·0	−19·8
13 14	+12·3	−20·0	13 24	+12·1	−19·7
13 35	+12·4	−19·9	13 45	+12·2	−19·6
13 56	+12·5	−19·8	14 07	+12·3	−19·5
14 18	+12·6	−19·7	14 30	+12·4	−19·4
14 42	+12·7	−19·6	14 54	+12·5	−19·3
15 06	+12·8	−19·5	15 19	+12·6	−19·2
15 32	+12·9	−19·4	15 46	+12·7	−19·1
15 59	+13·0	−19·3	16 14	+12·8	−19·0
16 28	+13·1	−19·2	16 44	+12·9	−18·9
16 59	+13·2	−19·1	17 15	+13·0	−18·8
17 32	+13·3	−19·0	17 48	+13·1	−18·7
18 06	+13·4	−18·9	18 24	+13·2	−18·6
18 42	+13·5	−18·8	19 01	+13·3	−18·5
19 21	+13·6	−18·7	19 42	+13·4	−18·4
20 03	+13·7	−18·6	20 25	+13·5	−18·3
20 48	+13·8	−18·5	21 11	+13·6	−18·2
21 35	+13·9	−18·4	22 00	+13·7	−18·1
22 26	+14·0	−18·3	22 54	+13·8	−18·0
23 22	+14·1	−18·2	23 51	+13·9	−17·9
24 21	+14·2	−18·1	24 53	+14·0	−17·8
25 26	+14·3	−18·0	26 00	+14·1	−17·7
26 36	+14·4	−17·9	27 13	+14·2	−17·6
27 52	+14·5	−17·8	28 33	+14·3	−17·5
29 15	+14·6	−17·7	30 00	+14·4	−17·4
30 46	+14·7	−17·6	31 35	+14·5	−17·3
32 26	+14·8	−17·5	33 20	+14·6	−17·2
34 17	+14·9	−17·4	35 17	+14·7	−17·1
36 20	+15·0	−17·3	37 26	+14·8	−17·0
38 36	+15·1	−17·2	39 50	+14·9	−16·9
41 08	+15·2	−17·1	42 31	+15·0	−16·8
43 59	+15·3	−17·0	45 31	+15·1	−16·7
47 10	+15·4	−16·9	48 55	+15·2	−16·6
50 46	+15·5	−16·8	52 44	+15·3	−16·5
54 49	+15·6	−16·7	57 02	+15·4	−16·4
59 23	+15·7	−16·6	61 51	+15·5	−16·3
64 30	+15·8	−16·5	67 17	+15·6	−16·2
70 12	+15·9	−16·4	73 16	+15·7	−16·1
76 26	+16·0	−16·3	79 43	+15·8	−16·0
83 05	+16·1	−16·2	86 32	+15·9	−15·9
90 00			90 00		

STARS AND PLANETS

App. Alt.	Corrⁿ
° ′	′
9 56	−5·3
10 08	−5·2
10 20	−5·1
10 33	−5·0
10 46	−4·9
11 00	−4·8
11 14	−4·7
11 29	−4·6
11 45	−4·5
12 01	−4·4
12 18	−4·3
12 35	−4·2
12 54	−4·1
13 13	−4·0
13 33	−3·9
13 54	−3·8
14 16	−3·7
14 40	−3·6
15 04	−3·5
15 30	−3·4
15 57	−3·3
16 26	−3·2
16 56	−3·1
17 28	−3·0
18 02	−2·9
18 38	−2·8
19 17	−2·7
19 58	−2·6
20 42	−2·5
21 28	−2·4
22 19	−2·3
23 13	−2·2
24 11	−2·1
25 14	−2·0
26 22	−1·9
27 36	−1·8
28 56	−1·7
30 24	−1·6
32 00	−1·5
33 45	−1·4
35 40	−1·3
37 48	−1·2
40 08	−1·1
42 44	−1·0
45 36	−0·9
48 47	−0·8
52 18	−0·7
56 11	−0·6
60 28	−0·5
65 08	−0·4
70 11	−0·3
75 34	−0·2
81 13	−0·1
87 03	0·0
90 00	

App. Alt.	Additional Corrⁿ
VENUS	
Jan. 1–Sept. 30	
0°	′
42	+0·1
Oct. 1–Nov. 15	
0°	′
47	+0·2
Nov. 16–Dec. 12	
0°	′
46	+0·3
Dec. 13–Dec. 28	
0°	′
11	+0·4
41	+0·5
Dec. 29–Dec. 31	
0°	′
6	+0·5
20	+0·6
31	+0·7
MARS	
Jan. 1–June 14	
0°	′
60	+0·1
June 15–Aug. 26	
0°	′
41	+0·2
75	+0·1
Aug. 27–Nov. 28	
0°	′
34	+0·3
60	+0·2
80	+0·1
Nov. 29–Dec. 31	
0°	′
41	+0·2
75	+0·1

DIP

Ht. of Eye (m)	Corrⁿ	Ht. of Eye (ft.)
m	′	ft.
2·4	−2·8	8·0
2·6	−2·9	8·6
2·8	−3·0	9·2
3·0	−3·1	9·8
3·2	−3·2	10·5
3·4	−3·3	11·2
3·6	−3·4	11·9
3·8	−3·5	12·6
4·0	−3·6	13·3
4·3	−3·7	14·1
4·5	−3·8	14·9
4·7	−3·9	15·7
5·0	−4·0	16·5
5·2	−4·1	17·4
5·5	−4·2	18·3
5·8	−4·3	19·1
6·1	−4·4	20·1
6·3	−4·5	21·0
6·6	−4·6	22·0
6·9	−4·7	22·9
7·2	−4·8	23·9
7·5	−4·9	24·9
7·9	−5·0	26·0
8·2	−5·1	27·1
8·5	−5·2	28·1
8·8	−5·3	29·2
9·2	−5·4	30·4
9·5	−5·5	31·5
9·9	−5·6	32·7
10·3	−5·7	33·9
10·6	−5·8	35·1
11·0	−5·9	36·3
11·4	−6·0	37·6
11·8	−6·1	38·9
12·2	−6·2	40·1
12·6	−6·3	41·5
13·0	−6·4	42·8
13·4	−6·5	44·2
13·8	−6·6	45·5
14·2	−6·7	46·9
14·7	−6·8	48·4
15·1	−6·9	49·8
15·5	−7·0	51·3
16·0	−7·1	52·8
16·5	−7·2	54·3
16·9	−7·3	55·8
17·4	−7·4	57·4
17·9	−7·5	58·9
18·4	−7·6	60·5
18·8	−7·7	62·1
19·3	−7·8	63·8
19·8	−7·9	65·4
20·4	−8·0	67·1
20·9	−8·1	68·8
21·4		70·5

Ht. of Eye	Corrⁿ
m	′
1·0	− 1·8
1·5	− 2·2
2·0	− 2·5
2·5	− 2·8
3·0	− 3·0
See table ←	
m	′
20	− 7·9
22	− 8·3
24	− 8·6
26	− 9·0
28	− 9·3
30	− 9·6
32	−10·0
34	−10·3
36	−10·6
38	−10·8
40	−11·1
42	−11·4
44	−11·7
46	−11·9
48	−12·2
ft.	′
2	− 1·4
4	− 1·9
6	− 2·4
8	− 2·7
10	− 3·1
See table ←	
ft.	′
70	− 8·1
75	− 8·4
80	− 8·7
85	− 8·9
90	− 9·2
95	− 9·5
100	− 9·7
105	− 9·9
110	−10·2
115	−10·4
120	−10·6
125	−10·8
130	−11·1
135	−11·3
140	−11·5
145	−11·7
150	−11·9
155	−12·1

App. Alt. = Apparent altitude = Sextant altitude corrected for index error and dip.

For daylight observations of Venus, see page 260.

Figure 4-13A. "Altitude Correction Tables," p. A2, Nautical Almanac.

ALTITUDE CORRECTION TABLES 0°–10°—SUN, STARS, PLANETS A3

App. Alt.	OCT.–MAR. SUN Lower Limb		APR.–SEPT. SUN Lower Limb		STARS PLANETS
	Lower Limb	Upper Limb	Lower Limb	Upper Limb	
° ′	′	′	′	′	′
0 00	−18·2	−50·5	−18·4	−50·2	−34·5
03	17·5	49·8	17·8	49·6	33·8
06	16·9	49·2	17·1	48·9	33·2
09	16·3	48·6	16·5	48·3	32·6
12	15·7	48·0	15·9	47·7	32·0
15	15·1	47·4	15·3	47·1	31·4
0 18	−14·5	−46·8	−14·8	−46·6	−30·8
21	14·0	46·3	14·2	46·0	30·3
24	13·5	45·8	13·7	45·5	29·8
27	12·9	45·2	13·2	45·0	29·2
30	12·4	44·7	12·7	44·5	28·7
33	11·9	44·2	12·2	44·0	28·2
0 36	−11·5	−43·8	−11·7	−43·5	−27·8
39	11·0	43·3	11·2	43·0	27·3
42	10·5	42·8	10·8	42·6	26·8
45	10·1	42·4	10·3	42·1	26·4
48	9·6	41·9	9·9	41·7	25·9
51	9·2	41·5	9·5	41·3	25·5
0 54	− 8·8	−41·1	− 9·1	−40·9	−25·1
0 57	8·4	40·7	8·7	40·5	24·7
1 00	8·0	40·3	8·3	40·1	24·3
03	7·7	40·0	7·9	39·7	24·0
06	7·3	39·6	7·5	39·3	23·6
09	6·9	39·2	7·2	39·0	23·2
1 12	− 6·6	−38·9	− 6·8	−38·6	−22·9
15	6·2	38·5	6·5	38·3	22·5
18	5·9	38·2	6·2	38·0	22·2
21	5·6	37·9	5·8	37·6	21·9
24	5·3	37·6	5·5	37·3	21·6
27	4·9	37·2	5·2	37·0	21·2
1 30	− 4·6	−36·9	− 4·9	−36·7	−20·9
35	4·2	36·5	4·4	36·2	20·5
40	3·7	36·0	4·0	35·8	20·0
45	3·2	35·5	3·5	35·3	19·5
50	2·8	35·1	3·1	34·9	19·1
1 55	2·4	34·7	2·6	34·4	18·7
2 00	− 2·0	−34·3	− 2·2	−34·0	−18·3
05	1·6	33·9	1·8	33·6	17·9
10	1·2	33·5	1·5	33·3	17·5
15	0·9	33·2	1·1	32·9	17·2
20	0·5	32·8	0·8	32·6	16·8
25	− 0·2	32·5	0·4	32·2	16·5
2 30	+ 0·2	−32·1	− 0·1	−31·9	−16·1
35	0·5	31·8	+ 0·2	31·6	15·8
40	0·8	31·5	0·5	31·3	15·5
45	1·1	31·2	0·8	31·0	15·2
50	1·4	30·9	1·1	30·7	14·9
2 55	1·6	30·7	1·4	30·4	14·7
3 00	+ 1·9	−30·4	+ 1·7	−30·1	−14·4
05	2·2	30·1	1·9	29·9	14·1
10	2·4	29·9	2·1	29·7	13·9
15	2·6	29·7	2·4	29·4	13·7
20	2·9	29·4	2·6	29·2	13·4
25	3·1	29·2	2·9	28·9	13·2
3 30	+ 3·3	−29·0	+ 3·1	−28·7	−13·0

App. Alt.	OCT.–MAR. SUN Lower Limb		APR.–SEPT. SUN Lower Limb		STARS PLANETS
	Lower Limb	Upper Limb	Lower Limb	Upper Limb	
° ′	′	′	′	′	′
3 30	+ 3·3	−29·0	+ 3·1	−28·7	−13·0
35	3·6	28·7	3·3	28·5	12·7
40	3·8	28·5	3·5	28·3	12·5
45	4·0	28·3	3·7	28·1	12·3
50	4·2	28·1	3·9	27·9	12·1
3 55	4·4	27·9	4·1	27·7	11·9
4 00	+ 4·5	−27·8	+ 4·3	−27·5	−11·8
05	4·7	27·6	4·5	27·3	11·6
10	4·9	27·4	4·6	27·2	11·4
15	5·1	27·2	4·8	27·0	11·2
20	5·2	27·1	5·0	26·8	11·1
25	5·4	26·9	5·1	26·7	10·9
4 30	+ 5·6	−26·7	+ 5·3	−26·5	−10·7
35	5·7	26·6	5·5	26·3	10·6
40	5·9	26·4	5·6	26·2	10·4
45	6·0	26·3	5·8	26·0	10·3
50	6·2	26·1	5·9	25·9	10·1
4 55	6·3	26·0	6·0	25·8	10·0
5 00	+ 6·4	−25·9	+ 6·2	−25·6	− 9·9
05	6·6	25·7	6·3	25·5	9·7
10	6·7	25·6	6·4	25·4	9·6
15	6·8	25·5	6·6	25·2	9·5
20	6·9	25·4	6·7	25·1	9·4
25	7·1	25·2	6·8	25·0	9·2
5 30	+ 7·2	−25·1	+ 6·9	−24·9	− 9·1
35	7·3	25·0	7·0	24·8	9·0
40	7·4	24·9	7·2	24·6	8·9
45	7·5	24·8	7·3	24·5	8·8
50	7·6	24·7	7·4	24·4	8·7
5 55	7·7	24·6	7·5	24·3	8·6
6 00	+ 7·8	−24·5	+ 7·6	−24·2	− 8·5
10	8·0	24·3	7·8	24·0	8·3
20	8·2	24·1	8·0	23·8	8·1
30	8·4	23·9	8·1	23·7	7·9
40	8·6	23·7	8·3	23·5	7·7
6 50	8·7	23·6	8·5	23·3	7·6
7 00	+ 8·9	−23·4	+ 8·6	−23·2	− 7·4
10	9·1	23·2	8·8	23·0	7·2
20	9·2	23·1	9·0	22·8	7·1
30	9·3	23·0	9·1	22·7	7·0
40	9·5	22·8	9·2	22·6	6·8
7 50	9·6	22·7	9·4	22·4	6·7
8 00	+ 9·7	−22·6	+ 9·5	−22·3	− 6·6
10	9·9	22·4	9·6	22·2	6·4
20	10·0	22·3	9·7	22·1	6·3
30	10·1	22·2	9·8	22·0	6·2
40	10·2	22·1	10·0	21·8	6·1
8 50	10·3	22·0	10·1	21·7	6·0
9 00	+10·4	−21·9	+10·2	−21·6	− 5·9
10	10·5	21·8	10·3	21·5	5·8
20	10·6	21·7	10·4	21·4	5·7
30	10·7	21·6	10·5	21·3	5·6
40	10·8	21·5	10·6	21·2	5·5
9 50	10·9	21·4	10·6	21·2	5·4
10 00	+11·0	−21·3	+10·7	−21·1	− 5·3

Additional corrections for temperature and pressure are given on the following page.
For bubble sextant observations ignore dip and use the star corrections for Sun, planets, and stars.

Figure 4-13B. "Altitude Correction Tables," p. A3, Nautical Almanac.

ALTITUDE CORRECTION TABLES 0°-35°—MOON

App. Alt.	0°-4° Corrⁿ	5°-9° Corrⁿ	10°-14° Corrⁿ	15°-19° Corrⁿ	20°-24° Corrⁿ	25°-29° Corrⁿ	30°-34° Corrⁿ	App. Alt.
′	° ′	° ′	° ′	° ′	° ′	° ′	° ′	′
00	0 33·8	5 58·2	10 62·1	15 62·8	20 62·2	25 60·8	30 58·9	00
10	35·9	58·5	62·2	62·8	62·1	60·8	58·8	10
20	37·8	58·7	62·2	62·8	62·1	60·7	58·8	20
30	39·6	58·9	62·3	62·8	62·1	60·7	58·7	30
40	41·2	59·1	62·3	62·8	62·0	60·6	58·6	40
50	42·6	59·3	62·4	62·7	62·0	60·6	58·5	50
00	1 44·0	6 59·5	11 62·4	16 62·7	21 62·0	26 60·5	31 58·5	00
10	45·2	59·7	62·4	62·7	61·9	60·4	58·4	10
20	46·3	59·9	62·5	62·7	61·9	60·4	58·3	20
30	47·3	60·0	62·5	62·7	61·9	60·3	58·2	30
40	48·3	60·2	62·5	62·7	61·8	60·3	58·2	40
50	49·2	60·3	62·6	62·7	61·8	60·2	58·1	50
00	2 50·0	7 60·5	12 62·6	17 62·7	22 61·7	27 60·1	32 58·0	00
10	50·8	60·6	62·6	62·6	61·7	60·1	57·9	10
20	51·4	60·7	62·6	62·6	61·6	60·0	57·8	20
30	52·1	60·9	62·7	62·6	61·6	59·9	57·8	30
40	52·7	61·0	62·7	62·6	61·5	59·9	57·7	40
50	53·3	61·1	62·7	62·6	61·5	59·8	57·6	50
00	3 53·8	8 61·2	13 62·7	18 62·5	23 61·5	28 59·7	33 57·5	00
10	54·3	61·3	62·7	62·5	61·4	59·7	57·4	10
20	54·8	61·4	62·7	62·5	61·4	59·6	57·4	20
30	55·2	61·5	62·8	62·5	61·3	59·6	57·3	30
40	55·6	61·6	62·8	62·4	61·3	59·5	57·2	40
50	56·0	61·6	62·8	62·4	61·2	59·4	57·1	50
00	4 56·4	9 61·7	14 62·8	19 62·4	24 61·2	29 59·3	34 57·0	00
10	56·7	61·8	62·8	62·3	61·1	59·3	56·9	10
20	57·1	61·9	62·8	62·3	61·1	59·2	56·9	20
30	57·4	61·9	62·8	62·3	61·0	59·1	56·8	30
40	57·7	62·0	62·8	62·2	60·9	59·1	56·7	40
50	57·9	62·1	62·8	62·2	60·9	59·0	56·6	50
H.P.	L U	L U	L U	L U	L U	L U	L U	H.P.
′	′ ′	′ ′	′ ′	′ ′	′ ′	′ ′	′ ′	′
54·0	0·3 0·9	0·3 0·9	0·4 1·0	0·5 1·1	0·6 1·2	0·7 1·3	0·9 1·5	54·0
54·3	0·7 1·1	0·7 1·2	0·7 1·2	0·8 1·3	0·9 1·4	1·1 1·5	1·2 1·7	54·3
54·6	1·1 1·4	1·1 1·4	1·1 1·4	1·2 1·5	1·3 1·6	1·4 1·7	1·5 1·8	54·6
54·9	1·4 1·6	1·5 1·6	1·5 1·6	1·6 1·7	1·6 1·8	1·8 1·9	1·9 2·0	54·9
55·2	1·8 1·8	1·8 1·8	1·9 1·9	1·9 1·9	2·0 2·0	2·1 2·1	2·2 2·2	55·2
55·5	2·2 2·0	2·2 2·0	2·3 2·1	2·3 2·1	2·4 2·2	2·4 2·3	2·5 2·4	55·5
55·8	2·6 2·2	2·6 2·2	2·6 2·3	2·7 2·3	2·7 2·4	2·8 2·4	2·9 2·5	55·8
56·1	3·0 2·4	3·0 2·5	3·0 2·5	3·0 2·5	3·1 2·6	3·1 2·6	3·2 2·7	56·1
56·4	3·4 2·7	3·4 2·7	3·4 2·7	3·4 2·7	3·4 2·8	3·5 2·8	3·5 2·9	56·4
56·7	3·7 2·9	3·7 2·9	3·8 2·9	3·8 2·9	3·8 3·0	3·8 3·0	3·9 3·0	56·7
57·0	4·1 3·1	4·1 3·1	4·1 3·1	4·1 3·1	4·2 3·1	4·2 3·2	4·2 3·2	57·0
57·3	4·5 3·3	4·5 3·3	4·5 3·3	4·5 3·3	4·5 3·3	4·5 3·4	4·6 3·4	57·3
57·6	4·9 3·5	4·9 3·5	4·9 3·5	4·9 3·5	4·9 3·5	4·9 3·5	4·9 3·6	57·6
57·9	5·3 3·8	5·3 3·8	5·2 3·8	5·2 3·7	5·2 3·7	5·2 3·7	5·2 3·7	57·9
58·2	5·6 4·0	5·6 4·0	5·6 4·0	5·6 4·0	5·6 3·9	5·6 3·9	5·6 3·9	58·2
58·5	6·0 4·2	6·0 4·2	6·0 4·2	6·0 4·2	6·0 4·1	5·9 4·1	5·9 4·1	58·5
58·8	6·4 4·4	6·4 4·4	6·4 4·4	6·3 4·4	6·3 4·3	6·3 4·3	6·2 4·2	58·8
59·1	6·8 4·6	6·8 4·6	6·7 4·6	6·7 4·6	6·7 4·5	6·6 4·5	6·6 4·4	59·1
59·4	7·2 4·8	7·1 4·8	7·1 4·8	7·1 4·8	7·0 4·7	7·0 4·7	6·9 4·6	59·4
59·7	7·5 5·1	7·5 5·0	7·5 5·0	7·5 5·0	7·4 4·9	7·3 4·8	7·2 4·7	59·7
60·0	7·9 5·3	7·9 5·3	7·9 5·2	7·8 5·2	7·8 5·1	7·7 5·0	7·6 4·9	60·0
60·3	8·3 5·5	8·3 5·5	8·2 5·4	8·2 5·4	8·1 5·3	8·0 5·2	7·9 5·1	60·3
60·6	8·7 5·7	8·7 5·7	8·6 5·7	8·6 5·6	8·5 5·5	8·4 5·4	8·2 5·3	60·6
60·9	9·1 5·9	9·0 5·9	9·0 5·9	8·9 5·8	8·8 5·7	8·7 5·6	8·6 5·4	60·9
61·2	9·5 6·2	9·4 6·1	9·4 6·1	9·3 6·0	9·2 5·9	9·1 5·8	8·9 5·6	61·2
61·5	9·8 6·4	9·8 6·3	9·7 6·3	9·7 6·2	9·5 6·1	9·4 5·9	9·2 5·8	61·5

DIP

Ht. of Eye	Corrⁿ	Ht. of Eye	Corrⁿ	Ht. of Eye	Corrⁿ
ft.	′	ft.	′	ft.	′
4·0	−2·0	24	−4·9	63	− 7·8
4·4	−2·1	26	−5·0	65	− 7·9
4·9	−2·2	27	−5·1	67	− 8·0
5·3	−2·3	28	−5·2	68	− 8·1
5·8	−2·4	29	−5·3	70	− 8·2
6·3	−2·5	30	−5·4	72	− 8·3
6·9	−2·6	31	−5·5	74	− 8·4
7·4	−2·7	32	−5·6	75	− 8·5
8·0	−2·8	33	−5·7	77	− 8·6
8·6	−2·9	35	−5·8	79	− 8·7
9·2	−3·0	36	−5·9	81	− 8·8
9·8	−3·1	37	−6·0	83	− 8·9
10·5	−3·2	38	−6·1	85	− 9·0
11·2	−3·3	40	−6·2	87	− 9·1
11·9	−3·4	41	−6·3	88	− 9·2
12·6	−3·5	42	−6·4	90	− 9·3
13·3	−3·6	44	−6·5	92	− 9·4
14·1	−3·7	45	−6·6	94	− 9·5
14·9	−3·8	47	−6·7	96	− 9·6
15·7	−3·9	48	−6·8	98	− 9·7
16·5	−4·0	49	−6·9	101	− 9·8
17·4	−4·1	51	−7·0	103	− 9·9
18·3	−4·2	52	−7·1	105	−10·0
19·1	−4·3	54	−7·2	107	−10·1
20·1	−4·4	55	−7·3	109	−10·2
21·0	−4·5	57	−7·4	111	−10·3
22·0	−4·6	58	−7·5	113	−10·4
22·9	−4·7	60	−7·6	116	−10·5
23·9	−4·8	62	−7·7	118	−10·6
24·9		63		120	

MOON CORRECTION TABLE

The correction is in two parts; the first correction is taken from the upper part of the table with argument apparent altitude, and the second from the lower part, with argument H.P., in the same column as that from which the first correction was taken. Separate corrections are given in the lower part for lower (L) and upper (U) limbs. All corrections are to be **added** to apparent altitude, *but 30′ is to be subtracted from the altitude of the upper limb.*

For corrections for pressure and temperature see page A4.

For bubble sextant observations ignore dip, take the mean of upper and lower limb corrections and subtract 15′ from the altitude.

App. Alt. = Apparent altitude = Sextant altitude corrected for index error and dip.

Figure 4-14A. "Altitude Correction Tables—Moon," p. xxxiv, Nautical Almanac.

ALTITUDE CORRECTION TABLES 35°-90°—MOON

App. Alt.	35°–39° Corrⁿ	40°–44° Corrⁿ	45°–49° Corrⁿ	50°–54° Corrⁿ	55°–59° Corrⁿ	60°–64° Corrⁿ	65°–69° Corrⁿ	70°–74° Corrⁿ	75°–79° Corrⁿ	80°–84° Corrⁿ	85°–89° Corrⁿ	App. Alt.
′	° ′	° ′	° ′	° ′	° ′	° ′	° ′	° ′	° ′	° ′	° ′	′
00	**35** 56·5	**40** 53·7	**45** 50·5	**50** 46·9	**55** 43·1	**60** 38·9	**65** 34·6	**70** 30·1	**75** 25·3	**80** 20·5	**85** 15·6	00
10	56·4	53·6	50·4	46·8	42·9	38·8	34·4	29·9	25·2	20·4	15·5	10
20	56·3	53·5	50·2	46·7	42·8	38·7	34·3	29·7	25·0	20·2	15·3	20
30	56·2	53·4	50·1	46·5	42·7	38·5	34·1	29·6	24·9	20·0	15·1	30
40	56·2	53·3	50·0	46·4	42·5	38·4	34·0	29·4	24·7	19·9	15·0	40
50	56·1	53·2	49·9	46·3	42·4	38·2	33·8	29·3	24·5	19·7	14·8	50
00	**36** 56·0	**41** 53·1	**46** 49·8	**51** 46·2	**56** 42·3	**61** 38·1	**66** 33·7	**71** 29·1	**76** 24·4	**81** 19·6	**86** 14·6	00
10	55·9	53·0	49·7	46·0	42·1	37·9	33·5	29·0	24·2	19·4	14·5	10
20	55·8	52·8	49·5	45·9	42·0	37·8	33·4	28·8	24·1	19·2	14·3	20
30	55·7	52·7	49·4	45·8	41·8	37·7	33·2	28·7	23·9	19·1	14·1	30
40	55·6	52·6	49·3	45·7	41·7	37·5	33·1	28·5	23·8	18·9	14·0	40
50	55·5	52·5	49·2	45·5	41·6	37·4	32·9	28·3	23·6	18·7	13·8	50
00	**37** 55·4	**42** 52·4	**47** 49·1	**52** 45·4	**57** 41·4	**62** 37·2	**67** 32·8	**72** 28·2	**77** 23·4	**82** 18·6	**87** 13·7	00
10	55·3	52·3	49·0	45·3	41·3	37·1	32·6	28·0	23·3	18·4	13·5	10
20	55·2	52·2	48·8	45·2	41·2	36·9	32·5	27·9	23·1	18·2	13·3	20
30	55·1	52·1	48·7	45·0	41·0	36·8	32·3	27·7	22·9	18·1	13·2	30
40	55·0	52·0	48·6	44·9	40·9	36·6	32·2	27·6	22·8	17·9	13·0	40
50	55·0	51·9	48·5	44·8	40·8	36·5	32·0	27·4	22·6	17·8	12·8	50
00	**38** 54·9	**43** 51·8	**48** 48·4	**53** 44·6	**58** 40·6	**63** 36·4	**68** 31·9	**73** 27·2	**78** 22·5	**83** 17·6	**88** 12·7	00
10	54·8	51·7	48·2	44·5	40·5	36·2	31·7	27·1	22·3	17·4	12·5	10
20	54·7	51·6	48·1	44·4	40·3	36·1	31·6	26·9	22·1	17·3	12·3	20
30	54·6	51·5	48·0	44·2	40·2	35·9	31·4	26·8	22·0	17·1	12·2	30
40	54·5	51·4	47·9	44·1	40·1	35·8	31·3	26·6	21·8	16·9	12·0	40
50	54·4	51·2	47·8	44·0	39·9	35·6	31·1	26·5	21·7	16·8	11·8	50
00	**39** 54·3	**44** 51·1	**49** 47·6	**54** 43·9	**59** 39·8	**64** 35·5	**69** 31·0	**74** 26·3	**79** 21·5	**84** 16·6	**89** 11·7	00
10	54·2	51·0	47·5	43·7	39·6	35·3	30·8	26·1	21·3	16·5	11·5	10
20	54·1	50·9	47·4	43·6	39·5	35·2	30·7	26·0	21·2	16·3	11·4	20
30	54·0	50·8	47·3	43·5	39·4	35·0	30·5	25·8	21·0	16·1	11·2	30
40	53·9	50·7	47·2	43·3	39·2	34·9	30·4	25·7	20·9	16·0	11·0	40
50	53·8	50·6	47·0	43·2	39·1	34·7	30·2	25·5	20·7	15·8	10·9	50
H.P.	**L U**	**L U**	**L U**	**L U**	**L U**	**L U**	**L U**	**L U**	**L U**	**L U**	**L U**	**H.P.**
′	′ ′	′ ′	′ ′	′ ′	′ ′	′ ′	′ ′	′ ′	′ ′	′ ′	′ ′	′
54·0	1·1 1·7	1·3 1·9	1·5 2·1	1·7 2·4	2·0 2·6	2·3 2·9	2·6 3·2	2·9 3·5	3·2 3·8	3·5 4·1	3·8 4·5	54·0
54·3	1·4 1·8	1·6 2·0	1·8 2·2	2·0 2·5	2·3 2·7	2·5 3·0	2·8 3·2	3·0 3·5	3·3 3·8	3·6 4·1	3·9 4·4	54·3
54·6	1·7 2·0	1·9 2·2	2·1 2·4	2·3 2·6	2·5 2·8	2·7 3·0	3·0 3·3	3·2 3·5	3·5 3·8	3·7 4·1	4·0 4·3	54·6
54·9	2·0 2·2	2·2 2·3	2·3 2·5	2·5 2·7	2·7 2·9	2·9 3·1	3·2 3·3	3·4 3·5	3·6 3·8	3·9 4·0	4·1 4·3	54·9
55·2	2·3 2·3	2·5 2·4	2·6 2·6	2·8 2·8	3·0 2·9	3·2 3·1	3·4 3·3	3·6 3·5	3·8 3·7	4·0 4·0	4·2 4·2	55·2
55·5	2·7 2·5	2·8 2·6	2·9 2·7	3·1 2·9	3·2 3·0	3·4 3·2	3·6 3·4	3·7 3·5	3·9 3·7	4·1 3·9	4·3 4·1	55·5
55·8	3·0 2·6	3·1 2·7	3·2 2·8	3·3 3·0	3·5 3·1	3·6 3·3	3·8 3·4	3·9 3·6	4·1 3·7	4·2 3·9	4·4 4·0	55·8
56·1	3·3 2·8	3·4 2·9	3·5 3·0	3·6 3·1	3·7 3·2	3·8 3·3	4·0 3·4	4·1 3·6	4·2 3·7	4·4 3·8	4·5 4·0	56·1
56·4	3·6 2·9	3·7 3·0	3·8 3·1	3·9 3·2	3·9 3·3	4·0 3·4	4·1 3·5	4·3 3·6	4·4 3·7	4·5 3·8	4·6 3·9	56·4
56·7	3·9 3·1	4·0 3·1	4·1 3·2	4·1 3·3	4·2 3·3	4·3 3·4	4·3 3·5	4·4 3·6	4·5 3·7	4·6 3·8	4·7 3·8	56·7
57·0	4·3 3·2	4·3 3·3	4·3 3·3	4·4 3·4	4·4 3·4	4·5 3·5	4·5 3·5	4·6 3·6	4·7 3·6	4·7 3·7	4·8 3·8	57·0
57·3	4·6 3·4	4·6 3·4	4·6 3·4	4·6 3·5	4·7 3·5	4·7 3·5	4·7 3·6	4·8 3·6	4·8 3·6	4·8 3·7	4·9 3·7	57·3
57·6	4·9 3·6	4·9 3·6	4·9 3·6	4·9 3·6	4·9 3·6	4·9 3·6	4·9 3·6	4·9 3·6	5·0 3·6	5·0 3·6	5·0 3·6	57·6
57·9	5·2 3·7	5·2 3·7	5·2 3·7	5·2 3·7	5·2 3·7	5·1 3·6	5·1 3·6	5·1 3·6	5·1 3·6	5·1 3·6	5·1 3·6	57·9
58·2	5·5 3·9	5·5 3·8	5·5 3·8	5·4 3·8	5·4 3·7	5·4 3·7	5·3 3·7	5·3 3·6	5·2 3·6	5·2 3·5	5·2 3·5	58·2
58·5	5·9 4·0	5·8 4·0	5·8 3·9	5·7 3·9	5·6 3·8	5·6 3·8	5·5 3·7	5·5 3·6	5·4 3·6	5·3 3·5	5·3 3·4	58·5
58·8	6·2 4·2	6·1 4·1	6·0 4·1	6·0 4·0	5·9 3·9	5·8 3·8	5·7 3·7	5·6 3·6	5·5 3·5	5·4 3·5	5·3 3·4	58·8
59·1	6·5 4·3	6·4 4·3	6·3 4·2	6·2 4·1	6·1 4·0	6·0 3·9	5·9 3·8	5·8 3·6	5·7 3·5	5·6 3·4	5·4 3·3	59·1
59·4	6·8 4·5	6·7 4·4	6·6 4·3	6·5 4·2	6·4 4·1	6·2 3·9	6·1 3·8	6·0 3·7	5·8 3·5	5·7 3·4	5·5 3·2	59·4
59·7	7·1 4·6	7·0 4·5	6·9 4·4	6·8 4·3	6·6 4·1	6·5 4·0	6·3 3·8	6·2 3·7	6·0 3·5	5·8 3·3	5·6 3·2	59·7
60·0	7·5 4·8	7·3 4·7	7·2 4·5	7·0 4·4	6·9 4·2	6·7 4·0	6·5 3·9	6·3 3·7	6·1 3·5	5·9 3·3	5·7 3·1	60·0
60·3	7·8 5·0	7·6 4·8	7·5 4·7	7·3 4·5	7·1 4·3	6·9 4·1	6·7 3·9	6·5 3·7	6·3 3·5	6·0 3·2	5·8 3·0	60·3
60·6	8·1 5·1	7·9 5·0	7·7 4·8	7·6 4·6	7·3 4·4	7·1 4·2	6·9 3·9	6·7 3·7	6·4 3·4	6·2 3·2	5·9 2·9	60·6
60·9	8·4 5·3	8·2 5·1	8·0 4·9	7·8 4·7	7·6 4·5	7·3 4·2	7·1 4·0	6·8 3·7	6·6 3·4	6·3 3·2	6·0 2·9	60·9
61·2	8·7 5·4	8·5 5·2	8·3 5·0	8·1 4·8	7·8 4·5	7·6 4·3	7·3 4·0	7·0 3·7	6·7 3·4	6·4 3·1	6·1 2·8	61·2
61·5	9·1 5·6	8·8 5·4	8·6 5·1	8·3 4·9	8·1 4·6	7·8 4·3	7·5 4·0	7·2 3·7	6·9 3·4	6·5 3·1	6·2 2·7	61·5

Figure 4-14B. "Altitude Correction Tables—Moon," p. xxxv, Nautical Almanac.

A4 ALTITUDE CORRECTION TABLES—ADDITIONAL CORRECTIONS

ADDITIONAL REFRACTION CORRECTIONS FOR NON-STANDARD CONDITIONS

App. Alt.	A	B	C	D	E	F	G	H	J	K	L	M	N	App. Alt.
° ′	′	′	′	′	′	′	′	′	′	′	′	′	′	° ′
0 00	−6·9	−5·7	−4·6	−3·4	−2·3	−1·1	0·0	+1·1	+2·3	+3·4	+4·6	+5·7	+6·9	0 00
0 30	5·2	4·4	3·5	2·6	1·7	0·9	0·0	0·9	1·7	2·6	3·5	4·4	5·2	0 30
1 00	4·3	3·5	2·8	2·1	1·4	0·7	0·0	0·7	1·4	2·1	2·8	3·5	4·3	1 00
1 30	3·5	2·9	2·4	1·8	1·2	0·6	0·0	0·6	1·2	1·8	2·4	2·9	3·5	1 30
2 00	3·0	2·5	2·0	1·5	1·0	0·5	0·0	0·5	1·0	1·5	2·0	2·5	3·0	2 00
2 30	−2·5	−2·1	−1·6	−1·2	−0·8	−0·4	0·0	+0·4	+0·8	+1·2	+1·6	+2·1	+2·5	2 30
3 00	2·2	1·8	1·5	1·1	0·7	0·4	0·0	0·4	0·7	1·1	1·5	1·8	2·2	3 00
3 30	2·0	1·6	1·3	1·0	0·7	0·3	0·0	0·3	0·7	1·0	1·3	1·6	2·0	3 30
4 00	1·8	1·5	1·2	0·9	0·6	0·3	0·0	0·3	0·6	0·9	1·2	1·5	1·8	4 00
4 30	1·6	1·4	1·1	0·8	0·5	0·3	0·0	0·3	0·5	0·8	1·1	1·4	1·6	4 30
5 00	−1·5	−1·3	−1·0	−0·8	−0·5	−0·2	0·0	+0·2	+0·5	+0·8	+1·0	+1·3	+1·5	5 00
6	1·3	1·1	0·9	0·6	0·4	0·2	0·0	0·2	0·4	0·6	0·9	1·1	1·3	6
7	1·1	0·9	0·7	0·6	0·4	0·2	0·0	0·2	0·4	0·6	0·7	0·9	1·1	7
8	1·0	0·8	0·7	0·5	0·3	0·2	0·0	0·2	0·3	0·5	0·7	0·8	1·0	8
9	0·9	0·7	0·6	0·4	0·3	0·1	0·0	0·1	0·3	0·4	0·6	0·7	0·9	9
10 00	−0·8	−0·7	−0·5	−0·4	−0·3	−0·1	0·0	+0·1	+0·3	+0·4	+0·5	+0·7	+0·8	10 00
12	0·7	0·6	0·5	0·3	0·2	0·1	0·0	0·1	0·2	0·3	0·5	0·6	0·7	12
14	0·6	0·5	0·4	0·3	0·2	0·1	0·0	0·1	0·2	0·3	0·4	0·5	0·6	14
16	0·5	0·4	0·3	0·3	0·2	0·1	0·0	0·1	0·2	0·3	0·3	0·4	0·5	16
18	0·4	0·4	0·3	0·2	0·2	0·1	0·0	0·1	0·2	0·2	0·3	0·4	0·4	18
20 00	−0·4	−0·3	−0·3	−0·2	−0·1	−0·1	0·0	+0·1	+0·1	+0·2	+0·3	+0·3	+0·4	20 00
25	0·3	0·3	0·2	0·2	0·1	−0·1	0·0	+0·1	0·1	0·2	0·2	0·3	0·3	25
30	0·3	0·2	0·2	0·1	0·1	0·0	0·0	0·0	0·1	0·1	0·2	0·2	0·3	30
35	0·2	0·2	0·1	0·1	0·1	0·0	0·0	0·0	0·1	0·1	0·1	0·2	0·2	35
40	0·2	0·1	0·1	0·1	−0·1	0·0	0·0	0·0	+0·1	0·1	0·1	0·1	0·2	40
50 00	−0·1	−0·1	−0·1	−0·1	0·0	0·0	0·0	0·0	0·0	+0·1	+0·1	+0·1	+0·1	50 00

The graph is entered with arguments temperature and pressure to find a zone letter; using as arguments this zone letter and apparent altitude (sextant altitude corrected for dip), a correction is taken from the table. This correction is to be applied to the sextant altitude in addition to the corrections for standard conditions (for the Sun, planets and stars from the inside front cover and for the Moon from the inside back cover).

242-578 O - 69

Figure 4-15. "Additional Corrections" table, p. A4, Nautical Almanac.

greater than 10°, as the amount of the correction for higher altitudes is so small as to be considered insignificant for most applications. Moreover, since atmospheric refraction of light from low-altitude bodies can vary unpredictably even in standard atmospheric conditions, and more so in nonstandard conditions, most navigators will normally not observe bodies with altitudes lower than 10°, unless there are no alternatives. Thus, neither Table A3 nor the additional corrections Table A4 are used very often in practical navigation.

Adjustments for Equivalent Reading for the Center of the Body

As was mentioned earlier in this chapter, the observed altitude Ho of a celestial body is defined as the angle that would be formed at the center of the earth between the observer's celestial horizon and the line of sight to the center of the body. Since a star is so far distant from earth, it always appears as a point source of light, having no measureable diameter. On the other hand, the sun, the moon, and at times the planets Venus and Mars do have significant diameters as viewed from earth. Although sextant observations could be made to the approximate centers of the sun and moon, it is the preferred procedure to bring either their upper or lower limbs tangent to the sea horizon, for increased accuracy. When this is done, there are three types of corrections that must be made to arrive at an equivalent reading for the center of the body: semidiameter, augmentation, and phase.

Semidiameter, abbreviated SD, is perhaps the most obvious of these three corrections. As can be seen from Figure 4-16, the apparent altitude as measured by the sextant must be decreased by the amount of one-half the diameter of the body as measured by the sextant when

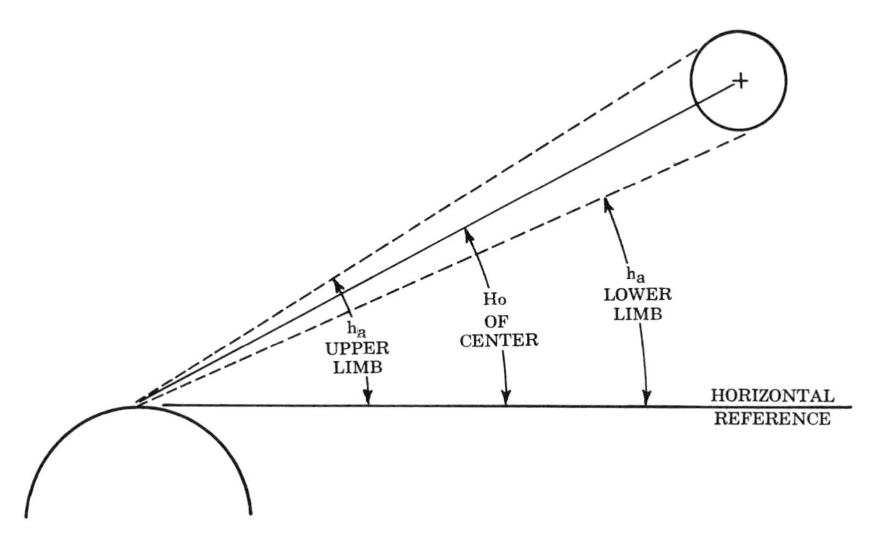

Figure 4-16. Effect of semidiameter on the altitude of a body.

the upper limb is observed, and increased by this amount if the lower limb is observed. The semidiameter of the sun varies from a little less than 15.8′ early in July, when the sun is at its greatest distance, to about 16.3′ early in January, when the earth is nearest the sun. The moon also varies in apparent size as it moves in its orbit around the earth; its semidiameter ranges from 14.7′ to 16.8′ over the period of each month. Although the navigational planets do have a measurable semidiameter at certain times of the year, it is customary to observe their centers at these times, thereby making any semidiameter correction for the planets unnecessary.

Although a correction for the semidiameter of the sun and moon could be made by using daily semidiameter figures for these bodies, which are given at the bottom of each daily page of the *Nautical Almanac*, the usual procedure is simply to use the *Nautical Almanac* "Altitude Correction Tables" for the sun and the moon. In the sun tables located inside the front cover, the semidiameter correction and the parallax correction to be discussed below are combined with the refraction correction to produce a single "altitude" correction to ha. In the moon tables, located inside the back cover, the semidiameter correction is combined with the augmentation correction discussed below as well as the refraction correction; a separate correction for parallax is also extracted. The sun "Altitude Correction Tables" are shown in Figures 4-13A and 4-13B. The table of Figure 4-13A, located on the left-hand side of the inside front cover of the *Almanac,* is, like the stars and planets table, a critical value table for ha between 10° and 90°, while the table on the right-hand side pictured in Figure 4-13B for ha between 0° and 10° requires interpolation. In each table, there are two columns of corrections; one is for upper or lower limb observations made from October through March, and the other is for observations made from April through September. The moon "Altitude Correction Tables" are shown in Figures 4-14A and 4-14B; their use will be described in the following section.

The increase in apparent size of the sun and moon as a result of increase in the apparent altitude of these bodies is called *augmentation.* The augmentation effect is illustrated in Figure 4-17.

If the celestial body is near the visible horizon of the observer on the earth's surface, its distance is about the same as it would be as viewed from the center of the earth. If the body is near the zenith, however, its distance is decreased by the radius of the earth. Hence, the body appears larger than it should when it is near the zenith of the observer, assuming its mean distance from earth is always the same. The augmentation correction for the sun is so small (a maximum of $\frac{1}{24}$ of 1″ of arc) that it can be ignored, while the augmentation correc-

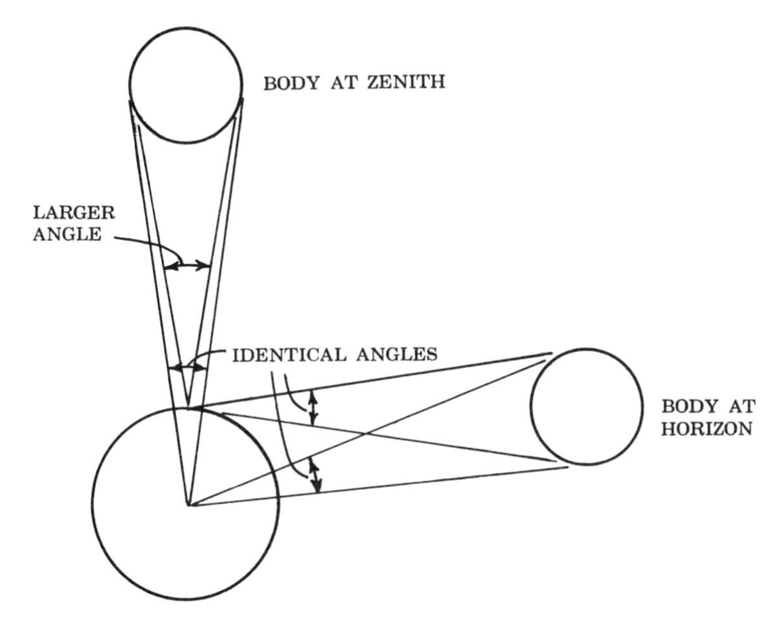

Figure 4-17. The augmentation effect.

tion for the moon as mentioned above is included in the moon "Altitude Correction Tables" on the inside back cover of the *Nautical Almanac* (Figures 4-14A and 4-14B).

The *phase* correction is necessary to compensate for the fact that the actual centers of the moon and the inferior planet Venus may differ from the apparent centers because of the phase of the body. A phase correction for the moon is unnecessary if its upper or lower limb is observed. A phase correction for Venus forms the main constituent of additional corrections for phase and parallax tabulated for that planet on the inside front cover of the *Nautical Almanac;* additional corrections for parallax for Mars are also tabulated. The other superior navigational planets Jupiter and Saturn, like Mars, have no phases, and they are so far distant that parallax corrections for them are too small to be significant. The entering arguments for the "Additional Correction" tables for Venus and Mars are the GMT date of the observation and the apparent altitude ha of the planet. If as occasionally happens these planets are observed by day, the additional corrections are not applied.

As an example of application of the corrections discussed in this section, suppose that the sextant altitude of Venus was observed on 1 December and recorded as 15° 13.5′. Height of eye is 15 feet and IC is −1.3′. Entering this information on the sight form, ha is first computed:

Body		VENUS
IC	+	-1.3' -
Dip (Ht 15 ')		-3.8'
Sum		-5.1'
hs		15° 13.5'
ha		15° 08.4'
Alt. Corr		
Add'l.		
H.P. ( )		
Corr. to ha		
Ho (Obs Alt)		

Using the resulting ha, 15° 08.4′, the altitude correction tables are entered for the basic altitude correction −3.5′. To complete the top of the sight form, the tables are entered once again for the additional correction for Venus. Entering arguments are the date of the observation and the ha.

If a correction for nonstandard temperature and barometric pressure conditions applied in this example, its value would be entered in the "Add'l" space on the sight form, along with the additional phase and parallax correction. These two additional corrections would then be summed with the basic altitude correction to form the total correction to ha.

Body		VENUS
IC	+	-1.3' -
Dip (Ht 15 ')		-3.8'
Sum		-5.1'
hs		15° 13.5'
ha		15° 08.4'
Alt. Corr		-3.5'
Add'l.	+.6	
H.P. ( )		
Corr. to ha		-2.9'
Ho (Obs Alt)		15° 05.5'

Adjustment to Equivalent Reading at the Center of the Earth

As was the case with semidiameter, the need to adjust the apparent altitude of a celestial body located within the solar system for the radius of the earth should be readily apparent. The difference in the apparent altitude of a body within the solar system as viewed from the surface of the earth and from its center is called *parallax.* The effects of parallax are greatest when the body is near the visible horizon, and least when the body is near the observer's zenith; the maximum value of parallax, occurring when a body is on the visible horizon, is called *horizontal parallax*, abbreviated *HP.*

In Figure 4-18, the angle formed at the center of the earth between the celestial horizon and the line of sight to the body, Ho, is greater than the corresponding angle ha formed at the earth's surface between the horizontal reference plane parallel to the celestial horizon and the line of sight to the body, because of the radius of the earth. The nearer the body is to the earth, the greater the difference in the two angles becomes. Hence, it follows that the effect of parallax for the moon is most pronounced, followed by the parallax of the sun and finally that of the planets and stars. A star is so far distant that incoming light rays from it are virtually parallel at all points between the earth's surface and center, thus rendering the effect of parallax negligible for a stellar observation. The effect for the moon, on the other hand, is so great that the values of horizontal parallax are tabulated for each hour in the daily pages of the *Nautical Almanac,* for use as described below in the altitude correction tables of the moon. The parallax correction for the sun is included in the altitude correction tables for the sun in the *Almanac,* and this correction forms the basis for the additional corrections for Mars and Venus mentioned earlier.

When using the altitude correction tables for the moon in the *Nautical Almanac,* the apparent altitude ha of either the upper or lower limb of the moon is the entering argument. The left-hand table of Figure 4-14A is used for values of ha between 0° and 35°, and the right-hand table of Figure 4-14B is for values between 35° and 90°. The total correction is in two parts, the first for the effects of refraction and augmentation, and the second for the effects of parallax. The first correction is obtained by entering the upper half of the appropriate

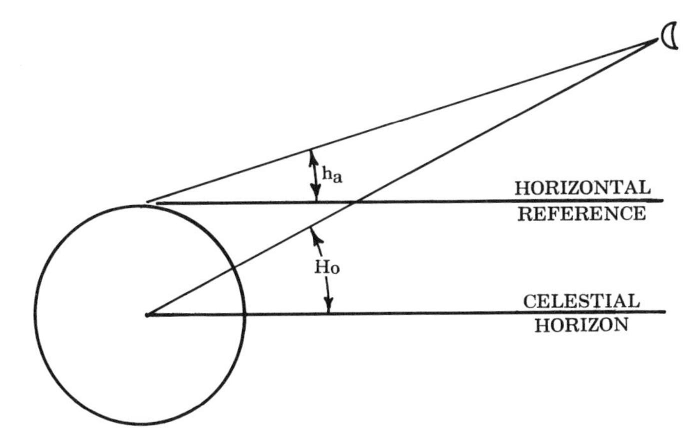

Figure 4-18. The effect of parallax.

table in the column containing values of ha; six values of the correction are given for each degree of the apparent altitude. The correction should be interpolated to the nearest tenth if an entering argument is between the tabulated values. To obtain the second correction, the column containing the entering ha is followed down to the lower half of the table. Here, two corrections are located in each column, one for upper limb observations (U) and the other for lower limb (L). The horizontal entering argument used in the lower half of the moon table is the horizontal parallax (HP) of the moon at the GMT of the observation, obtained from the daily pages of the *Almanac;* the HP extracted from the *Almanac* is the value tabulated for the whole hour of GMT closest to the GMT of the observation. Interpolation is performed in the lower table if the actual value of HP extracted from the *Almanac* lies between two tabulated values. After the HP correction has been obtained, it is added to the altitude correction obtained from the top half of the table. Finally, if the upper limb of the moon was observed, 30′ is then subtracted from this combined correction to yield the total altitude and parallax correction for the moon.

As an example of finding the Ho of the moon, suppose that the sextant altitude of the lower limb of the moon were obtained at GMT 02-45-04 on 15 December and found to be 49° 45.6′. Height of eye is 16 feet and IC is +1.5′. Using the *Nautical Almanac* and the sight form, an ha of 49° 43.2′ is computed. To enter the altitude correction tables for the moon shown in Figure 4-14B, the value of

DECEMBER 15, 16, 17 (SAT., SUN., MON.) 243

G.M.T.	SUN		MOON					Lat.	Twilight		Sun-rise	Moonrise			
	G.H.A.	Dec.	G.H.A.	v	Dec.	d	H.P.		Naut.	Civil		15	16	17	18
d h	° ′	° ′	° ′	′	° ′	′	′	°	h m	h m	h m	h m	h m	h m	h m
								N 72	08 21	10 49	■	23 01	25 03	01 03	03 07
15 00	181 16·4	S23 15·0	292 09·8	10·9	N 6 43·8	13·7	59·1	N 70	08 01	09 49	■	23 03	24 56	00 56	02 50
01	196 16·1	15·2	306 39·7	10·9	6 30·1	13·7	59·1	68	07 46	09 14	■	23 04	24 51	00 51	02 36
02	211 15·8	15·3	321 09·6	11·0	6 16·4	13·8	59·0	66	07 33	08 49	10 29	23 05	24 46	00 46	02 26
03	226 15·5	· · 15·4	335 39·6	11·1	6 02·6	13·8	59·0	64	07 22	08 30	09 47	23 06	24 42	00 42	02 17
04	241 15·2	15·6	350 09·7	11·2	5 48·8	13·8	59·0	62	07 12	08 14	09 19	23 07	24 39	00 39	02 09
05	256 14·9	15·7	4 39·9	11·2	5 35·0	13·8	58·9	60	07 04	08 00	08 58	23 08	24 36	00 36	02 02
06	271 14·6	S23 15·8	19 10·1	11·3	N 5 21·2	13·8	58·9	N 58	06 56	07 49	08 41	23 09	24 34	00 34	01 57
07	286 14·2	16·0	33 40·4	11·3	5 07·4	13·8	58·9	56	06 50	07 39	08 26	23 10	24 32	00 32	01 51
S 08	301 13·9	16·1	48 10·7	11·4	4 53·6	13·9	58·8	54	06 44	07 30	08 13	23 10	24 30	00 30	01 47
A 09	316 13·6	· · 16·2	62 41·1	11·5	4 39·7	13·8	58·8	52	06 38	07 22	08 02	23 11	24 28	00 28	01 43
T 10	331 13·3	16·4	77 11·6	11·5	4 25·9	13·9	58·8	50	06 33	07 14	07 52	23 11	24 26	00 26	01 39
U 11	346 13·0	16·5	91 42·1	11·6	4 12·0	13·8	58·7	45	06 21	06 58	07 32	23 12	24 22	00 22	01 31
R 12	1 12·7	S23 16·6	106 12·7	11·7	N 3 58·2	13·9	58·7	N 40	06 11	06 45	07 15	23 13	24 20	00 20	01 24
D 13	16 12·4	16·8	120 43·4	11·7	3 44·3	13·8	58·6	35	06 01	06 33	07 01	23 14	24 17	00 17	01 19
A 14	31 12·1	16·9	135 14·1	11·7	3 30·5	13·9	58·6	30	05 53	06 22	06 49	23 14	24 15	00 15	01 14
Y 15	46 11·8	· · 17·0	149 44·8	11·8	3 16·6	13·8	58·6	20	05 36	06 04	06 28	23 16	24 11	00 11	01 05
16	61 11·5	17·1	164 15·6	11·9	3 02·8	13·9	58·5	N 10	05 20	05 46	06 09	23 17	24 08	00 08	00 57
17	76 11·2	17·2	178 46·5	11·9	2 48·9	13·9	58·5	0	05 03	05 29	05 52	23 18	24 04	00 04	00 50
18	91 10·9	S23 17·4	193 17·4	12·0	N 2 35·0	13·8	58·5	S 10	04 44	05 11	05 34	23 19	24 01	00 01	00 44
19	106 10·6	17·5	207 48·4	12·0	2 21·2	13·8	58·4	20	04 22	04 51	05 16	23 20	23 58	24 36	00 36
20	121 10·3	17·6	222 19·4	12·1	2 07·4	13·9	58·4	30	03 53	04 26	04 54	23 21	23 54	24 28	00 28

Figure 4-19. Excerpt from right-hand daily page, Nautical Almanac, *15 December.*

the HP of the moon for the time and date of the observation must first be found. This information is obtained from the daily pages of the *Almanac*, a portion of which appears in Figure 4-19 for the day in question.

The HP of the moon listed for GMT 0300 on 15 December, the value closest to the GMT of the observation, is 59.0. Entering the moon altitude table of Figure 4-14B with ha, the 49° portion locates the observer in the third column. The tabulated correction for 49° 40.0′ is 47.2′; and for 49° 50.0′, 47.0′. Interpolation between these values for 49° 43.2′ yields an altitude correction of 47.1′. Next, the third column is followed down into the lower half of the table. Since a lower limb was shot, the figures under the column heading "L" will be used to obtain the parallax correction. Tabulated values for an HP of 58.8 is 6.0′; and for an HP of 59.1, 6.3′. Interpolation between these values for an HP of 59.0 results in an HP correction of 6.2′. Since the lower limb was observed, it remains only to enter these values in the sight form, and add them together for the total correction to ha:

Body	Moon (LL)	
IC	+1.5′ −	+
Dip (Ht 16 ′)	−3.9′	
Sum	−2.4′	
hs	49° 45.6′	
ha	49° 43.2′	
Alt. Corr	47.1′	
Add'l.		
H.P. (59.0)	6.2′	
Corr. to ha	53.3′	
Ho (Obs Att)	50° 36.5′	

The augmentation effect correction for lunar observations discussed in the preceding section can also be considered an adjustment of the apparent altitude at the earth's surface to obtain the desired observed altitude at the center of the earth.

Sextant Altitude Corrections Using the *Air Almanac*

As was stated at the beginning of the section on sextant altitude corrections, the *Air Almanac* can also be used to obtain corrections for each of the categories of possible error discussed in the preceding sections, but the *Nautical Almanac* is generally preferred in marine navigation because of the increased precision of the tabulated corrections. In the *Air Almanac* the refraction correction for all bodies is extracted from the same refraction table, using ha as the entering argument. When applicable, the effects of semidiameter, augmentation, phase, and parallax must be separately reckoned using data in the daily pages, and combined with the extracted refraction correction to form the total correction to ha for each body observed. An additional adjustment to the resulting line of position necessitated by the Coriolis effect on a fast-moving aircraft is also required in air navigation; tables for this correction are also contained in the *Air Almanac*. The use of the *Air Almanac* in determining sextant altitude corrections will be explained in Chapter 6.

Summary

This chapter has discussed the marine sextant and its role in the practice of celestial navigation at sea. While there are many different models of sextants available, all are designed to measure the angle between a celestial body and the visible horizon with great precision. Even the most precise instrument must be handled with skill, which develops only with experience, in order to produce observed sextant altitudes that yield highly accurate celestial LOPs. The care of his sextant and the skill with which he uses it are matters of great pride to the professional navigator.

After the sextant altitude of a celestial body relative to the visible horizon has been obtained, certain corrections must always be applied to account for various errors inherent in any sextant altitude observation. Certain adjustments must then be made to the resulting apparent altitude to obtain the observed altitude relative to the celestial horizon, upon which the solution of the celestial or navigational triangle is based. These corrections and adjustments are tabulated in several tables in both the *Nautical Almanac* and the *Air Almanac*. The corrections and adjustments applying to each of the bodies normally observed in celestial navigation when the *Nautical Almanac* is used are summarized below:

> Corrections for *IC* and *Dip* must be applied to all sextant altitude observations to obtain the apparent altitude; the apparent altitude is then used as an entering argument in the *Nautical Almanac* to find the appropriate corrections and adjustments for the body:
>
> > *Venus* and *Mars*—Corrections must be made for refraction, parallax, and phase by night, and refraction only by day. All are combined in the Altitude Correction Table A2 of the *Nautical Almanac*.
> >
> > Other *planets* and *stars*—A single correction must be made for refraction; it is obtained from Table A2 of the *Nautical Almanac*.

The *sun*—Corrections must be made for refraction, parallax, and semidiameter. All are combined in Table A2 of the *Nautical Almanac.*

The *moon*—Corrections must be made for refraction, parallax, semidiameter, and augmentation. Two combined corrections for all of these effects are obtained from the Moon Tables of the *Nautical Almanac.*

If the apparent altitude of the planets, stars, or the sun is less than 10°, Table A3 instead of A2 of the *Nautical Almanac* is used; if this low-altitude observation was made under non-standard temperature and barometric pressure conditions, an additional correction from Table A4 of the *Nautical Almanac* is applied to the apparent altitude.

In the following chapters, the procedures used to convert the observed altitude into a celestial line of position, and the combination of several such LOPs to form a celestial fix or running fix, will be presented. Chapter 5 explains the use of the *Nautical Almanac* in conjunction with the *Sight Reduction Tables for Marine Navigation* to obtain the quantities required by the altitude-intercept method; Chapter 6 demonstrates the use of the *Air Almanac* with the *Sight Reduction Tables for Air Navigation* as an alternative method of determining the same quantities; and Chapter 7 sets forth the procedures for plotting the celestial fix and running fix. Regardless of the publications employed, the basic techniques of sextant observation and the application of sextant altitude corrections described herein remain unchanged.

5

The Complete Solution by the *Nautical Almanac* and *Tables No. 229*

In Chapter 2, the technique of plotting the celestial LOP by the altitude-intercept method was demonstrated. It was pointed out that the method is based on the solution of the navigational triangle for the computed altitude and true azimuth of the body from an assumed position of the observer.

The process of deriving the information needed to plot a celestial line of position from an observation of a selected celestial body is called *sight reduction.* A complete set of calculations performed during the sight reduction process is referred to as the *complete solution* for the body observed. The *Sight Reduction Tables for Marine Navigation, No. 229,* yield a solution to the navigational triangle in the form of a computed altitude Hc and an azimuth angle Z for an observed celestial body from an assumed position at a given time of observation; the azimuth angle Z is then immediately converted to a true azimuth Zn to plot the celestial LOP by the altitude-intercept method. Entering arguments for *Tables No. 229* are three quantities from which two legs and an included vertex of the navigational triangle are derived—the assumed latitude (colatitude), the exact declination of the GP of the body (polar distance), and the local hour angle (meridian angle *t*). Thus, *Tables No. 229* are in reality a set of precalculated solutions for the third leg (coaltitude) and an interior angle (azimuth angle) of the navigational triangle, given the other two sides and the included angle between them.

The *Nautical Almanac* (or the *Air Almanac*) is used during sight reduction to provide a set of coordinates for the GP of the observed celestial body at the time of the observation—the exact declination and Greenwich hour angle (GHA). These coordinates, in combination with the terrestrial coordinates of the assumed position, form the three entering arguments for *Tables No. 229:* integral (whole) degrees of declination, local hour angle (from the GHA and assumed longitude), and assumed latitude.

In this chapter, the use of the *Nautical Almanac* in conjunction with the *Sight Reduction Tables for Marine Navigation, No. 229,* to obtain the complete solution of a celestial observation will be demonstrated. The first few sections will deal with the layout of the *Almanac* and examples of how to use it to determine the GHA and declination for the bodies observed in celestial navigation. The determination of the local hour angle will then be explained, followed by a discussion of the layout and use of *Tables No. 229.* In Chapter 6, the use of the *Air Almanac* with the *Sight Reduction Tables for Air Navigation, No. 249,* will be examined. While it is possible to use either almanac with either set of sight reduction tables, it has been found that the introduction of these publications to the student navigator is usually best accomplished by presenting them in this manner.

The *Nautical Almanac*

As has been mentioned, the *Nautical Almanac* is used during sight reduction to obtain the exact declination and GHA of a celestial body at the moment of its observation, for further use with the assumed position as entering arguments for the sight reduction tables. The *Nautical Almanac* is published annually in a single volume by the Naval Observatory in the United States, and by Her Majesty's Stationery Office in England. The data in each edition of the almanac is compiled for the sun, moon, navigational planets, and stars for the entire year for which the almanac is printed. The data pertaining to the sun and navigational stars may be used for the succeeding year as well, by following instructions given for this purpose in the "Explanation" section.

In the white daily pages comprising the body of the *Nautical Almanac,* the values of GHA and declination for each hour of Greenwich mean time (GMT) are given for a three-day period for Aries and the navigational planets on the left-hand pages, and for the sun and the moon on the right-hand pages. The left-hand pages also contain a listing of the sidereal hour angle (SHA) and declination of each of 57 so-called *navigational stars,* selected primarily on the basis of their magnitude. The listed values of SHA and declination change slightly from one three-day period to the next because of the combined effects of precession and nutation, described in Chapter 1, and *aberration.* This latter effect is caused by bending of the incoming light rays from the stars as a result of the earth's orbital velocity; the effect of aberration is cyclical, varying as the orbital velocity of the earth varies over the period of a year. Various tables used for the prediction of certain rising and setting phenomena appear on the right-hand pages in addition to the sun and moon data; the equation of time discussed in Chapter 3, and the times of the meridian passage and phase of the moon appear at the bottom of the right-hand pages as well. The use of these tables will be covered later in this text. A typical pair of daily pages appear in Figures 5-1A and 5-1B on the following pages.

G.M.T.		ARIES	VENUS −4·4		MARS +1·8		JUPITER −1·3		SATURN +0·1	
d	h	G.H.A.	G.H.A.	Dec.	G.H.A.	Dec.	G.H.A.	Dec.	G.H.A.	Dec.
		° ′	° ′	° ′	° ′	° ′	° ′	° ′	° ′	° ′
15	00	83 14·2	221 28·5	S13 14·2	229 56·4	S12 25·1	211 11·1	S17 58·5	38 16·8	N14 29·8
	01	98 16·7	236 29·7	14·3	244 57·4	25·7	226 13·0	58·7	53 19·4	29·7
	02	113 19·1	251 30·9	14·4	259 58·3	26·2	241 15·0	58·8	68 22·0	29·7
	03	128 21·6	266 32·1	· · 14·5	274 59·2	· · 26·8	256 16·9	· · 58·9	83 24·6	· · 29·7
	04	143 24·1	281 33·3	14·6	290 00·2	27·3	271 18·8	59·0	98 27·2	29·6
	05	158 26·5	296 34·5	14·7	305 01·1	27·8	286 20·8	59·2	113 29·8	29·6
	06	173 29·0	311 35·7	S13 14·8	320 02·0	S12 28·4	301 22·7	S17 59·3	128 32·4	N14 29·6
	07	188 31·5	326 36·8	14·9	335 03·0	28·9	316 24·7	59·4	143 35·0	29·5
T	08	203 33·9	341 38·0	15·0	350 03·9	29·5	331 26·6	59·5	158 37·6	29·5
U	09	218 36·4	356 39·2	· · 15·1	5 04·9	· · 30·0	346 28·5	· · 59·7	173 40·2	· · 29·5
E	10	233 38·9	11 40·4	15·2	20 05·8	30·5	1 30·5	59·8	188 42·8	29·4
S	11	248 41·3	26 41·6	15·4	35 06·7	31·1	16 32·4	17 59·9	203 45·4	29·4
D	12	263 43·8	41 42·7	S13 15·5	50 07·7	S12 31·6	31 34·3	S18 00·0	218 48·0	N14 29·4
A	13	278 46·3	56 43·9	15·6	65 08·6	32·1	46 36·3	00·2	233 50·7	29·4
Y	14	293 48·7	71 45·1	15·7	80 09·5	32·7	61 38·2	00·3	248 53·3	29·3
	15	308 51·2	86 46·2	· · 15·8	95 10·5	· · 33·2	76 40·2	· · 00·4	263 55·9	· · 29·3
	16	323 53·6	101 47·4	16·0	110 11·4	33·8	91 42·1	00·5	278 58·5	29·3
	17	338 56·1	116 48·5	16·1	125 12·3	34·3	106 44·0	00·7	294 01·1	29·2
	18	353 58·6	131 49·7	S13 16·2	140 13·3	S12 34·8	121 46·0	S18 00·8	309 03·7	N14 29·2
	19	9 01·0	146 50·8	16·3	155 14·2	35·4	136 47·9	00·9	324 06·3	29·2
	20	24 03·5	161 52·0	16·4	170 15·1	35·9	151 49·8	01·0	339 08·9	29·1
	21	39 06·0	176 53·1	· · 16·6	185 16·1	· · 36·5	166 51·8	· · 01·2	354 11·5	· · 29·1
	22	54 08·4	191 54·3	16·7	200 17·0	37·0	181 53·7	01·3	9 14·1	29·1
	23	69 10·9	206 55·4	16·8	215 17·9	37·5	196 55·7	01·4	24 16·7	29·0
16	00	84 13·4	221 56·6	S13 17·0	230 18·9	S12 38·1	211 57·6	S18 01·5	39 19·3	N14 29·0
	01	99 15·8	236 57·7	17·1	245 19·8	38·6	226 59·5	01·7	54 21·9	29·0
	02	114 18·3	251 58·8	17·2	260 20·7	39·1	242 01·5	01·8	69 24·5	29·0
	03	129 20·8	267 00·0	· · 17·4	275 21·7	· · 39·7	257 03·4	· · 01·9	84 27·1	· · 28·9
	04	144 23·2	282 01·1	17·5	290 22·6	40·2	272 05·4	02·0	99 29·7	28·9
	05	159 25·7	297 02·2	17·6	305 23·5	40·8	287 07·3	02·2	114 32·4	28·9
	06	174 28·1	312 03·4	S13 17·8	320 24·5	S12 41·3	302 09·2	S18 02·3	129 35·0	N14 28·8
W	07	189 30·6	327 04·5	17·9	335 25·4	41·8	317 11·2	02·4	144 37·6	28·8
E	08	204 33·1	342 05·6	18·1	350 26·3	42·4	332 13·1	02·5	159 40·2	28·8
D	09	219 35·5	357 06·7	· · 18·2	5 27·3	· · 42·9	347 15·0	· · 02·7	174 42·8	· · 28·7
N	10	234 38·0	12 07·8	18·4	20 28·2	43·4	2 17·0	02·8	189 45·4	28·7
E	11	249 40·5	27 09·0	18·5	35 29·1	44·0	17 18·9	02·9	204 48·0	28·7
S	12	264 42·9	42 10·1	S13 18·6	50 30·1	S12 44·5	32 20·9	S18 03·0	219 50·6	N14 28·7
D	13	279 45·4	57 11·2	18·8	65 31·0	45·0	47 22·8	03·2	234 53·2	28·6
A	14	294 47·9	72 12·3	18·9	80 31·9	45·6	62 24·7	03·3	249 55·8	28·6
Y	15	309 50·3	87 13·4	· · 19·1	95 32·9	· · 46·1	77 26·7	· · 03·4	264 58·4	· · 28·6
	16	324 52·8	102 14·5	19·2	110 33·8	46·6	92 28·6	03·5	280 01·0	28·5
	17	339 55·3	117 15·6	19·4	125 34·7	47·2	107 30·6	03·7	295 03·6	28·5
	18	354 57·7	132 16·7	S13 19·6	140 35·7	S12 47·7	122 32·5	S18 03·8	310 06·2	N14 28·5
	19	10 00·2	147 17·8	19·7	155 36·6	48·3	137 34·4	03·9	325 08·8	28·4
	20	25 02·6	162 18·8	19·9	170 37·5	48·8	152 36·4	04·0	340 11·4	28·4
	21	40 05·1	177 19·9	· · 20·0	185 38·5	· · 49·3	167 38·3	· · 04·2	355 14·0	· · 28·4
	22	55 07·6	192 21·0	20·2	200 39·4	49·9	182 40·3	04·3	10 16·6	28·4
	23	70 10·0	207 22·1	20·3	215 40·3	50·4	197 42·2	04·4	25 19·2	28·3
17	00	85 12·5	222 23·2	S13 20·5	230 41·3	S12 50·9	212 44·1	S18 04·5	40 21·8	N14 28·3
	01	100 15·0	237 24·3	20·7	245 42·2	51·5	227 46·1	04·7	55 24·4	28·3
	02	115 17·4	252 25·3	20·8	260 43·1	52·0	242 48·0	04·8	70 27·0	28·2
	03	130 19·9	267 26·4	· · 21·0	275 44·1	· · 52·5	257 50·0	· · 04·9	85 29·6	· · 28·2
	04	145 22·4	282 27·5	21·2	290 45·0	53·1	272 51·9	05·0	100 32·2	28·2
	05	160 24·8	297 28·5	21·3	305 45·9	53·6	287 53·8	05·2	115 34·8	28·2
	06	175 27·3	312 29·6	S13 21·5	320 46·9	S12 54·1	302 55·8	S18 05·3	130 37·4	N14 28·1
	07	190 29·7	327 30·7	21·7	335 47·8	54·7	317 57·7	05·4	145 40·0	28·1
T	08	205 32·2	342 31·7	21·8	350 48·7	55·2	332 59·7	05·5	160 42·6	28·1
H	09	220 34·7	357 32·8	· · 22·0	5 49·6	· · 55·7	348 01·6	· · 05·7	175 45·2	· · 28·0
U	10	235 37·1	12 33·8	22·2	20 50·6	56·3	3 03·5	05·8	190 47·8	28·0
R	11	250 39·6	27 34·9	22·4	35 51·5	56·8	18 05·5	05·9	205 50·4	28·0
S	12	265 42·1	42 36·0	S13 22·5	50 52·4	S12 57·3	33 07·4	S18 06·0	220 53·0	N14 27·9
D	13	280 44·5	57 37·0	22·7	65 53·4	57·8	48 09·4	06·2	235 55·6	27·9
A	14	295 47·0	72 38·0	22·9	80 54·3	58·4	63 11·3	06·3	250 58·2	27·9
Y	15	310 49·5	87 39·1	· · 23·1	95 55·2	· · 58·9	78 13·2	· · 06·4	266 00·8	· · 27·9
	16	325 51·9	102 40·1	23·3	110 56·2	12 59·4	93 15·2	06·5	281 03·4	27·8
	17	340 54·4	117 41·2	23·4	125 57·1	13 00·0	108 17·1	06·6	296 06·0	27·8
	18	355 56·9	132 42·2	S13 23·6	140 58·0	S13 00·5	123 19·1	S18 06·8	311 08·6	N14 27·8
	19	10 59·3	147 43·2	23·8	155 58·9	01·0	138 21·0	06·9	326 11·2	27·7
	20	26 01·8	162 44·3	24·0	170 59·9	01·6	153 22·9	07·0	341 13·8	27·7
	21	41 04·2	177 45·3	· · 24·2	186 00·8	· · 02·1	168 24·9	· · 07·1	356 16·4	· · 27·7
	22	56 06·7	192 46·3	24·4	201 01·7	02·6	183 26·8	07·3	11 19·0	27·7
	23	71 09·2	207 47·4	24·6	216 02·7	03·2	198 28·8	07·4	26 21·6	27·6
Mer. Pass.		h m 18 20·1	v 1·1	d 0·1	v 0·9	d 0·5	v 1·9	d 0·1	v 2·6	d 0·0

STARS

Name	S.H.A.	Dec.
	° ′	° ′
Acamar	315 42·1	S 40 25·2
Achernar	335 49·9	S 57 23·2
Acrux	173 45·6	S 62 56·1
Adhara	255 37·3	S 28 55·7
Aldebaran	291 25·7	N 16 27·3
Alioth	166 48·6	N 56 06·7
Alkaid	153 24·1	N 49 27·1
Al Na'ir	28 23·8	S 47 06·4
Alnilam	276 18·4	S 1 13·0
Alphard	218 27·3	S 8 31·8
Alphecca	126 38·3	N 26 48·5
Alpheratz	358 16·7	N 28 56·1
Altair	62 39·7	N 8 47·4
Ankaa	353 46·9	S 42 28·0
Antares	113 05·9	S 26 22·2
Arcturus	146 25·0	N 19 19·8
Atria	108 37·1	S 68 58·7
Avior	234 30·8	S 59 24·7
Bellatrix	279 05·9	N 6 19·6
Betelgeuse	271 35·5	N 7 24·3
Canopus	264 09·8	S 52 40·6
Capella	281 21·2	N 45 58·4
Deneb	49 53·7	N 45 10·7
Denebola	183 06·2	N 14 43·9
Diphda	349 27·8	S 18 08·8
Dubhe	194 30·3	N 61 54·2
Elnath	278 52·6	N 28 35·2
Eltanin	91 01·5	N 51 29·4
Enif	34 18·7	N 9 44·5
Fomalhaut	15 59·1	S 29 46·7
Gacrux	172 37·0	S 56 56·8
Gienah	176 25·3	S 17 22·8
Hadar	149 34·0	S 60 13·9
Hamal	328 36·7	N 23 19·8
Kaus Aust.	84 26·6	S 34 24·1
Kochab	137 19·1	N 74 16·1
Markab	14 10·3	N 15 03·1
Menkar	314 48·2	N 3 58·8
Menkent	148 45·7	S 36 13·6
Miaplacidus	221 46·3	S 69 35·5
Mirfak	309 25·9	N 49 45·8
Nunki	76 38·2	S 26 20·2
Peacock	54 09·7	S 56 50·0
Pollux	244 06·4	N 28 05·9
Procyon	245 32·8	N 5 18·1
Rasalhague	96 36·5	N 12 34·7
Regulus	208 17·3	N 12 06·5
Rigel	281 42·4	S 8 13·9
Rigil Kent.	140 36·2	S 60 42·8
Sabik	102 49·6	S 15 41·5
Schedar	350 17·1	N 56 23·1
Shaula	97 05·7	S 37 05·2
Sirius	259 01·6	S 16 40·4
Spica	159 05·1	S 11 00·6
Suhail	223 15·8	S 43 18·6
Vega	81 01·0	N 38 45·3
Zuben'ubi	137 41·1	S 15 55·4

	S.H.A.	Mer. Pass.
	° ′	h m
Venus	137 43·2	9 12
Mars	146 05·5	8 38
Jupiter	127 44·2	9 51
Saturn	315 06·0	21 19

Figure 5-1A. Left-hand daily page, Nautical Almanac, *15, 16, 17 December.*

G.M.T. d	h	SUN G.H.A. ° ′	SUN Dec. ° ′	MOON G.H.A. ° ′	v ′	Dec. ° ′	d ′	H.P. ′
15	00	181 17·8	S23 14·3	333 25·4	9·4	N25 19·4	6·5	55·8
	01	196 17·5	14·4	347 53·8	9·4	25 12·9	6·7	55·7
	02	211 17·2	14·5	2 22·2	9·5	25 06·2	6·7	55·7
	03	226 16·9	.. 14·7	16 50·7	9·7	24 59·5	6·9	55·7
	04	241 16·7	14·8	31 19·4	9·7	24 52·6	7·0	55·7
	05	256 16·4	15·0	45 48·1	9·8	24 45·6	7·1	55·6
	06	271 16·1	S23 15·1	60 16·9	9·9	N24 38·5	7·2	55·6
	07	286 15·8	15·2	74 45·8	10·0	24 31·3	7·3	55·6
TUESDAY	08	301 15·5	15·4	89 14·8	10·1	24 24·0	7·5	55·6
	09	316 15·2	.. 15·5	103 43·9	10·2	24 16·5	7·6	55·6
	10	331 14·9	15·6	118 13·1	10·3	24 08·9	7·7	55·5
	11	346 14·6	15·8	132 42·4	10·4	24 01·2	7·8	55·5
	12	1 14·3	S23 15·9	147 11·8	10·5	N23 53·4	7·9	55·5
	13	16 14·0	16·0	161 41·3	10·5	23 45·5	8·0	55·5
	14	31 13·7	16·2	176 10·8	10·7	23 37·5	8·1	55·4
	15	46 13·4	.. 16·3	190 40·5	10·8	23 29·4	8·3	55·4
	16	61 13·1	16·4	205 10·3	10·9	23 21·1	8·3	55·4
	17	76 12·8	16·6	219 40·2	10·9	23 12·8	8·5	55·4
	18	91 12·5	S23 16·7	234 10·1	11·1	N23 04·3	8·5	55·4
	19	106 12·1	16·8	248 40·2	11·1	22 55·8	8·7	55·3
	20	121 11·8	16·9	263 10·3	11·3	22 47·1	8·7	55·3
	21	136 11·5	.. 17·1	277 40·6	11·3	22 38·4	8·9	55·3
	22	151 11·2	17·2	292 10·9	11·5	22 29·5	8·9	55·3
	23	166 10·9	17·3	306 41·4	11·5	22 20·6	9·1	55·2
16	00	181 10·6	S23 17·4	321 11·9	11·7	N22 11·5	9·1	55·2
	01	196 10·3	17·6	335 42·6	11·7	22 02·4	9·3	55·2
	02	211 10·0	17·7	350 13·3	11·8	21 53·1	9·3	55·2
	03	226 09·7	.. 17·8	4 44·1	11·9	21 43·8	9·4	55·2
	04	241 09·4	17·9	19 15·0	12·1	21 34·4	9·6	55·1
	05	256 09·1	18·0	33 46·1	12·1	21 24·8	9·6	55·1
	06	271 08·8	S23 18·2	48 17·2	12·2	N21 15·2	9·7	55·1
WEDNESDAY	07	286 08·5	18·3	62 48·4	12·3	21 05·5	9·8	55·1
	08	301 08·2	18·4	77 19·7	12·4	20 55·7	9·8	55·1
	09	316 07·9	.. 18·5	91 51·1	12·5	20 45·9	10·0	55·0
	10	331 07·6	18·6	106 22·6	12·5	20 35·9	10·0	55·0
	11	346 07·3	18·7	120 54·1	12·7	20 25·9	10·2	55·0
	12	1 07·0	S23 18·9	135 25·8	12·8	N20 15·7	10·2	55·0
	13	16 06·7	19·0	149 57·6	12·8	20 05·5	10·3	55·0
	14	31 06·4	19·1	164 29·4	12·9	19 55·2	10·3	55·0
	15	46 06·1	.. 19·2	179 01·3	13·1	19 44·9	10·5	54·9
	16	61 05·8	19·3	193 33·4	13·1	19 34·4	10·5	54·9
	17	76 05·5	19·4	208 05·5	13·2	19 23·9	10·6	54·9
	18	91 05·2	S23 19·5	222 37·7	13·3	N19 13·3	10·7	54·9
	19	106 04·9	19·6	237 10·0	13·4	19 02·6	10·7	54·9
	20	121 04·6	19·7	251 42·4	13·4	18 51·9	10·8	54·8
	21	136 04·3	.. 19·8	266 14·8	13·6	18 41·1	10·9	54·8
	22	151 04·0	19·9	280 47·4	13·6	18 30·2	11·0	54·8
	23	166 03·7	20·0	295 20·0	13·7	18 19·2	11·0	54·8
17	00	181 03·4	S23 20·2	309 52·7	13·8	N18 08·2	11·1	54·8
	01	196 03·1	20·3	324 25·5	13·9	17 57·1	11·2	54·8
	02	211 02·8	20·4	338 58·4	14·0	17 45·9	11·2	54·7
	03	226 02·5	.. 20·5	353 31·4	14·0	17 34·7	11·3	54·7
	04	241 02·2	20·6	8 04·4	14·2	17 23·4	11·4	54·7
	05	256 01·9	20·7	22 37·6	14·2	17 12·0	11·4	54·7
	06	271 01·6	S23 20·8	37 10·8	14·3	N17 00·6	11·5	54·7
THURSDAY	07	286 01·2	20·9	51 44·1	14·3	16 49·1	11·5	54·7
	08	301 00·9	21·0	66 17·4	14·5	16 37·6	11·7	54·7
	09	316 00·6	.. 21·0	80 50·9	14·5	16 25·9	11·6	54·6
	10	331 00·3	21·1	95 24·4	14·6	16 14·3	11·8	54·6
	11	346 00·0	21·2	109 58·0	14·7	16 02·5	11·7	54·6
	12	0 59·7	S23 21·3	124 31·7	14·7	N15 50·8	11·9	54·6
	13	15 59·4	21·4	139 05·4	14·8	15 38·9	11·9	54·6
	14	30 59·1	21·5	153 39·2	14·9	15 27·0	11·9	54·6
	15	45 58·8	.. 21·6	168 13·1	15·0	15 15·1	12·0	54·6
	16	60 58·5	21·7	182 47·1	15·0	15 03·1	12·1	54·5
	17	75 58·2	21·8	197 21·1	15·1	14 51·0	12·1	54·5
	18	90 57·9	S23 21·9	211 55·2	15·1	N14 38·9	12·2	54·5
	19	105 57·6	22·0	226 29·3	15·3	14 26·7	12·2	54·5
	20	120 57·3	22·1	241 03·6	15·3	14 14·5	12·2	54·5
	21	135 57·0	.. 22·1	255 37·9	15·3	14 02·3	12·4	54·5
	22	150 56·7	22·2	270 12·2	15·5	13 49·9	12·3	54·5
	23	165 56·4	22·3	284 46·7	15·4	13 37·6	12·4	54·5
		S.D. 16·3	d 0·1	S.D. 15·1		15·0		14·9

Lat.	Twilight Naut. h m	Twilight Civil h m	Sunrise h m	Moonrise 15 h m	16 h m	17 h m	18 h m
N 72	08 20	10 48	■	□	□	18 22	20 42
N 70	08 01	09 48	■	□	16 01	18 55	20 56
68	07 45	09 14	■	□	17 08	19 18	21 08
66	07 32	08 49	10 28	15 14	17 44	19 36	21 17
64	07 21	08 29	09 47	16 14	18 09	19 51	21 25
62	07 12	08 13	09 19	16 47	18 29	20 03	21 31
60	07 03	08 00	08 58	17 11	18 45	20 13	21 37
N 58	06 56	07 48	08 40	17 31	18 58	20 22	21 42
56	06 49	07 38	08 26	17 47	19 10	20 30	21 47
54	06 43	07 29	08 13	18 01	19 20	20 36	21 51
52	06 38	07 21	08 02	18 13	19 29	20 43	21 54
50	06 32	07 14	07 52	18 23	19 37	20 48	21 58
45	06 21	06 58	07 32	18 45	19 54	21 00	22 05
N 40	06 10	06 44	07 15	19 03	20 07	21 10	22 10
35	06 01	06 33	07 01	19 18	20 19	21 18	22 16
30	05 52	06 22	06 49	19 31	20 29	21 26	22 20
20	05 36	06 04	06 28	19 52	20 47	21 38	22 28
N 10	05 20	05 46	06 09	20 11	21 02	21 49	22 34
0	05 03	05 29	05 52	20 29	21 16	22 00	22 41
S 10	04 44	05 11	05 34	20 46	21 30	22 10	22 47
20	04 22	04 51	05 15	21 05	21 45	22 21	22 53
30	03 53	04 26	04 53	21 26	22 02	22 33	23 01
35	03 34	04 11	04 41	21 39	22 12	22 40	23 05
40	03 11	03 53	04 26	21 53	22 23	22 48	23 10
45	02 41	03 30	04 08	22 10	22 36	22 58	23 16
S 50	01 56	03 01	03 45	22 31	22 52	23 09	23 22
52	01 28	02 46	03 34	22 41	23 00	23 14	23 26
54	00 44	02 28	03 22	22 52	23 08	23 20	23 29
56	////	02 06	03 08	23 04	23 17	23 26	23 33
58	////	01 37	02 52	23 19	23 28	23 33	23 37
S 60	////	00 49	02 31	23 36	23 40	23 41	23 41

Lat.	Sunset h m	Twilight Civil h m	Twilight Naut. h m	Moonset 15 h m	16 h m	17 h m	18 h m
N 72	■	13 02	15 30	□	□	13 37	12 45
N 70	■	14 02	15 49	□	14 25	13 03	12 28
68	■	14 36	16 05	□	13 16	12 38	12 15
66	13 22	15 01	16 18	13 29	12 40	12 19	12 04
64	14 04	15 21	16 29	12 29	12 13	12 03	11 55
62	14 31	15 37	16 39	11 55	11 53	11 50	11 47
60	14 53	15 50	16 47	11 30	11 36	11 39	11 40
N 58	15 10	16 02	16 54	11 10	11 22	11 29	11 34
56	15 25	16 12	17 01	10 53	11 10	11 20	11 28
54	15 38	16 21	17 07	10 39	10 59	11 13	11 23
52	15 49	16 29	17 13	10 27	10 49	11 06	11 19
50	15 58	16 37	17 18	10 16	10 41	11 00	11 15
45	16 19	16 53	17 30	09 53	10 23	10 46	11 06
N 40	16 36	17 06	17 41	09 34	10 08	10 35	10 58
35	16 50	17 18	17 50	09 19	09 55	10 25	10 52
30	17 02	17 29	17 59	09 05	09 44	10 17	10 46
20	17 23	17 48	18 15	08 42	09 25	10 02	10 36
N 10	17 42	18 05	18 31	08 22	09 08	09 49	10 28
0	17 59	18 22	18 48	08 03	08 52	09 37	10 20
S 10	18 17	18 40	19 07	07 44	08 36	09 25	10 11
20	18 36	19 00	19 29	07 23	08 19	09 12	10 02
30	18 58	19 25	19 58	06 59	07 59	08 57	09 52
35	19 11	19 40	20 17	06 45	07 47	08 48	09 46
40	19 26	19 58	20 40	06 29	07 34	08 38	09 39
45	19 44	20 21	21 10	06 09	07 18	08 26	09 31
S 50	20 06	20 50	21 56	05 45	06 58	08 11	09 22
52	20 17	21 06	22 23	05 33	06 49	08 04	09 17
54	20 29	21 23	23 08	05 19	06 39	07 57	09 12
56	20 43	21 45	////	05 03	06 26	07 48	09 07
58	21 00	22 15	////	04 44	06 13	07 39	09 01
S 60	21 20	23 04	////	04 21	05 56	07 27	08 54

Day	SUN Eqn. of Time 00ʰ m s	12ʰ m s	Mer. Pass. h m	MOON Mer. Pass. Upper h m	Lower h m	Age d	Phase
15	05 12	04 58	11 55	01 50	14 16	17	
16	04 43	04 29	11 56	02 40	15 04	18	
17	04 14	04 00	11 56	03 27	15 49	19	

Figure 5-1B. Right-hand daily page, Nautical Almanac, *15, 16, 17 December.*

The extraction of the GHA and declination from the *Nautical Almanac* varies somewhat for each of the four types of bodies observed in celestial navigation. These variations are best illustrated by means of examples. Consequently, the following four sections of this chapter will each present an example of the use of the *Nautical Almanac* to find the GHA and declination of one of the four types of bodies normally observed—a star, a planet, the sun, and the moon. In each of the following examples, the observed altitude Ho of the body will be considered already determined by the methods discussed in Chapter 4.

Because of the many additions, subtractions, and interpolations necessary during sight reduction, it is customary to use a standardized form called a *slight reduction form* on which to record the various calculations. The use of the form developed by the U.S. Naval Academy for the *Nautical Almanac* and *Tables No. 229* will be demonstrated herein; the use of the top portion of this form to assist in the determination of the observed altitude Ho has already been presented in Chapter 4.

Determining the GHA and Declination of a Star

As an example of the use of the *Nautical Almanac* to obtain the GHA and declination of a star for a given time, suppose that the star Canopus was observed and was found to have an Ho of 50° 39.4′ at a watch time of 04-11-33 on 16 December. The watch has a watch error (WE) of 36 seconds slow (S), and the ship's DR position at the time of the observation was L 34° 19.0′S, λ 163° 05.7′ E.

In order to enter the *Almanac*, the exact GMT and date of the time of observation must first be calculated, since this is the basis by which all hourly values of coordinates in the daily pages of the *Almanac* are tabulated. To assist in the conversion from local zone time (ZT) to GMT, the time diagram on the top of the sight reduction form may be used. Later, the time diagram will again be of assistance in combining the GHA of the body with the assumed longitude to form the LHA.

Figure 5-2 depicts the Canopus sight form as it appears after the computation of the GMT of the observation. The slow watch error (WE) was added to the watch time of the observation (Obs. Time) to obtain the zone time (ZT) of the observation, and the zone time was then converted to the GMT time and date. Note the use of dashes in place of degree (°) signs in the DR latitude and longitude; the minutes sign (′) has also been omitted. In practice, positions and angles are always written on the sight form in this matter, to facilitate the mathematic operations performed during sight reduction.

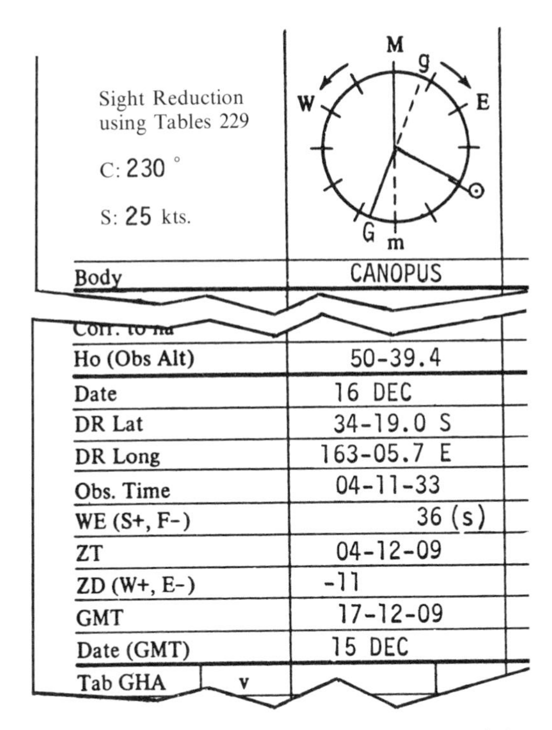

Sight Reduction using Tables 229

C: 230 °

S: 25 kts.

M, g, W, E, G, m

Body	CANOPUS
Corr. to ha	
Ho (Obs Alt)	50-39.4
Date	16 DEC
DR Lat	34-19.0 S
DR Long	163-05.7 E
Obs. Time	04-11-33
WE (S+, F-)	36 (s)
ZT	04-12-09
ZD (W+, E-)	-11
GMT	17-12-09
Date (GMT)	15 DEC
Tab GHA v	

Figure 5-2. Portion of Canopus sight form, GMT time and date computed.

After the GMT time and date of the observation have been determined, the GHA of Canopus can be found. Since Canopus is a star, its GHA is not separately tabulated, but rather it must be computed by the use of the formula

$$\text{GHA}☆ = \text{GHA}♈ + \text{SHA}☆$$

previously introduced in Chapter 1.

To find the GHA♈ for the time of the observation, its tabulated value for the whole hour of GMT immediately preceding the observation is first obtained and recorded on the sight form as the "Tabulated (Tab.) GHA." From Figure 5-1A, the tabulation for 1700 GMT on 15 December is 338° 56.1′. The "v" on the form alongside the Tab. GHA applies only to GHA calculations for the planets and the moon, and will be explained later. Next, the amount of the additional movement of Aries during the remaining 12 minutes 9 seconds until 07-12-09 GMT must be determined and added to the tabulated GHA. This additional movement of Aries is referred to as the "GHA increment." A set of "Increments and Corrections" tables designed for obtaining this and certain other related increments and corrections are printed on yellow pages *ii* through *xxxi* located in the back of the *Almanac*.

There is a separate table for each additional minute of GMT, and for each minute, the various increments and corrections corresponding with each of the 60 seconds comprising the minute are listed in vertical columns. Figure 5-3 pictures the "Increments and Corrections" table for 12 minutes. From this table the GHA Aries increment for 12 minutes 9 seconds is found to be 3° 02.7′; this figure is inserted on the form.

The SHA of Canopus listed on the daily page shown in Figure 5-1A is 264° 09.8′. After this value has been entered on the form, the GHA of Canopus is computed:

$$\text{GHA Canopus} = \underbrace{338^\circ\ 56.1' + 3^\circ\ 02.7'}_{\text{GHA}\,\Upsilon} + \underbrace{264^\circ\ 09.8'}_{\text{SHA}\,\star} = 606^\circ\ 08.6'$$

Since this result is greater than 360°, 360° is subtracted to yield the equivalent GHA of Canopus of 246° 08.6′.

With the GHA computed, the hour circle of Canopus can now be located relative to the Greenwich meridian (G) on the time diagram, for later use in determining the LHA.

To complete the use of the *Almanac* for Canopus, the declination of Canopus is obtained from its listing alongside the SHA on the daily page of Figure 5-1A and entered on the form as the "True Declination (True Dec.)." The form now appears as in Figure 5-4, page 98. After the LHA has been computed by methods to be discussed shortly, the form will be ready for use with *Sight Reduction Tables No. 229.*

Determining the GHA and Declination of a Planet

The use of the *Nautical Almanac* to find the GHA and declination of a navigational planet differs somewhat from the procedure just described for a star, because both the GHA and declination for the navigational planets are tabulated in the *Almanac* for each whole hour of GMT. These tabulations are necessitated by the fact that the planets, being relatively close to earth in the solar system, seem to move across the unchanging patterns of the stars from one hour to the next.

Suppose that the planet Venus was observed at watch time 04-03-36 on the same occasion as the star Canopus, and its observed altitude was determined to be 16° 38.6′. The DR position at the time of the sight was L 34° 17.0′S, λ 163° 09.1′E. After the GMT time and date are computed, the sight form appears as in Figure 5-5A, page 99.

The *Nautical Almanac* is first entered to compute the GHA of Venus for the time of the observation. The tabulated value of the GHA of Venus for the whole hour of GMT immediately preceding the observation is recorded on the form. In addition, the *v* value at the bottom

12ᵐ

12ᵐ	SUN PLANETS	ARIES	MOON	v or d	Corrⁿ	v or d	Corrⁿ	v or d	Corrⁿ
s	° ′	° ′	° ′	′	′	′	′	′	′
00	3 00·0	3 00·5	2 51·8	0·0	0·0	6·0	1·3	12·0	2·5
01	3 00·3	3 00·7	2 52·0	0·1	0·0	6·1	1·3	12·1	2·5
02	3 00·5	3 01·0	2 52·3	0·2	0·0	6·2	1·3	12·2	2·5
03	3 00·8	3 01·2	2 52·5	0·3	0·1	6·3	1·3	12·3	2·6
04	3 01·0	3 01·5	2 52·8	0·4	0·1	6·4	1·3	12·4	2·6
05	3 01·3	3 01·7	2 53·0	0·5	0·1	6·5	1·4	12·5	2·6
06	3 01·5	3 02·0	2 53·2	0·6	0·1	6·6	1·4	12·6	2·6
07	3 01·8	3 02·2	2 53·5	0·7	0·1	6·7	1·4	12·7	2·6
08	3 02·0	3 02·5	2 53·7	0·8	0·2	6·8	1·4	12·8	2·7
09	3 02·3	3 02·7	2 53·9	0·9	0·2	6·9	1·4	12·9	2·7
10	3 02·5	3 03·0	2 54·2	1·0	0·2	7·0	1·5	13·0	2·7
11	3 02·8	3 03·3	2 54·4	1·1	0·2	7·1	1·5	13·1	2·7
12	3 03·0	3 03·5	2 54·7	1·2	0·3	7·2	1·5	13·2	2·8
13	3 03·3	3 03·8	2 54·9	1·3	0·3	7·3	1·5	13·3	2·8
14	3 03·5	3 04·0	2 55·1	1·4	0·3	7·4	1·5	13·4	2·8
15	3 03·8	3 04·3	2 55·4	1·5	0·3	7·5	1·6	13·5	2·8
16	3 04·0	3 04·5	2 55·6	1·6	0·3	7·6	1·6	13·6	2·8
17	3 04·3	3 04·8	2 55·9	1·7	0·4	7·7	1·6	13·7	2·9
18	3 04·5	3 05·0	2 56·1	1·8	0·4	7·8	1·6	13·8	2·9
19	3 04·8	3 05·3	2 56·3	1·9	0·4	7·9	1·6	13·9	2·9
20	3 05·0	3 05·5	2 56·6	2·0	0·4	8·0	1·7	14·0	2·9
21	3 05·3	3 05·8	2 56·8	2·1	0·4	8·1	1·7	14·1	2·9
22	3 05·5	3 06·0	2 57·0	2·2	0·5	8·2	1·7	14·2	3·0
23	3 05·8	3 06·3	2 57·3	2·3	0·5	8·3	1·7	14·3	3·0
24	3 06·0	3 06·5	2 57·5	2·4	0·5	8·4	1·8	14·4	3·0
25	3 06·3	3 06·8	2 57·8	2·5	0·5	8·5	1·8	14·5	3·0
26	3 06·5	3 07·0	2 58·0	2·6	0·5	8·6	1·8	14·6	3·0
27	3 06·8	3 07·3	2 58·2	2·7	0·6	8·7	1·8	14·7	3·1
28	3 07·0	3 07·5	2 58·5	2·8	0·6	8·8	1·8	14·8	3·1
29	3 07·3	3 07·8	2 58·7	2·9	0·6	8·9	1·9	14·9	3·1
30	3 07·5	3 08·0	2 59·0	3·0	0·6	9·0	1·9	15·0	3·1
31	3 07·8	3 08·3	2 59·2	3·1	0·6	9·1	1·9	15·1	3·1
32	3 08·0	3 08·5	2 59·4	3·2	0·7	9·2	1·9	15·2	3·2
33	3 08·3	3 08·8	2 59·7	3·3	0·7	9·3	1·9	15·3	3·2
34	3 08·5	3 09·0	2 59·9	3·4	0·7	9·4	2·0	15·4	3·2
35	3 08·8	3 09·3	3 00·2	3·5	0·7	9·5	2·0	15·5	3·2
36	3 09·0	3 09·5	3 00·4	3·6	0·8	9·6	2·0	15·6	3·3
37	3 09·3	3 09·8	3 00·6	3·7	0·8	9·7	2·0	15·7	3·3
38	3 09·5	3 10·0	3 00·9	3·8	0·8	9·8	2·0	15·8	3·3
39	3 09·8	3 10·3	3 01·1	3·9	0·8	9·9	2·1	15·9	3·3
40	3 10·0	3 10·5	3 01·3	4·0	0·8	10·0	2·1	16·0	3·3
41	3 10·3	3 10·8	3 01·6	4·1	0·9	10·1	2·1	16·1	3·4
42	3 10·5	3 11·0	3 01·8	4·2	0·9	10·2	2·1	16·2	3·4
43	3 10·8	3 11·3	3 02·1	4·3	0·9	10·3	2·1	16·3	3·4
44	3 11·0	3 11·5	3 02·3	4·4	0·9	10·4	2·2	16·4	3·4
45	3 11·3	3 11·8	3 02·5	4·5	0·9	10·5	2·2	16·5	3·4
46	3 11·5	3 12·0	3 02·8	4·6	1·0	10·6	2·2	16·6	3·5
47	3 11·8	3 12·3	3 03·0	4·7	1·0	10·7	2·2	16·7	3·5
48	3 12·0	3 12·5	3 03·3	4·8	1·0	10·8	2·3	16·8	3·5
49	3 12·3	3 12·8	3 03·5	4·9	1·0	10·9	2·3	16·9	3·5
50	3 12·5	3 13·0	3 03·7	5·0	1·0	11·0	2·3	17·0	3·5
51	3 12·8	3 13·3	3 04·0	5·1	1·1	11·1	2·3	17·1	3·6
52	3 13·0	3 13·5	3 04·2	5·2	1·1	11·2	2·3	17·2	3·6
53	3 13·3	3 13·8	3 04·4	5·3	1·1	11·3	2·4	17·3	3·6
54	3 13·5	3 14·0	3 04·7	5·4	1·1	11·4	2·4	17·4	3·6
55	3 13·8	3 14·3	3 04·9	5·5	1·1	11·5	2·4	17·5	3·6
56	3 14·0	3 14·5	3 05·2	5·6	1·2	11·6	2·4	17·6	3·7
57	3 14·3	3 14·8	3 05·4	5·7	1·2	11·7	2·4	17·7	3·7
58	3 14·5	3 15·0	3 05·6	5·8	1·2	11·8	2·5	17·8	3·7
59	3 14·8	3 15·3	3 05·9	5·9	1·2	11·9	2·5	17·9	3·7
60	3 15·0	3 15·5	3 06·1	6·0	1·3	12·0	2·5	18·0	3·8

Figure 5-3. "Increments and Corrections," 12 minutes, Nautical Almanac.

Sight Reduction using Tables 229

C: 230 °

S: 25 kts.

M g W E G m

Body		CANOPUS
Corr. to ha		
Ho (Obs Alt)		50-39.4
Date		16 DEC
DR Lat		34-19.0 S
DR Long		163-05.7 E
Obs. Time		04-11-33
WE (S+, F-)		36 (s)
ZT		04-12-09
ZD (W+, E-)		-11
GMT		17-12-09
Date (GMT)		15 DEC
Tab GHA	v	338-56.1
GHA incr'mt.		3-02.7
SHA or v Corr.		264-09.8
GHA		606-08.6
±360 if needed		246-08.6
aλ (-W, +E)		
LHA		
Tab Dec	d	
d Corr (+ or −)		
True Dec		S 52-40.6
a Lat (N or S)		Same Cont.

Figure 5-4. Canopus sight reduction form, GHA and declination determined.

of the column containing the Venus tabulations is noted and recorde on the form alongside the tabulated GHA. This *v* value represents th average irregularity in the rate of increase of the GHA of Venus fror one hour to the next over the three-day period covered by the dail page, as a result of the motion of the planet in its orbit. Because th rate of increase of the GHA is not constant, a *v* correction derive from this *v* value must be applied to the tabulated GHA of the plane in addition to the GHA increment. The *v* correction, like the GH increment, is taken from the appropriate "Increments and Corre tions" table in the back of the *Almanac.*

It should be mentioned here that all bodies in the solar system ar characterized by this irregular rate of increase of GHA. Thus,

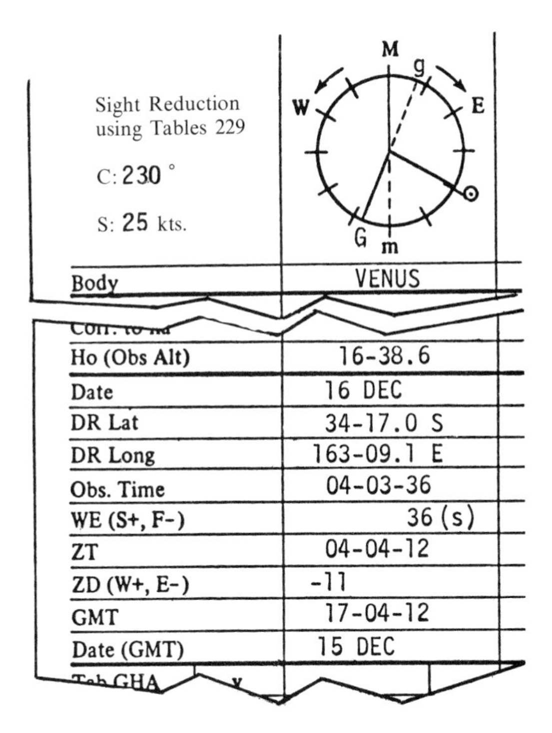

Figure 5-5A. Portion of Venus sight form, GMT time and date computed.

v correction is applied to the GHA increments of all the navigational planets and the moon; the *v* correction for the sun is so small that it is ignored when using the *Nautical Almanac.* The *v* corrections for Mars, Jupiter, Saturn, and the moon are always positive; at certain times of the year, the Venus *v* correction can be negative.

Returning to the example at hand, a GHA increment for 4 minutes 12 seconds is found from the 4-minute "Increments and Corrections" table and entered on the sight form; a portion of the 4-minute table is shown in Figure 5-5B on the following page. To find the *v* correction, the columns in the appropriate table with the headings

v
"or Corr^{n}"
d

are used (the *d* correction will be explained below). The left side of each column contains *v* or *d* values, and the right side, the corresponding *v* or *d* correction. In this case, a *v* correction of +0.1′ corresponding to a *v* value of +1.1 is obtained. The tabulated GHA, GHA increment, and *v* correction are all summed to yield the

4m	SUN PLANETS	ARIES	MOON	v or d	Corrⁿ	v or d	Corrⁿ	v or d	Corrⁿ
s	° ′	° ′	° ′	′	′	′	′	′	′
00	1 00·0	1 00·2	0 57·3	0·0	0·0	6·0	0·5	12·0	0·9
01	1 00·3	1 00·4	0 57·5	0·1	0·0	6·1	0·5	12·1	0·9
02	1 00·5	1 00·7	0 57·7	0·2	0·0	6·2	0·5	12·2	0·9
03	1 00·8	1 00·9	0 58·0	0·3	0·0	6·3	0·5	12·3	0·9
04	1 01·0	1 01·2	0 58·2	0·4	0·0	6·4	0·5	12·4	0·9
05	1 01·3	1 01·4	0 58·5	0·5	0·0	6·5	0·5	12·5	0·9
06	1 01·5	1 01·7	0 58·7	0·6	0·0	6·6	0·5	12·6	0·9
07	1 01·8	1 01·9	0 58·9	0·7	0·1	6·7	0·5	12·7	1·0
08	1 02·0	1 02·2	0 59·2	0·8	0·1	6·8	0·5	12·8	1·0
09	1 02·3	1 02·4	0 59·4	0·9	0·1	6·9	0·5	12·9	1·0
10	1 02·5	1 02·7	0 59·7	1·0	0·1	7·0	0·5	13·0	1·0
11	1 02·8	1 02·9	0 59·9	1·1	0·1	7·1	0·5	13·1	1·0
12	1 03·0	1 03·2	1 00·1	1·2	0·1	7·2	0·5	13·2	1·0
13	1 03·3	1 03·4	1 00·4	1·3	0·1	7·3	0·5	13·3	1·0
14	1 03·5	1 03·7	1 00·6	1·4	0·1	7·4	0·6	13·4	1·0
15	1 03·8	1 03·9	1 00·8	1·5	0·1	7·5	0·6	13·5	1·0
16	1 04·0	1 04·2	1 01·1	1·6	0·1	7·6	0·6	13·6	1·0
17	1 04·3	1 04·4	1 01·3	1·7	0·1	7·7	0·6	13·7	1·0
18	1 04·5	1 04·7	1 01·6	1·8	0·1	7·8	0·6	13·8	1·0
19	1 04·8	1 04·9	1 01·8	1·9	0·1	7·9	0·6	13·9	1·0
20	1 05·0	1 05·2	[illegible]	[illegible]	0·2	8·0	0·6	[illegible]	[illegible]
						8·1	[illegible]		

Figure 5-5B. Portion of 4-minute "Increments and Corrections" table, Nautical Almanac.

GHA of Venus of 117° 51.6′. The hour circle of Venus can now be located on the time diagram, for later use in determining the LHA.

To find the declination of Venus for the time of the observation, the tabulated declination for the whole hour of GMT preceding the sight is first extracted and entered on the form as the tabulated declination (Tab. Dec.). Just as the GHA of a body in the solar system is continually changing from one hour to the next, the declination of these bodies is also changing. In contrast to the irregular rate of change of GHA, however, the rate of change in declination is nearly constant. The *d* value appearing at the bottom of the Venus column represents the average hourly rate of change of its declination over the three-day period covered by the daily page. A *d* correction (actually analogous to the GHA increment) derived from this *d* value is found from the appropriate "Increments and Corrections" table and applied to the tabulated declination to form the true declination of Venus at the time of the observation. The sign of the *d* value and hence the sign of the *d* correction is positive if the trend of the tabulated declination is increasing. If the tabulated declination is decreasing, the sign is negative. In this case, a *d* correction of 0.0 is obtained from the 4-minute table partially shown in Figure 5-5B, using a *d* value of .1 as entering

Sight Reduction using Tables 229 C:230 ° S: 25 kts.	M g W E ♀ ⊙ G m
Body	VENUS
Ho (Obs Alt)	16-38.6
Date	16 DEC
DR Lat	34-17.0 S
DR Long	163-09.1 E
Obs. Time	04-03-36
WE (S+, F-)	36 (s)
ZT	04-04-12
ZD (W+, E-)	-11
GMT	17-04-12
Date (GMT)	15 DEC
Tab GHA / v	116-48.5 / 1.1
GHA incr'mt.	1-03.0
SHA or v Corr.	0.1
GHA	117-51.6
±360 if needed	
aλ (-W, +E)	
LHA	
Tab Dec / d	S 13-16.1 / 0.1
d Corr (+ or -)	0.0
True Dec	S 13-16.1
a Lat (N or S)	Same Cont.

Figure 5-5C. Venus sight reduction form, GHA and declination determined.

argument in the "*v* or *d* correction" column. Thus, the true declination of Venus at GMT 17-04-12 is S 13° 16.1′.

The use of the *Nautical Almanac* for Venus is now complete; the form at this point is shown in Figure 5-5C.

Determining the GHA and Declination of the Moon

The use of the *Nautical Almanac* to find the GHA and declination of the moon is virtually identical to the procedure just described for the planets, with the exception of the *v* and *d* corrections. Because the moon is much closer to the earth than the planets, the rate of irregular change of its GHA and the rate of change of its declination are both so rapid as to necessitate the tabulation of a *v* and *d* value for each hour of GMT, rather than to average values for the three-day period. After recording the tabulated GHA and declination for the whole hour of GMT immediately preceding the time of the sight, the *v* and *d* values for this same time are also recorded. The GHA increment, *v* correction, and *d* correction are then determined as before, using the appropriate "Increments and Corrections" page. The sign of the *v* value, and hence the *v* correction for the moon, is always positive; the sign of the

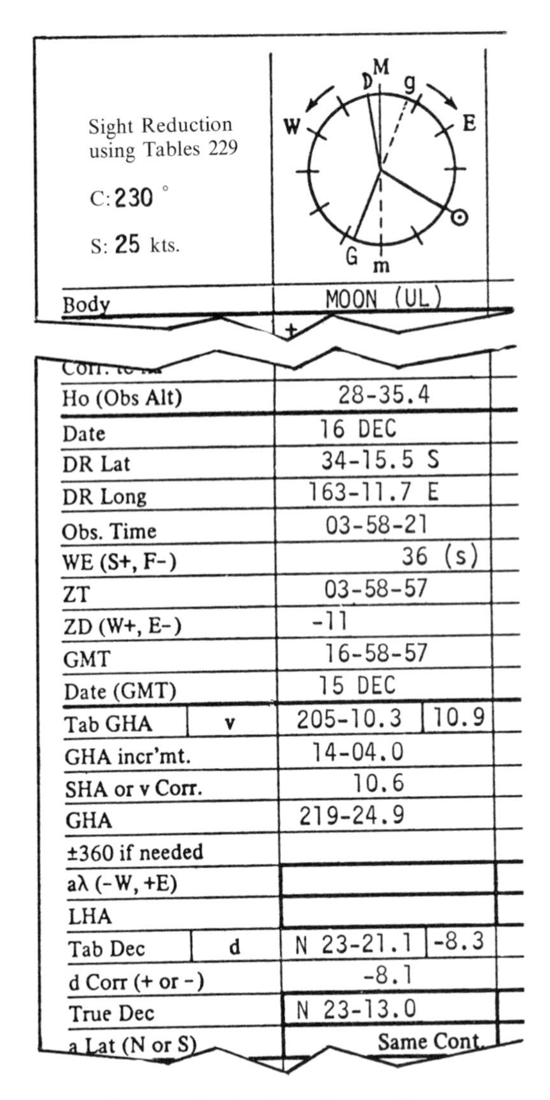

Sight Reduction using Tables 229

C: 230 °

S: 25 kts.

Body		MOON (UL)	
Ho (Obs Alt)		28-35.4	
Date		16 DEC	
DR Lat		34-15.5 S	
DR Long		163-11.7 E	
Obs. Time		03-58-21	
WE (S+, F-)		36 (s)	
ZT		03-58-57	
ZD (W+, E-)		-11	
GMT		16-58-57	
Date (GMT)		15 DEC	
Tab GHA	v	205-10.3	10.9
GHA incr'mt.		14-04.0	
SHA or v Corr.		10.6	
GHA		219-24.9	
±360 if needed			
aλ (-W, +E)			
LHA			
Tab Dec	d	N 23-21.1	-8.3
d Corr (+ or -)		-8.1	
True Dec		N 23-13.0	
a Lat (N or S)		Same Cont.	

Figure 5-6. Portion of a moon(UL)sight reduction form.

d value and the resulting *d* correction is determined by inspection of the trend of increase or decrease in the tabulated declination near the time of the observation.

Determining the GHA and Declination of the Sun

To complete the examples of the use of the *Nautical Almanac* during sight reduction to determine the GHA and declination of the various types of celestial bodies normally observed, the following example for a lower limb observation of the sun is included. The computation is essentially the same as that for a planet, with the difference being that no *v* correction is applied to the sum of the tabulated GHA and the GHA increment; the irregularity in the hourly rate of change of the GHA of the sun is so small as to be disregarded for most practical purposes. The sign of the *d* correction is determined by inspection of the trend of the hourly tabulated declination values. A partially completed sun sight form appears in Figure 5-7, next page.

Determining the Local Hour Angle

In the preceding four sections, examples of the use of the *Nautical Almanac* to find the GHA and declination of each of the four types of bodies observed in the practice of celestial navigation were given. The declination so determined forms one of the three entering arguments for *Sight Reduction Tables No. 229*. In this section, the methods by which GHA is combined with the assumed longitude to produce a second entering argument, LHA, will be set forth.

It might be recalled from Chapter 1 that the local hour angle (LHA) of a celestial body is defined as the hour angle measured from the observer's celestial meridian to the hour circle of the body; it is always measured in a westerly direction, and increases from 0° through 360° as the earth rotates beneath the celestial sphere. For the purpose of sight reduction, LHA is measured from the meridian of the assumed position. Since the assumed longitude is defined as the horizontal angle between the Greenwich meridian and the location of the AP of the observer, the following relationship exists between the local hour angle (LHA), the GHA of the body, and the assumed longitude (aλ):

$$\text{LHA} = \text{GHA} \begin{array}{l} + a\lambda(\text{E}) \\ - a\lambda(\text{W}) \end{array}$$

This relationship is depicted in Figure 5-8 on page 105. Occasionally, when the assumed longitude is west its value may be greater than the GHA of the body; this would result in a negative value for the LHA. In

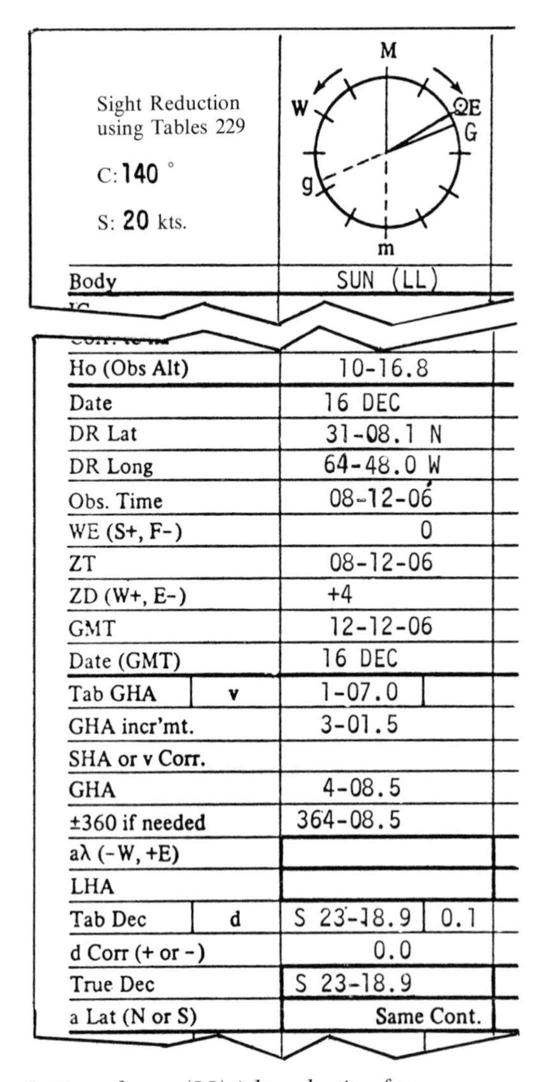
Sight Reduction using Tables 229

C: 140 °

S: 20 kts.

Body		SUN (LL)	
Ho (Obs Alt)		10-16.8	
Date		16 DEC	
DR Lat		31-08.1 N	
DR Long		64-48.0 W	
Obs. Time		08-12-06	
WE (S+, F-)		0	
ZT		08-12-06	
ZD (W+, E-)		+4	
GMT		12-12-06	
Date (GMT)		16 DEC	
Tab GHA	v	1-07.0	
GHA incr'mt.		3-01.5	
SHA or v Corr.			
GHA		4-08.5	
±360 if needed		364-08.5	
aλ (-W, +E)			
LHA			
Tab Dec	d	S 23-18.9	0.1
d Corr (+ or -)		0.0	
True Dec		S 23-18.9	
a Lat (N or S)		Same Cont.	

Figure 5-7. Portion of a sun(LL)sight reduction form.

these cases, 360° is added to the GHA prior to subtracting the westerly assumed longitude. An equivalent positive LHA results.

Since the entering argument for *Tables No. 229* must be an integral LHA, the assumed longitude is chosen such that when it is combined with the GHA determined from the *Nautical* (or *Air*) *Almanac*, an integral LHA results. Thus, if the observer is located in east longitude at the time of the observation, the minutes of the assumed longitude are selected in such a manner that when they are added to the minutes of the GHA a whole degree results. If the observer is in west longitude, the minutes of the assumed longitude chosen should be equal to the

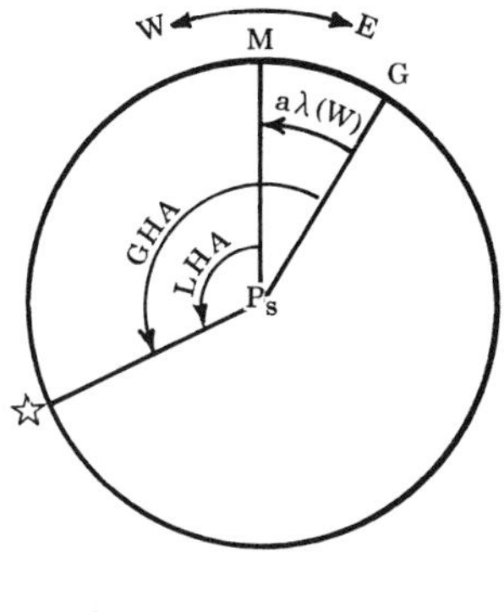

LHA = GHA − aλ (W)

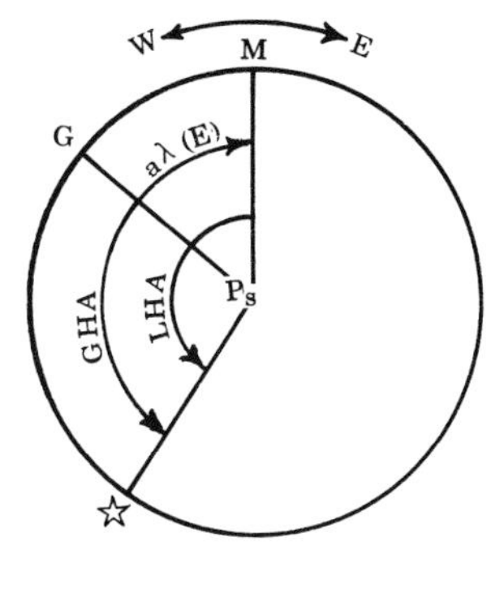

LHA = GHA + aλ (E)

Figure 5-8. Two possible relationships between a star and the Greenwich and local meridians.

minutes of the GHA, so that when the assumed longitude is subtracted from the GHA an integral LHA results. In order that the intercept distance from the AP of the observer to the LOP produced from an observation will not be excessive, the following rule has been developed:

> The assumed longitude chosen should not be more than 30 minutes of longitude from the DR longitude for the time of the observation.

As examples of the determination of local hour angles for use as entering arguments in *Sight Reduction Tables No. 229*, the integral LHA for each of the four examples of the use of the *Nautical Almanac* given previously will be determined, as outlined in the following paragraphs.

In the first example, the GHA of the star Canopus at the time of the observation was determined to be 246° 08.6′; the DR longitude for this time was λ 163° 05.7′ E. From an inspection of the time diagram in Figure 5-4, it appears as though the LHA of Canopus at the time of the observation should be about 50°. To compute the exact integral value of the LHA from an assumed position close to the DR position at the time of the observation, the formula

$$\text{LHA} = \text{GHA} + a\lambda(E)$$

applies, since the DR longitude is east. Since the GHA and assumed longitude are to be added to form the LHA, it follows that to obtain an integral LHA the minutes of the selected assumed longitude should be chosen such that when they are added to the minutes of the GHA the result will be 60 minutes or one degree. The minutes (m) of the assumed longitude must therefore satisfy the equation

$$60 = 08.6 + m$$

Obviously, the minutes chosen must be 51.4. Now, care must be taken to choose the degrees of the assumed longitude so that the result will be within 30 minutes of the DR longitude. An assumed longitude 163° 51.4′ would exceed this limit, so the value 162° 51.4′ is selected. Adding this value to the computed GHA yields the integral LHA 409°:

$$246° \ 08.6' + 162° \ 51.4' = 409°$$

Since this result is greater than 360°, 360° is subtracted to form the equivalent LHA 49°.

In the second example, the GHA of the planet Venus was found to be 117° 51.6′, and the DR longitude was 163° 09.1′ E at the time of the observation. From an inspection of the hour circle of Venus in the time diagram of Figure 5-5C, it appears as though the LHA should be about 280°. The formula LHA = GHA + aλ(E) again applies. To obtain an integral LHA, the minutes of the GHA when added to the minutes of the assumed longitude must equal 60. Hence, the minutes of the assumed longitude chosen must be (60′ – 51.6′) or 08.4°. Inasmuch as the DR longitude is 163° 09.1′ E, the complete assumed longitude is 163° 08.4′. The LHA is then 117° 51.6′ + 163° 08.4′ or 281°.

In the third example for the moon, the GHA was 219° 24.9′ and the DR longitude 163° 11.7′ E. The selected assumed longitude is 163° 35.1′, resulting in an integral LHA of 383°:

$$219° \ 24.9' + 163° \ 35.1' = 383°$$

Since the result is larger than 360°, the equivalent LHA of 23° is formed by subtraction of 360°.

In the sun example, the GHA was 4° 08.5′ and the DR longitude was 64° 48.0′W. In this situation the formula

$$\text{LHA} = \text{GHA} - a\lambda(\text{W})$$

applies, and the minutes of the GHA and assumed longitude must be identical to produce an integral LHA. Thus, the minutes of the assumed longitude must be 08.5′. Considering the DR position, care must again be taken to select the proper degrees of the assumed longitude. In accordance with the "30-minute" rule, the assumed longitude chosen is 65° 08.5′. Had a value of 64° 08.5′ been chosen, the assumed longitude would have been more than 30 minutes from the DR longitude (48.0′ − 08.5′ = 39.5′). Since the GHA is smaller than the assumed longitude, 360° is added to the GHA, and the resulting LHA is given by the following expression:

$$(360^\circ + 4^\circ\ 08.5') - 65^\circ\ 08.5' = 299^\circ$$

Certain of the LHAs calculated as examples above will be used as entering arguments for *Tables No. 229* later in this chapter.

Determining the Assumed Latitude

The remaining entering argument for *Tables No. 229* is an integral value of assumed latitude. The assumed latitude, similar to the assumed longitude, should not be more than 30 minutes of latitude from the DR latitude. The following rule is used for determining the assumed latitude:

> The assumed latitude chosen should be the closest whole degree of latitude to the DR latitude for the time of the observation.

Thus, the assumed latitudes for the Canopus, Venus, and moon observations used as examples earlier are all 34° S, as the DR latitudes for all three of these observations varied between 34° 15.5° S and 34° 19.0′ S. The assumed latitude for the sun observation, based on the DR latitude 31° 08.1′ N, is 31° N.

The *Sight Reduction Tables for Marine Navigation, No. 229*

The remainder of this chapter will be concerned with the use of the *Sight Reduction Tables for Marine Navigation, No. 229,* to obtain the complete solution for a celestial observation. As was mentioned earlier, *Tables No. 229* are, in effect, a set of precalculated solutions for the computed altitude Hc and the azimuth angle Z of the navigational triangle, given two other sides and the included angle between them. The tables themselves are divided into six volumes, each covering a basic 15° band of latitude. A 1° overlap occurs between the volumes, so that Volume 1 covers latitudes between 0° and 15°, Volume 2 covers 15° to 30°, Volume 3 covers 30° to 45°, and so on. Entering arguments for the tables are integral (whole number) degrees of local hour angle (LHA), assumed latitude, and declination. Values of Hc and Z are tabulated for each whole degree of each of the entering arguments, and interpolation tables are included inside the front and back covers of each volume for interpolating both Hc and Z for the exact declination. No interpolation for LHA or assumed latitude is necessary since, as has been explained, the assumed position from which the intercept distance is laid down is selected to yield an integral LHA and assumed latitude.

Each volume of *Tables No. 229* contains two sets of tabulations for all integral degrees of LHA between 0° and 360°. One set, comprising the front half of the volume, is for the first eight degrees of latitude covered by that volume, and the other set, comprising the second half of the volume, is for the remaining eight degrees. The values of the local hour angle, considered to be the primary entering argument within each of the eight-degree latitude "zones" thus established, are prominently displayed both at the top and bottom of each page. The eight degrees of latitude within each zone form the horizontal argument for each LHA page, while the vertical argument is declination. The tabulations are arranged in columns, with one column for each of the eight degrees of latitude covered.

Instructions printed across the top and bottom of each page indicate whether the tabulations on that page are for assumed latitudes on the same side of the celestial equator as the declination (latitude *same* name as declination) or for assumed latitudes on the opposite side of the equator (latitude *contrary* name to declination). If same name and contrary name tabulations appear on the same page, they are separated in each column by a horizontal line that forms a configuration across the page resembling the profile of a staircase. Each horizontal line indicates the degree of declination for that particular LHA and assumed latitude combination in which the visible sea horizon occurs.

Figure 5-9A on page 110 depicts a left-hand page of Volume 3 of *Tables No. 229* for LHAs of 49° and 311° for the 30° to 37° latitude zone with latitude the same name as declination. Figure 5-9B shows a right-hand page for the same latitude zone for LHAs of 61°, 299°, 119° and 241°; the top of the page is used if the latitude and the declination are of contrary name, and the bottom of the page is for use if the latitude and declination are of the same name.

As was the case with the *Nautical Almanac*, the use of the *Sight Reduction Tables No. 229* is best explained by means of examples. Because the procedures for the use of the tables during sight reduction are the same for any type of celestial body observed, only two of the four sample problems introduced earlier in this chapter will be completed here—Canopus and the sun. The complete solutions for the planet Venus and the moon will appear in Chapter 7 in connection with the plot of a celestial fix.

Completing the Sight Reduction by *No. 229*—First Example

As the first example of the completion of the sight reduction process by use of *Sight Reduction Tables No. 229*, the Canopus form shown in Figure 5-4 will be completed. The entering arguments, determined in previous sections of this chapter, are reproduced below:

LHA	49
…	…
True Dec	S 52–40.6
a Lat (N or S)	34 S (Same) Cont.

The first step in entering *Tables No. 229* is to select the proper volume and page number. Since the assumed latitude in this example is 34° S, Volume 3 containing tabulations for latitudes from 30° to 45° is selected. The volume is opened to the page containing tabulations for LHA of 49° with latitude the same name as declination, in the 30° to 37° latitude zone. This is the page appearing in Figure 5-9A.

The first quantity extracted from the table is the tabulated value of computed altitude Hc in the column of figures corresponding to the assumed latitude for the nearest whole degree of declination less than the exact declination. In this example an Hc of 50° 51.1′ is extracted, using as a horizontal argument 34° of latitude, and as a vertical argument 52° of declination. In practice, to save time the azimuth angle Z would also be recorded at this time for future use, but to simplify this initial explanation, this will be omitted here.

Next, the exact value of Hc corresponding to the exact declination of the body at the time of observation must be determined by interpolation. To simplify this interpolation process, a set of interpolation tables is included inside the front and back cover of each volume.

The entering arguments for the interpolation tables are the declination increment (Dec. Inc.)—the remaining minutes and tenths of the exact declination—and the altitude difference d between the two tabulated Hc's bracketing the exact declination. The value of d between successive tabulated Hc's has been precalculated and appears in the center of each column of tabulations. Both the declination increment and altitude difference d are recorded in the appropriate spaces on the form; in this case, the declination increment is 40.6′, and the altitude difference d is −8.7′. If the Hc decreases in value with increasing declination, as is the case in this instance, the sign of the altitude difference d is negative; this is indicated in the tables by the placement of a minus sign adjacent to the initial negative value

49°, 311° L.H.A. LATITUDE SAME NAME AS DECLINATION

N. Lat. { L.H.A. greater than 180°......Zn=Z ; L.H.A. less than 180°............Zn=360°−Z }

Dec.	30° Hc	d	Z	31° Hc	d	Z	32° Hc	d	Z	33° Hc	d	Z
°	° ′	′	°	° ′	′	°	° ′	′	°	° ′	′	°
0	34 37.3	+36.3	113.5	34 13.1	+37.2	114.1	33 48.3	+38.1	114.7	33 22.9	+38.9	115.3
1	35 13.6	35.7	112.5	34 50.3	36.7	113.2	34 26.4	37.6	113.8	34 01.8	38.6	114.4
2	35 49.3	35.3	111.5	35 27.0	36.2	112.2	35 04.0	37.2	112.9	34 40.4	38.1	113.5
3	36 24.6	34.7	110.5	36 03.2	35.7	111.2	35 41.2	36.7	111.9	35 18.5	37.6	112.5
4	36 59.3	34.2	109.5	36 38.9	35.3	110.2	36 17.9	36.2	110.9	35 56.1	37.2	111.6
5	37 33.5	+33.7	108.5	37 14.2	+34.7	109.2	36 54.1	+35.7	109.9	36 33.3	+36.7	110.6
6	38 07.2	33.1	107.4	37 48.9	34.1	108.2	37 29.8	35.2	108.9	37 10.0	36.2	109.6
7	38 40.3	32.6	106.4	38 23.0	33.6	107.1	38 05.0	34.6	107.9	37 46.2	35.7	108.6
8	39 12.9	31.9	105.3	38 56.6	33.0	106.1	38 39.6	34.1	106.8	38 21.9	35.1	107.6
9	39 44.8	31.2	104.2	39 29.6	32.5	105.0	39 13.7	33.5	105.8	38 57.0	34.6	106.6
10	40 16.0	+30.7	103.1	40 02.1	+31.7	103.9	39 47.2	+32.9	104.7	39 31.6	+34.0	105.5
11	40 46.7	29.9	101.9	40 33.8	31.2	102.8	40 20.1	32.3	103.6	40 05.6	33.4	104.4
12	41 16.6	29.3	100.8	41 05.0	30.4	101.7	40 52.4	31.6	102.5	40 39.0	32.7	103.3
13	41 45.9	28.5	99.6	41 35.4	29.8	100.5	41 24.0	31.0	101.4	41 11.7	32.2	102.2
14	42 14.4	27.9	98.4	42 05.2	29.0	99.3	41 55.0	30.2	100.2	41 43.9	31.4	101.1
15	42 42.3	+27.0	97.2	42 34.2	+28.3	98.2	42 25.2	+29.6	99.1	42 15.3	+30.8	100.0
16	43 09.3	26.2	96.0	43 02.5	27.5	97.0	42 54.8	28.7	97.9	42 46.1	30.0	98.8
17	43 35.5	25.5	94.8	43 30.0	26.8	95.7	43 23.5	28.1	96.7	43 16.1	29.3	97.6
18	44 01.0	24.6	93.5	43 56.8	25.9	94.5	43 51.6	27.2	95.5	43 45.4	28.5	96.4
19	44 25.6	23.7	92.3	44 22.7	25.1	93.2	44 18.8	26.4	94.2	44 13.9	27.7	95.2
20	44 49.3	+22.9	91.0	44 47.8	+24.2	92.0	44 45.2	+25.6	93.0	44 41.6	+26.9	93.9
21	45 12.2	21.9	89.7	45 12.0	23.4	90.7	45 10.8	24.7	91.7	45 08.5	26.1	92.7
22	45 34.1	21.1	88.3	45 35.4	22.4	89.4	45 35.5	23.8	90.4	45 34.6	25.1	91.4
23	45 55.2	20.0	87.0	45 57.8	21.5	88.0	45 59.3	22.9	89.1	45 59.7	24.3	90.1
24	46 15.2	19.1	85.6	46 19.3	20.5	86.7	46 22.2	21.9	87.7	46 24.0	23.4	88.8
25	46 34.3	+18.1	84.3	46 39.8	+19.5	85.3	46 44.1	+21.0	86.4	46 47.4	+22.4	87.4
26	46 52.4	17.1	82.9	46 59.3	18.5	83.9	47 05.1	20.0	85.0	47 09.8	21.4	86.1
27	47 09.5	16.0	81.5	47 17.8	17.5	82.5	47 25.1	18.9	83.6	47 31.2	20.4	84.7
28	47 25.5	15.0	80.0	47 35.3	16.5	81.1	47 44.0	17.9	82.2	47 51.6	19.4	83.3
29	47 40.5	13.9	78.6	47 51.8	15.3	79.7	48 01.9	16.9	80.8	48 11.0	18.3	81.9
30	47 54.4	+12.7	77.2	48 07.1	+14.3	78.2	48 18.8	+15.7	79.3	48 29.3	+17.3	80.5
31	48 07.1	11.7	75.7	48 21.4	13.1	76.8	48 34.5	14.7	77.9	48 46.6	16.1	79.0
32	48 18.8	10.5	74.2	48 34.5	12.1	75.3	48 49.2	13.5	76.4	49 02.7	15.0	77.5
33	48 29.3	9.4	72.7	48 46.6	10.8	73.8	49 02.7	12.4	74.9	49 17.7	13.9	76.1
34	48 38.7	8.2	71.3	48 57.4	9.7	72.3	49 15.1	11.2	73.4	49 31.6	12.7	74.6
35	48 46.9	+7.0	69.8	49 07.1	+8.5	70.8	49 26.3	+10.0	71.9	49 44.3	+11.5	73.1
36	48 53.9	5.9	68.2	49 15.6	7.3	69.3	49 36.3	8.8	70.4	49 55.8	10.4	71.5
37	48 59.8	4.6	66.7	49 22.9	6.2	67.8	49 45.1	7.6	68.9	50 06.2	9.1	70.0
38	49 04.4	3.5	65.2	49 29.1	4.9	66.3	49 52.7	6.4	67.4	50 15.3	7.8	68.5
39	49 07.9	2.2	63.7	49 34.0	3.6	64.7	49 59.1	5.1	65.8	50 23.1	6.6	66.9
40	49 10.1	+1.0	62.2	49 37.6	+2.5	63.2	50 04.2	+3.9	64.3	50 29.7	+5.4	65.3
41	49 11.1	−0.2	60.6	49 40.1	+1.2	61.6	50 08.1	2.6	62.7	50 35.1	4.1	63.8
42	49 10.9	1.4	59.1	49 41.3	0.0	60.1	50 10.7	1.4	61.1	50 39.2	2.8	62.2
43	49 09.5	2.6	57.6	49 41.3	−1.3	58.6	50 12.1	+0.2	59.6	50 42.0	1.6	60.6
44	49 06.9	3.8	56.0	49 40.0	2.5	57.0	50 12.3	−1.2	58.0	50 43.6	+0.2	59.0
45	49 03.1	−5.0	54.5	49 37.5	−3.7	55.5	50 11.1	−2.3	56.5	50 43.8	−1.0	57.5
46	48 58.1	6.2	53.0	49 33.8	4.9	53.9	50 08.8	3.7	54.9	50 42.8	2.3	55.9
47	48 51.9	7.4	51.5	49 28.9	6.1	52.4	50 05.1	4.9	53.3	50 40.5	3.5	54.3
48	48 44.5	8.5	50.0	49 22.8	7.4	50.9	50 00.2	6.1	51.8	50 37.0	4.9	52.7
49	48 36.0	9.8	48.5	49 15.4	8.6	49.3	49 54.1	7.3	50.2	50 32.1	6.1	51.2
50	48 26.2	−10.8	47.0	49 06.8	−9.7	47.8	49 46.8	−8.6	48.7	50 26.0	−7.3	49.6
51	48 15.4	12.0	45.5	48 57.1	10.9	46.3	49 38.2	9.7	47.2	50 18.7	8.6	48.0
52	48 03.4	13.1	44.0	48 46.2	12.0	44.8	49 28.5	11.0	45.6	50 10.1	9.8	46.5
53	47 50.3	14.2	42.6	48 34.2	13.2	43.3	49 17.5	12.1	44.1	50 00.3	11.0	45.0
54	47 36.1	15.3	41.1	48 21.0	14.3	41.9	49 05.4	13.3	42.6	49 49.3	12.3	43.4
55	47 20.8	−16.4	39.7	48 06.7	−15.4	40.4	48 52.1	−14.4	41.2	49 37.0	−13.4	41.9
56	47 04.4	17.3	38.3	47 51.3	16.5	39.0	48 37.7	15.5	39.7	49 23.6	14.5	40.4
57	46 47.1	18.4	36.9	47 34.8	17.5	37.5	48 22.2	16.6	38.2	49 09.1	15.7	38.9
58	46 28.7	19.4	35.5	47 17.3	18.5	36.1	48 05.6	17.7	36.8	48 53.4	16.8	37.5
59	46 09.3	20.4	34.1	46 58.8	19.6	34.7	47 47.9	18.8	35.4	48 36.6	17.9	36.0
60	45 48.9	−21.3	32.8	46 39.2	−20.6	33.3	47 29.1	−19.7	33.9	48 18.7	−18.9	34.6
61	45 27.6	22.2	31.4	46 18.6	21.5	32.0	47 09.4	20.8	32.6	47 59.8	20.0	33.1
62	45 05.4	23.2	30.1	45 57.1	22.4	30.6	46 48.6	21.7	31.2	47 39.8	21.0	31.7
63	44 42.2	24.0	28.8	45 34.7	23.4	29.3	46 26.9	22.7	29.8	47 18.8	22.0	30.4
64	44 18.2	24.8	27.5	45 11.3	24.2	28.0	46 04.2	23.6	28.5	46 56.8	23.0	29.0
65	43 53.4	−25.7	26.3	44 47.1	−25.1	26.7	45 40.6	−24.5	27.2	46 33.8	−23.9	27.6
66	43 27.7	26.5	25.0	44 22.0	26.0	25.4	45 16.1	25.4	25.9	46 09.9	24.8	26.3
67	43 01.2	27.3	23.8	43 56.0	26.8	24.2	44 50.7	26.3	24.6	45 45.1	25.7	25.0
68	42 33.9	28.0	22.6	43 29.2	27.5	22.9	44 24.4	27.0	23.3	45 19.4	26.5	23.7
69	42 05.9	28.8	21.4	43 01.7	28.3	21.7	43 57.4	27.9	22.1	44 52.9	27.4	22.4
70	41 37.1	−29.5	20.2	42 33.4	−29.1	20.5	43 29.5	−28.6	20.8	44 25.5	−28.2	21.2
71	41 07.6	30.1	19.0	42 04.3	29.8	19.3	43 00.9	29.4	19.6	43 57.3	28.9	20.0
72	40 37.5	30.9	17.9	41 34.5	30.4	18.2	42 31.5	30.1	18.4	43 28.4	29.8	18.7
73	40 06.6	31.4	16.8	41 04.1	31.2	17.0	42 01.4	30.8	17.3	42 58.6	30.4	17.6
74	39 35.2	32.1	15.7	40 32.9	31.8	15.9	41 30.6	31.5	16.1	42 28.2	31.2	16.4
75	39 03.1	−32.7	14.6	40 01.1	−32.4	14.8	40 59.1	−32.1	15.0	41 57.0	−31.8	15.2
76	38 30.4	33.3	13.5	39 28.7	33.0	13.7	40 27.0	32.8	13.9	41 25.2	32.5	14.1
77	37 57.1	33.9	12.4	38 55.7	33.7	12.6	39 54.2	33.4	12.8	40 52.7	33.1	13.0
78	37 23.2	34.3	11.4	38 22.0	34.1	11.5	39 20.8	33.9	11.7	40 19.6	33.8	11.9
79	36 48.9	34.9	10.4	37 47.9	34.8	10.5	38 46.9	34.6	10.6	39 45.8	34.3	10.8
80	36 14.0	−35.5	9.4	37 13.1	−35.2	9.5	38 12.3	−35.1	9.6	39 11.5	−34.9	9.7
81	35 38.5	35.9	8.4	36 37.9	35.8	8.5	37 37.2	35.6	8.6	38 36.6	35.5	8.7
82	35 02.6	36.3	7.4	36 02.1	36.2	7.5	37 01.6	36.1	7.6	38 01.1	36.0	7.7
83	34 26.3	36.8	6.4	35 25.9	36.7	6.5	36 25.5	36.6	6.6	37 25.1	36.5	6.7
84	33 49.5	37.3	5.4	34 49.2	37.2	5.5	35 48.9	37.1	5.6	36 48.6	37.0	5.7
85	33 12.2	−37.7	4.5	34 12.0	−37.6	4.6	35 11.8	−37.5	4.6	36 11.6	−37.4	4.7
86	32 34.5	38.0	3.6	33 34.4	38.0	3.6	34 34.3	38.0	3.7	35 34.2	37.9	3.7
87	31 56.5	38.5	2.7	32 56.4	38.4	2.7	33 56.3	38.3	2.7	34 56.3	38.4	2.8
88	31 18.0	38.8	1.8	32 18.0	38.8	1.8	33 18.0	38.8	1.8	34 17.9	38.7	1.8
89	30 39.2	39.2	0.9	31 39.2	39.2	0.9	32 39.2	39.2	0.9	33 39.2	39.2	0.9
90	30 00.0	−39.5	0.0	31 00.0	−39.5	0.0	32 00.0	−39.5	0.0	33 00.0	−39.6	0.0

34° Hc	d	Z	35° Hc	d	Z	36° Hc	d	Z	37° Hc	d	Z	Dec.
° ′	′	°	° ′	′	°	° ′	′	°	° ′	′	°	°
32 57.0	+39.8	115.9	32 30.5	+40.6	116.5	32 03.4	+41.5	117.1	31 35.9	+42.2	117.6	0
33 36.8	39.4	115.0	33 11.1	40.2	115.6	32 44.9	41.0	116.2	32 18.1	41.9	116.8	1
34 16.2	38.9	114.1	33 51.3	39.9	114.7	33 25.9	40.8	115.3	33 00.0	41.5	115.9	2
34 55.1	38.6	113.2	34 31.2	39.5	113.8	34 06.7	40.3	114.5	33 41.5	41.2	115.1	3
35 33.7	38.1	112.3	35 10.7	39.0	112.9	34 47.0	39.9	113.6	34 22.7	40.8	114.2	4
36 11.8	+37.7	111.3	35 49.7	+38.6	112.0	35 26.9	+39.5	112.6	35 03.5	+40.4	113.3	5
36 49.5	37.2	110.3	36 28.3	38.1	111.0	36 06.4	39.1	111.7	35 43.9	40.0	112.4	6
37 26.7	36.6	109.4	37 06.4	37.7	110.1	36 45.5	38.6	110.8	36 23.9	39.5	111.5	7
38 03.3	36.2	108.4	37 44.1	37.2	109.1	37 24.1	38.2	109.8	37 03.4	39.1	110.5	8
38 39.5	35.6	107.3	38 21.3	36.6	108.1	38 02.3	37.6	108.8	37 42.5	38.6	109.6	9
39 15.1	+35.1	106.3	38 57.9	+36.1	107.1	38 39.9	+37.1	107.8	38 21.1	+38.2	108.6	10
39 50.2	34.5	105.2	39 34.0	35.6	106.0	39 17.0	36.7	106.8	38 59.3	37.6	107.6	11
40 24.7	33.9	104.2	40 09.6	35.0	105.0	39 53.7	36.0	105.8	39 36.9	37.1	106.6	12
40 58.6	33.2	103.1	40 44.6	34.4	103.9	40 29.7	35.5	104.8	40 14.0	36.5	105.6	13
41 31.8	32.7	102.0	41 19.0	33.7	102.8	41 05.2	34.9	103.7	40 50.5	36.0	104.5	14
42 04.5	+31.9	100.9	41 52.7	+33.1	101.7	41 40.1	+34.2	102.6	41 26.5	+35.4	103.5	15
42 36.4	31.2	99.7	42 25.8	32.5	100.6	42 14.3	33.6	101.5	42 01.9	34.7	102.4	16
43 07.6	30.6	98.5	42 58.3	31.7	99.5	42 47.9	32.9	100.4	42 36.6	34.1	101.3	17
43 38.2	29.8	97.4	43 30.0	31.0	98.3	43 20.8	32.3	99.2	43 10.7	33.5	100.2	18
44 08.0	29.0	96.2	44 01.0	30.3	97.1	43 53.1	31.5	98.1	43 44.2	32.7	99.0	19
44 37.0	+28.2	94.9	44 31.3	+29.5	95.9	44 24.6	+30.8	96.9	44 16.9	+32.0	97.9	20
45 05.2	27.4	93.7	45 00.8	28.7	94.7	44 55.4	30.0	95.7	44 48.9	31.3	96.7	21
45 32.6	26.5	92.4	45 29.5	27.9	93.4	45 25.4	29.1	94.5	45 20.2	30.5	95.5	22
45 59.1	25.6	91.1	45 57.4	27.0	92.2	45 54.5	28.4	93.2	45 50.7	29.6	94.2	23
46 24.7	24.8	89.8	46 24.4	26.1	90.9	46 22.9	27.5	91.9	46 20.3	28.9	93.0	24
46 49.5	+23.8	88.5	46 50.5	+25.2	89.6	46 50.4	+26.6	90.6	46 49.2	+27.9	91.7	25
47 13.3	22.8	87.2	47 15.7	24.3	88.2	47 17.0	25.7	89.3	47 17.1	27.1	90.4	26
47 36.1	21.9	85.8	47 40.0	23.3	86.9	47 42.7	24.7	88.0	47 44.2	26.2	89.1	27
47 58.0	20.9	84.4	48 03.3	22.3	85.5	48 07.4	23.8	86.6	48 10.4	25.1	87.7	28
48 18.9	19.8	83.0	48 25.6	21.3	84.1	48 31.2	22.7	85.2	48 35.5	24.3	86.4	29
48 38.7	+18.7	81.6	48 46.9	+20.2	82.7	48 53.9	+21.7	83.8	48 59.8	+23.1	85.0	30
48 57.4	17.7	80.1	49 07.1	19.2	81.3	49 15.6	20.7	82.4	49 22.9	22.2	83.6	31
49 15.1	16.5	78.7	49 26.3	18.0	79.8	49 36.3	19.5	81.0	49 45.1	21.1	82.1	32
49 31.6	15.4	77.2	49 44.3	16.9	78.3	49 55.8	18.5	79.5	50 06.2	19.9	80.7	33
49 47.0	14.2	75.7	50 01.2	15.8	76.9	50 14.3	17.3	78.0	50 26.1	18.8	79.2	34
50 01.2	+13.1	74.2	50 17.0	+14.6	75.4	50 31.6	+16.1	76.5	50 44.9	+17.7	77.7	35
50 14.3	11.8	72.7	50 31.6	13.3	73.8	50 47.7	14.9	75.0	51 02.6	16.5	76.2	36
50 26.1	10.6	71.1	50 44.9	12.2	72.3	51 02.6	13.7	73.5	51 19.1	15.2	74.7	37
50 36.7	9.4	69.6	50 57.1	10.9	70.7	51 16.3	12.5	71.9	51 34.3	14.1	73.1	38
50 46.1	8.1	68.0	51 08.0	9.7	69.2	51 28.8	11.2	70.3	51 48.4	12.7	71.5	39
50 54.2	+6.9	66.5	51 17.7	+8.3	67.6	51 40.0	+9.9	68.8	52 01.1	+11.5	70.0	40
51 01.1	5.6	64.9	51 26.0	7.1	66.0	51 49.9	8.6	67.2	52 12.6	10.1	68.4	41
51 06.7	4.3	63.3	51 33.1	5.8	64.4	51 58.5	7.3	65.6	52 22.7	8.9	66.7	42
51 11.0	3.0	61.7	51 38.9	4.5	62.8	52 05.8	5.9	64.0	52 31.6	7.5	65.1	43
51 14.0	1.6	60.1	51 43.4	3.1	61.2	52 11.7	4.7	62.3	52 39.1	6.1	63.5	44
51 15.6	+0.4	58.5	51 46.5	+1.8	59.6	52 16.4	+3.2	60.7	52 45.2	+4.8	61.9	45
51 16.0	−0.9	56.9	51 48.3	+0.5	58.0	52 19.6	2.0	59.1	52 50.0	3.4	60.2	46
51 15.1	2.2	55.3	51 48.8	−0.8	56.4	52 21.6	+0.5	57.4	52 53.4	2.0	58.5	47
51 12.9	3.5	53.7	51 48.0	2.2	54.7	52 22.1	−0.8	55.8	52 55.4	+0.6	56.9	48
51 09.4	4.9	52.1	51 45.8	3.5	53.1	52 21.3	2.1	54.2	52 56.0	−0.7	55.2	49
51 04.5	−6.1	50.5	51 42.3	−4.8	51.5	52 19.2	−3.5	52.5	52 55.3	−2.2	53.6	50
50 58.4	7.3	49.0	51 37.5	6.2	49.9	52 15.7	4.8	50.9	52 53.1	3.5	51.9	51
50 51.1	8.7	47.4	51 31.3	7.4	48.3	52 10.9	6.2	49.3	52 49.6	4.9	50.3	52
50 42.4	9.9	45.8	51 23.9	8.7	46.7	52 04.7	7.5	47.6	52 44.7	6.2	48.6	53
50 32.5	11.1	44.3	51 15.2	10.0	45.1	51 57.2	8.9	46.0	52 38.5	7.6	47.0	54
50 21.4	−12.3	42.7	51 05.2	−11.3	43.6	51 48.3	−10.1	44.4	52 30.9	−9.0	45.3	55
50 09.1	13.6	41.2	50 53.9	12.5	42.0	51 38.2	11.4	42.8	52 21.9	10.3	43.7	56
49 55.5	14.7	39.7	50 41.4	13.7	40.5	51 26.8	12.6	41.3	52 11.6	11.5	42.1	57
49 40.8	15.9	38.2	50 27.7	14.9	38.9	51 14.2	13.9	39.7	52 00.1	12.9	40.5	58
49 24.9	17.0	36.7	50 12.8	16.0	37.4	51 00.3	15.2	38.1	51 47.2	14.1	38.9	59
49 07.9	−18.1	35.2	49 56.8	−17.3	35.9	50 45.1	−16.3	36.6	51 33.1	−15.4	37.4	60
48 49.8	19.1	33.8	49 39.5	18.3	34.4	50 28.8	17.5	35.1	51 17.7	16.6	35.8	61
48 30.7	20.3	32.3	49 21.2	19.5	33.0	50 11.3	18.6	33.6	51 01.1	17.7	34.3	62
48 10.4	21.3	30.9	49 01.7	20.5	31.5	49 52.7	19.7	32.1	50 43.4	19.0	32.8	63
47 49.1	22.2	29.5	48 41.2	21.6	30.1	49 33.0	20.8	30.7	50 24.4	20.0	31.3	64
47 26.9	−23.3	28.1	48 19.6	−22.5	28.7	49 12.2	−21.9	29.2	50 04.4	−21.2	29.8	65
47 03.6	24.2	26.8	47 57.1	23.6	27.3	48 50.3	22.9	27.8	49 43.2	22.2	28.3	66
46 39.4	25.1	25.4	47 33.5	24.5	25.9	48 27.4	24.0	26.4	49 21.0	23.3	26.9	67
46 14.3	26.0	24.1	47 09.0	25.5	24.6	48 03.4	24.9	25.0	48 57.7	24.3	25.5	68
45 48.3	26.9	22.8	46 43.5	26.4	23.2	47 38.5	25.8	23.7	48 33.4	25.3	24.1	69
45 21.4	−27.7	21.6	46 17.1	−27.2	21.9	47 12.7	−26.7	22.3	48 08.1	−26.2	22.8	70
44 53.7	28.6	20.3	45 49.9	28.1	20.6	46 46.0	27.7	21.0	47 41.9	27.2	21.4	71
44 25.1	29.3	19.1	45 21.8	28.9	19.4	46 18.3	28.5	19.7	47 14.7	28.0	20.1	72
43 55.8	30.1	17.8	44 52.9	29.7	18.1	45 49.8	29.3	18.5	46 46.7	28.9	18.8	73
43 25.7	30.8	16.6	44 23.2	30.5	16.9	45 20.5	30.1	17.2	46 17.8	29.8	17.5	74
42 54.9	−31.5	15.5	43 52.7	−31.2	15.7	44 50.4	−30.9	16.0	45 48.0	−30.5	16.3	75
42 23.4	32.3	14.3	43 21.5	32.0	14.5	44 19.5	31.6	14.8	45 17.5	31.3	15.0	76
41 51.1	32.9	13.2	42 49.5	32.6	13.4	43 47.9	32.4	13.6	44 46.2	32.1	13.8	77
41 18.2	33.5	12.1	42 16.9	33.3	12.2	43 15.5	33.0	12.4	44 14.1	32.8	12.7	78
40 44.7	34.1	11.0	41 43.6	33.9	11.1	42 42.5	33.7	11.3	43 41.3	33.5	11.5	79
40 10.6	−34.7	9.9	41 09.7	−34.6	10.0	42 08.8	−34.4	10.2	43 07.8	−34.2	10.3	80
39 35.9	35.3	8.8	40 35.1	35.1	8.9	41 34.4	35.0	9.1	42 33.6	34.8	9.2	81
39 00.6	35.9	7.8	40 00.0	35.7	7.9	40 59.4	35.5	8.0	41 58.8	35.4	8.1	82
38 24.7	36.4	6.7	39 24.3	36.3	6.8	40 23.9	36.2	6.9	41 23.4	36.0	7.0	83
37 48.3	36.9	5.7	38 48.0	36.8	5.8	39 47.7	36.7	5.9	40 47.4	36.6	6.0	84
37 11.4	−37.3	4.7	38 11.2	−37.3	4.8	39 11.0	−37.2	4.9	40 10.8	−37.1	4.9	85
36 34.1	37.9	3.8	37 33.9	37.8	3.8	38 33.8	37.7	3.9	39 33.7	37.7	3.9	86
35 56.2	38.3	2.8	36 56.1	38.2	2.8	37 56.1	38.3	2.9	38 56.0	38.2	2.9	87
35 17.9	38.7	1.8	36 17.9	38.7	1.9	37 17.8	38.7	1.9	38 17.8	38.7	1.9	88
34 39.2	39.2	0.9	35 39.2	39.2	0.9	36 39.1	39.1	0.9	37 39.1	39.1	1.0	89
34 00.0	39.6	0.0	35 00.0	−39.6	0.0	36 00.0	39.6	0.0	37 00.0	−39.6	0.0	90

30° 31° 32° 33° 34° 35° 36° 37°

49°, 311° L.H.A. LATITUDE SAME NAME AS DECLINATION

Figure 5-9A. Same-name page, Tables No. 229.

LATITUDE CONTRARY NAME TO DECLINATION — L.H.A. 61°, 299°

Dec.	30° Hc	d	Z	31° Hc	d	Z	32° Hc	d	Z	33° Hc	d	Z
°	° ′	′	°	° ′	′	°	° ′	′	°	° ′	′	°
0	24 49.5	-33.2	105.5	24 33.3	-34.1	105.9	24 16.6	-35.0	106.4	23 59.5	-35.9	106.8
1	24 16.3	33.5	106.4	23 59.2	34.5	106.8	23 41.6	35.4	107.3	23 23.6	36.3	107.7
2	23 42.8	33.9	107.3	23 24.7	34.8	107.7	23 06.2	35.6	108.1	22 47.3	36.4	108.5
3	23 08.9	34.2	108.2	22 49.9	35.0	108.6	22 30.6	35.9	109.0	22 10.9	36.8	109.4
4	22 34.7	34.4	109.1	22 14.9	35.3	109.5	21 54.7	36.2	109.9	21 34.1	37.0	110.3
5	22 00.3	-34.8	110.0	21 39.6	-35.6	110.4	21 18.5	-36.4	110.7	20 57.1	-37.3	111.1
6	21 25.5	35.0	110.9	21 04.0	35.9	111.2	20 42.1	36.7	111.6	20 19.8	37.5	111.9
7	20 50.5	35.3	111.7	20 28.1	36.1	112.1	20 05.4	36.9	112.4	19 42.3	37.7	112.8
8	20 15.2	35.5	112.6	19 52.0	36.3	112.9	19 28.5	37.2	113.3	19 04.6	37.9	113.6
9	19 39.7	35.7	113.5	19 15.7	36.6	113.8	18 51.3	37.3	114.1	18 26.7	38.2	114.4
10	19 04.0	-36.1	114.3	18 39.1	-36.8	114.6	18 14.0	-37.6	114.9	17 48.5	-38.3	115.2
11	18 27.9	36.2	115.2	18 02.3	37.0	115.5	17 36.4	37.8	115.7	17 10.2	38.6	116.0
12	17 51.7	36.4	116.0	17 25.3	37.2	116.3	16 58.6	38.0	116.6	16 31.6	38.7	116.8
13	17 15.3	36.7	116.8	16 48.1	37.4	117.1	16 20.6	38.1	117.4	15 52.9	38.9	117.6
14	16 38.6	36.8	117.7	16 10.7	37.6	117.9	15 42.5	38.4	118.2	15 14.0	39.0	118.4
15	16 01.8	-37.0	118.5	15 33.1	-37.8	118.7	15 04.1	-38.5	119.0	14 35.0	-39.3	119.2
16	15 24.8	37.2	119.3	14 55.3	37.9	119.5	14 25.6	38.6	119.8	13 55.7	39.3	120.0
17	14 47.6	37.4	120.1	14 17.4	38.1	120.3	13 47.0	38.8	120.5	13 16.4	39.5	120.8
18	14 10.2	37.5	120.9	13 39.3	38.3	121.1	13 08.2	39.0	121.3	12 36.9	39.7	121.5
19	13 32.7	37.7	121.7	13 01.0	38.4	121.9	12 29.2	39.1	122.1	11 57.2	39.7	122.3
20	12 55.0	-37.9	122.5	12 22.6	-38.5	122.7	11 50.1	-39.2	122.9	11 17.5	-39.9	123.1
21	12 17.1	37.9	123.3	11 44.1	38.6	123.5	11 10.9	39.3	123.7	10 37.6	40.0	123.8
22	11 39.2	38.1	124.1	11 05.5	38.8	124.3	10 31.6	39.4	124.4	9 57.6	40.1	124.6
23	11 01.1	38.2	124.9	10 26.7	38.9	125.0	9 52.2	39.6	125.2	9 17.5	40.2	125.3
24	10 22.9	38.4	125.7	9 47.8	39.0	125.8	9 12.6	39.6	126.0	8 37.3	40.2	126.1
25	9 44.5	-38.4	126.5	9 08.8	-39.1	126.6	8 33.0	-39.7	126.7	7 57.1	-40.4	126.8
26	9 06.1	38.6	127.2	8 29.7	39.2	127.4	7 53.3	39.8	127.5	7 16.7	40.4	127.6
27	8 27.5	38.6	128.0	7 50.5	39.2	128.1	7 13.5	39.9	128.2	6 36.3	40.5	128.3
28	7 48.9	38.7	128.8	7 11.3	39.4	128.9	6 33.6	40.0	129.0	5 55.8	40.6	129.1
29	7 10.2	38.8	129.6	6 31.9	39.4	129.6	5 53.6	40.0	129.7	5 15.2	40.6	129.8
30	6 31.4	-38.9	130.3	5 52.5	-39.4	130.4	5 13.6	-40.0	130.5	4 34.6	-40.6	130.5
31	5 52.5	38.9	131.1	5 13.1	39.5	131.2	4 33.6	40.2	131.2	3 54.0	40.7	131.3
32	5 13.6	39.0	131.9	4 33.6	39.6	131.9	3 53.4	40.1	132.0	3 13.3	40.7	132.0
33	4 34.6	39.0	132.6	3 54.0	39.6	132.7	3 13.3	40.2	132.7	2 32.6	40.8	132.8
34	3 55.6	39.1	133.4	3 14.4	39.7	133.4	2 33.1	40.2	133.5	1 51.8	40.7	133.5
35	3 16.5	-39.1	134.1	2 34.7	-39.6	134.2	1 52.9	-40.2	134.2	1 11.1	-40.8	134.2
36	2 37.4	39.1	134.9	1 55.1	39.7	134.9	1 12.7	40.3	134.9	0 30.3	-40.8	135.0
37	1 58.3	39.2	135.7	1 15.4	39.7	135.7	0 32.4	-40.2	135.7	0 10.5	+40.8	44.3
38	1 19.1	39.1	136.4	0 35.7	-39.7	136.4	0 07.8	+40.2	43.6	0 51.3	40.7	43.6
39	0 40.0	39.2	137.2	0 04.0	+39.7	42.8	0 48.0	40.3	42.8	1 32.0	40.8	42.8
40	0 00.8	-39.2	137.9	0 43.7	+39.7	42.1	1 28.3	+40.2	42.1	2 12.8	+40.7	42.1
41	0 38.4	+39.1	41.3	1 23.4	39.7	41.3	2 08.5	40.2	41.3	2 53.5	40.7	41.4
42	1 17.5	39.2	40.6	2 03.1	39.7	40.6	2 48.7	40.1	40.6	3 34.2	40.7	40.6
43	1 56.7	39.1	39.8	2 42.8	39.6	39.8	3 28.8	40.2	39.9	4 14.9	40.6	39.9
44	2 35.8	39.1	39.0	3 22.4	39.6	39.1	4 09.0	40.1	39.1	4 55.5	40.6	39.2
45	3 14.9	+39.1	38.3	4 02.0	+39.6	38.3	4 49.1	+40.0	38.4	5 36.1	+40.5	38.4
46	3 54.0	39.0	37.5	4 41.6	39.5	37.6	5 29.1	40.0	37.6	6 16.6	40.5	37.7
47	4 33.0	39.0	36.8	5 21.1	39.4	36.8	6 09.1	39.9	36.9	6 57.1	40.4	36.9
48	5 12.0	38.9	36.0	6 00.5	39.4	36.0	6 49.0	39.9	36.1	7 37.5	40.3	36.2
49	5 50.9	38.9	35.2	6 39.9	39.3	35.3	7 28.9	39.8	35.4	8 17.8	40.2	35.4
50	6 29.8	+38.8	34.5	7 19.2	+39.3	34.5	8 08.7	+39.6	34.6	8 58.0	+40.1	34.7
51	7 08.6	38.7	33.7	7 58.5	39.1	33.8	8 48.3	39.6	33.8	9 38.1	40.1	33.9
52	7 47.3	38.6	32.9	8 37.6	39.1	33.0	9 27.9	39.5	33.1	10 18.2	39.9	33.2
53	8 25.9	38.6	32.1	9 16.7	39.0	32.2	10 07.4	39.4	32.3	10 58.1	39.8	32.4
54	9 04.5	38.4	31.4	9 55.7	38.9	31.5	10 46.8	39.3	31.6	11 37.9	39.7	31.7
55	9 42.9	+38.4	30.6	10 34.6	+38.7	30.7	11 26.1	+39.2	30.8	12 17.6	+39.6	30.9
56	10 21.3	38.2	29.8	11 13.3	38.6	29.9	12 05.3	39.0	30.0	12 57.2	39.4	30.1
57	10 59.5	38.1	29.0	11 51.9	38.5	29.1	12 44.3	38.9	29.2	13 36.6	39.3	29.3
58	11 37.6	38.0	28.2	12 30.4	38.4	28.3	13 23.2	38.8	28.5	14 15.9	39.2	28.6
59	12 15.6	37.8	27.5	13 08.8	38.2	27.6	14 02.0	38.6	27.7	14 55.1	39.0	27.8
60	12 53.4	+37.7	26.7	13 47.0	+38.1	26.8	14 40.6	+38.4	26.9	15 34.1	+38.8	27.0
61	13 31.1	37.6	25.9	14 25.1	37.9	26.0	15 19.0	38.3	26.1	16 12.9	38.6	26.2
62	14 08.7	37.3	25.1	15 03.0	37.7	25.2	15 57.3	38.0	25.3	16 51.5	38.4	25.4
63	14 46.0	37.2	24.2	15 40.7	37.6	24.4	16 35.3	38.0	24.5	17 29.9	38.3	24.6
64	15 23.2	37.1	23.4	16 18.3	37.3	23.5	17 13.3	37.7	23.7	18 08.2	38.0	23.8
65	16 00.3	+36.8	22.6	16 55.6	+37.2	22.7	17 51.0	+37.5	22.9	18 46.2	+37.9	23.0
66	16 37.1	36.7	21.8	17 32.8	37.0	21.9	18 28.5	37.2	22.0	19 24.1	37.6	22.2
67	17 13.8	36.4	21.0	18 09.8	36.7	21.1	19 05.7	37.1	21.2	20 01.7	37.3	21.3
68	17 50.2	36.3	20.1	18 46.5	36.6	20.2	19 42.8	36.8	20.4	20 39.0	37.2	20.5
69	18 26.5	36.0	19.3	19 23.1	36.3	19.4	20 19.6	36.6	19.5	21 16.2	36.8	19.7
70	19 02.5	+35.7	18.4	19 59.4	+36.0	18.6	20 56.2	+36.4	18.7	21 53.0	+36.7	18.8
71	19 38.2	35.6	17.6	20 35.4	35.8	17.7	21 32.6	36.0	17.8	22 29.7	36.3	18.0
72	20 13.8	35.3	16.7	21 11.2	35.6	16.8	22 08.6	35.8	17.0	23 06.0	36.1	17.1
73	20 49.1	35.0	15.9	21 46.8	35.2	16.0	22 44.4	35.5	16.1	23 42.1	35.7	16.2
74	21 24.1	34.7	15.0	22 22.0	35.0	15.1	23 19.9	35.3	15.2	24 17.8	35.5	15.3
75	21 58.8	+34.5	14.1	22 57.0	+34.7	14.2	23 55.2	+34.9	14.3	24 53.3	+35.1	14.5
76	22 33.3	34.2	13.2	23 31.7	34.4	13.3	24 30.1	34.6	13.4	25 28.4	34.8	13.6
77	23 07.5	33.9	12.4	24 06.1	34.1	12.4	25 04.7	34.2	12.5	26 03.2	34.5	12.7
78	23 41.4	33.5	11.5	24 40.2	33.7	11.5	25 38.9	34.0	11.6	26 37.7	34.1	11.7
79	24 14.9	33.3	10.5	25 13.9	33.4	10.6	26 12.9	33.5	10.7	27 11.8	33.8	10.8
80	24 48.2	+32.9	9.6	25 47.3	+33.1	9.7	26 46.4	+33.3	9.8	27 45.6	+33.3	9.9
81	25 21.1	32.5	8.7	26 20.4	32.7	8.8	27 19.7	32.8	8.9	28 18.9	33.0	8.9
82	25 53.6	32.2	7.8	26 53.1	32.3	7.8	27 52.5	32.4	7.9	28 51.9	32.6	8.0
83	26 25.8	31.8	6.8	27 25.4	31.9	6.9	28 24.9	32.1	7.0	29 24.5	32.1	7.0
84	26 57.6	31.5	5.9	27 57.3	31.5	5.9	28 57.0	31.6	6.0	29 56.6	31.8	6.1
85	27 29.1	+31.0	4.9	28 28.8	+31.1	5.0	29 28.6	+31.2	5.0	30 28.4	+31.2	5.1
86	28 00.1	30.6	4.0	28 59.9	30.7	4.0	29 59.8	30.7	4.0	30 59.6	30.9	4.1
87	28 30.7	30.2	3.0	29 30.6	30.3	3.0	30 30.5	30.3	3.0	31 30.5	30.3	3.1
88	29 00.9	29.8	2.0	30 00.9	29.8	2.0	31 00.8	29.9	2.0	32 00.8	29.9	2.1
89	29 30.7	29.3	1.0	30 30.7	29.3	1.0	31 30.7	29.3	1.0	32 30.7	29.3	1.0
90	30 00.0	+28.9	0.0	31 00.0	+28.8	0.0	32 00.0	+28.8	0.0	33 00.0	+28.8	0.0
	30°			31°			32°			33°		

Dec.	34° Hc	d	Z	35° Hc	d	Z	36° Hc	d	Z	37° Hc	d	Z
°	° ′	′	°	° ′	′	°	° ′	′	°	° ′	′	°
0	23 41.9	-36.8	107.2	23 23.9	-37.6	107.6	23 05.6	-38.5	108.0	22 46.8	-39.3	108.4
1	23 05.1	37.0	108.1	22 46.3	37.9	108.5	22 27.1	38.7	108.9	22 07.5	39.5	109.3
2	22 28.1	37.4	108.9	22 08.4	38.1	109.3	21 48.4	39.0	109.7	21 28.0	39.8	110.1
3	21 50.7	37.5	109.8	21 30.3	38.4	110.2	21 09.4	39.2	110.5	20 48.2	40.0	110.9
4	21 13.2	37.9	110.6	20 51.9	38.7	111.0	20 30.2	39.4	111.3	20 08.2	40.2	111.7
5	20 35.3	-38.0	111.4	20 13.2	-38.8	111.8	19 50.8	-39.7	112.1	19 28.0	-40.4	112.5
6	19 57.3	38.3	112.3	19 34.4	39.1	112.6	19 11.1	39.8	112.9	18 47.6	40.6	113.2
7	19 19.0	38.5	113.1	18 55.3	39.3	113.4	18 31.3	40.0	113.7	18 07.0	40.8	114.0
8	18 40.5	38.8	113.9	18 16.0	39.5	114.2	17 51.3	40.3	114.5	17 26.2	40.9	114.8
9	18 01.7	38.9	114.7	17 36.5	39.6	115.0	17 11.0	40.4	115.3	16 45.3	41.1	115.6
10	17 22.8	-39.1	115.5	16 56.9	-39.9	115.8	16 30.6	-40.5	116.1	16 04.2	-41.3	116.3
11	16 43.7	39.3	116.3	16 17.0	40.0	116.6	15 50.1	40.8	116.8	15 22.9	41.5	117.1
12	16 04.4	39.4	117.1	15 37.0	40.2	117.3	15 09.3	40.8	117.6	14 41.4	41.5	117.8
13	15 25.0	39.6	117.9	14 56.8	40.3	118.1	14 28.5	41.1	118.3	13 59.9	41.7	118.6
14	14 45.4	39.8	118.6	14 16.5	40.5	118.9	13 47.4	41.1	119.1	13 18.2	41.9	119.3
15	14 05.6	-39.9	119.4	13 36.0	-40.6	119.6	13 06.3	-41.3	119.8	12 36.3	-41.9	120.0
16	13 25.7	40.1	120.2	12 55.4	40.7	120.4	12 25.0	41.4	120.6	11 54.4	42.1	120.8
17	12 45.6	40.2	121.0	12 14.7	40.9	121.1	11 43.6	41.6	121.3	11 12.3	42.2	121.5
18	12 05.4	40.3	121.7	11 33.8	41.0	121.9	11 02.0	41.6	122.1	10 30.1	42.3	122.2
19	11 25.1	40.4	122.5	10 52.8	41.1	122.6	10 20.4	41.7	122.8	9 47.8	42.3	122.9
20	10 44.7	-40.6	123.2	10 11.7	-41.2	123.4	9 38.7	-41.9	123.5	9 05.5	-42.5	123.7
21	10 04.1	40.6	124.0	9 30.5	41.2	124.1	8 56.8	41.9	124.3	8 23.0	42.5	124.4
22	9 23.5	40.7	124.7	8 49.3	41.4	124.9	8 14.9	42.0	125.0	7 40.5	42.6	125.1
23	8 42.8	40.9	125.5	8 07.9	41.4	125.6	7 32.9	42.0	125.7	6 57.9	42.7	125.8
24	8 01.9	40.9	126.2	7 26.5	41.6	126.3	6 50.9	42.1	126.4	6 15.2	42.7	126.5
25	7 21.0	-40.9	126.9	6 44.9	-41.5	127.0	6 08.8	-42.2	127.1	5 32.5	-42.8	127.2
26	6 40.1	41.1	127.7	6 03.4	41.7	127.8	5 26.6	42.2	127.8	4 49.7	42.8	127.9
27	5 59.0	41.1	128.4	5 21.7	41.7	128.5	4 44.4	42.3	128.6	4 06.9	42.8	128.6
28	5 17.9	41.1	129.1	4 40.0	41.7	129.2	4 02.1	42.3	129.3	3 24.1	42.9	129.3
29	4 36.8	41.2	129.9	3 58.3	41.8	129.9	3 19.8	42.4	130.0	2 41.2	42.9	130.0
30	3 55.6	-41.2	130.6	3 16.5	-41.8	130.7	2 37.4	-42.3	130.7	1 58.3	-42.9	130.7
31	3 14.4	41.3	131.3	2 34.7	41.8	131.4	1 55.1	42.4	131.4	1 15.4	43.0	131.4
32	2 33.1	41.3	132.1	1 52.9	41.8	132.1	1 12.7	42.4	132.1	0 32.4	-42.9	132.1
33	1 51.8	41.3	132.8	1 11.1	41.9	132.8	0 30.3	-42.4	132.8	0 10.5	+42.9	47.2
34	1 10.5	41.3	133.5	0 29.2	-41.8	133.5	0 12.1	+42.4	46.5	0 53.4	42.9	46.5
35	0 29.2	-41.3	134.2	0 12.6	+41.9	45.8	0 54.5	+42.4	45.8	1 36.3	+43.0	45.8
36	0 12.1	+41.3	45.0	0 54.5	41.8	45.0	1 36.9	42.4	45.1	2 19.3	42.9	45.1
37	0 53.4	41.3	44.3	1 36.3	41.9	44.3	2 19.3	42.3	44.4	3 02.2	42.8	44.4
38	1 34.7	41.3	43.6	2 18.2	41.8	43.6	3 01.6	42.3	43.6	3 45.0	42.8	43.7
39	2 16.0	41.3	42.9	3 00.0	41.8	42.9	3 43.9	42.3	42.9	4 27.8	42.8	43.0
40	2 57.3	+41.2	42.1	3 41.8	+41.7	42.2	4 26.2	+42.3	42.2	5 10.6	+42.8	42.3
41	3 38.5	41.2	41.4	4 23.5	41.7	41.5	5 08.5	42.2	41.5	5 53.4	42.7	41.6
42	4 19.7	41.2	40.7	5 05.2	41.7	40.7	5 50.7	42.1	40.8	6 36.1	42.6	40.9
43	5 00.9	41.1	39.9	5 46.9	41.6	40.0	6 32.8	42.1	40.1	7 18.7	42.6	40.2
44	5 42.0	41.1	39.2	6 28.5	41.5	39.3	7 14.9	42.0	39.4	8 01.3	42.5	39.4
45	6 23.1	+41.0	38.5	7 10.0	+41.5	38.6	7 56.9	+42.0	38.6	8 43.8	+42.4	38.7
46	7 04.1	40.9	37.7	7 51.5	41.4	37.8	8 38.9	41.8	37.9	9 26.2	42.3	38.0
47	7 45.0	40.9	37.0	8 32.9	41.3	37.1	9 20.7	41.8	37.2	10 08.5	42.2	37.3
48	8 25.9	40.7	36.3	9 14.2	41.2	36.4	10 02.5	41.7	36.5	10 50.7	42.1	36.6
49	9 06.6	40.7	35.5	9 55.4	41.2	35.6	10 44.2	41.5	35.7	11 32.8	42.1	35.8
50	9 47.3	+40.6	34.8	10 36.6	+41.0	34.9	11 25.7	+41.5	35.0	12 14.9	+41.9	35.1
51	10 27.9	40.5	34.0	11 17.6	40.9	34.1	12 07.2	41.4	34.3	12 56.8	41.7	34.4
52	11 08.4	40.3	33.3	11 58.5	40.8	33.4	12 48.6	41.2	33.5	13 38.5	41.7	33.6
53	11 48.7	40.3	32.5	12 39.3	40.7	32.6	13 29.8	41.1	32.8	14 20.2	41.5	32.9
54	12 29.0	40.1	31.8	13 20.0	40.5	31.9	14 10.9	40.9	32.0	15 01.7	41.4	32.2
55	13 09.1	+40.0	31.0	14 00.5	+40.4	31.1	14 51.8	+40.8	31.3	15 43.1	+41.2	31.4
56	13 49.1	39.8	30.2	14 40.9	40.2	30.4	15 32.6	40.7	30.5	16 24.3	41.0	30.7
57	14 28.9	39.7	29.5	15 21.1	40.1	29.6	16 13.3	40.4	29.7	17 05.3	40.9	29.9
58	15 08.6	39.5	28.7	16 01.2	39.9	28.8	16 53.7	40.3	29.0	17 46.2	40.7	29.1
59	15 48.1	39.4	27.9	16 41.1	39.7	28.1	17 34.0	40.1	28.2	18 26.9	40.5	28.4
60	16 27.5	+39.2	27.1	17 20.8	+39.6	27.3	18 14.1	+40.0	27.4	19 07.4	+40.3	27.6
61	17 06.7	39.0	26.3	18 00.4	39.4	26.5	18 54.1	39.7	26.6	19 47.7	40.1	26.8
62	17 45.7	38.8	25.5	18 39.8	39.1	25.7	19 33.8	39.5	25.8	20 27.8	39.8	26.0
63	18 24.5	38.5	24.7	19 18.9	39.0	24.9	20 13.3	39.3	25.0	21 07.6	39.7	25.2
64	19 03.0	38.4	23.9	19 57.9	38.7	24.1	20 52.6	39.1	24.2	21 47.3	39.4	24.4
65	19 41.4	+38.2	23.1	20 36.6	+38.5	23.3	21 31.7	+38.8	23.4	22 26.7	+39.2	23.6
66	20 19.6	37.9	22.3	21 15.1	38.2	22.4	22 10.5	38.6	22.6	23 05.9	38.9	22.8
67	20 57.5	37.7	21.5	21 53.3	38.0	21.6	22 49.1	38.3	21.8	23 44.8	38.6	21.9
68	21 35.2	37.4	20.6	22 31.3	37.8	20.8	23 27.4	38.1	20.9	24 23.4	38.4	21.1
69	22 12.6	37.2	19.8	23 09.1	37.4	19.9	24 05.5	37.7	20.1	25 01.8	38.1	20.2
70	22 49.8	+36.9	18.9	23 46.5	+37.2	19.1	24 43.2	+37.5	19.2	25 39.9	+37.7	19.4
71	23 26.7	36.6	18.1	24 23.7	36.9	18.2	25 20.7	37.2	18.4	26 17.6	37.5	18.5
72	24 03.3	36.4	17.2	25 00.6	36.6	17.4	25 57.9	36.8	17.5	26 55.1	37.1	17.6
73	24 39.7	36.0	16.3	25 37.2	36.3	16.5	26 34.7	36.6	16.6	27 32.2	36.8	16.8
74	25 15.7	35.7	15.5	26 13.5	35.9	15.6	27 11.3	36.1	15.7	28 09.0	36.4	15.9
75	25 51.4	+35.3	14.6	26 49.4	+35.6	14.7	27 47.4	+35.9	14.8	28 45.4	+36.1	15.0
76	26 26.7	35.1	13.7	27 25.0	35.3	13.8	28 23.3	35.4	13.9	29 21.5	35.7	14.0
77	27 01.8	34.6	12.8	28 00.3	34.8	12.9	28 58.7	35.1	13.0	29 57.2	35.3	13.1
78	27 36.4	34.3	11.8	28 35.1	34.5	12.0	29 33.8	34.7	12.1	30 32.5	34.9	12.2
79	28 10.7	34.0	10.9	29 09.6	34.2	11.0	30 08.5	34.3	11.1	31 07.4	34.5	11.2
80	28 44.7	+33.5	10.0	29 43.8	+33.7	10.1	30 42.8	+33.9	10.2	31 41.9	+34.0	10.3
81	29 18.2	33.1	9.0	30 17.5	33.2	9.1	31 16.7	33.4	9.2	32 15.9	33.6	9.3
82	29 51.3	32.7	8.1	30 50.7	32.9	8.2	31 50.1	33.0	8.2	32 49.5	33.1	8.3
83	30 24.0	32.3	7.1	31 23.6	32.4	7.2	32 23.1	32.5	7.3	33 22.6	32.6	7.3
84	30 56.3	31.8	6.1	31 56.0	31.9	6.2	32 55.6	32.0	6.3	33 55.2	32.2	6.3
85	31 28.1	+31.4	5.1	32 27.9	+31.4	5.2	33 27.6	+31.6	5.2	34 27.4	+31.6	5.3
86	31 59.5	30.9	4.1	32 59.3	31.0	4.2	33 59.2	31.0	4.2	34 59.0	31.1	4.3
87	32 30.4	30.4	3.1	33 30.3	30.4	3.1	34 30.2	30.5	3.2	35 30.1	30.5	3.2
88	33 00.8	29.8	2.1	34 00.7	29.9	2.1	35 00.7	29.9	2.1	36 00.6	30.0	2.2
89	33 30.6	29.4	1.0	34 30.6	29.4	1.1	35 30.6	29.4	1.1	36 30.6	29.4	1.1
90	34 00.0	+28.8	0.0	35 00.0	+28.8	0.0	36 00.0	+28.8	0.0	37 00.0	+28.8	0.0
	34°			35°			36°			37°		

S. Lat. {L.H.A. greater than 180°......Zn=180°−Z; L.H.A. less than 180°............Zn=180°+Z}

LATITUDE SAME NAME AS DECLINATION — L.H.A. 119°, 241°

Figure 5-9B. Contrary-name page, Tables No. 229.

and every fifth value thereafter. If the Hc increases with increasing declination, the sign of the altitude difference d is positive, indicated by a plus sign.

In almost all cases, two increments are extracted from the interpolation table—one for the tens of minutes of the altitude difference d and the other for the remaining units and tenths. Adding the two parts together yields the total interpolation correction (Tot. Corr.), which in turn is added algebraically to the tabulated Hc to obtain the final computed altitude. In about one percent of all cases, a third increment called a "double second difference" correction (DS Corr.) must also be found. Occasions for which this is necessitated are indicated in the tables by the d value being printed in italic type followed by a dot.

The interpolation tables inside the front cover are used for declination increments in the range 0.0′ to 31.9′, and those inside the back cover for the range 28.0′ to 59.9′. Since the declination increment in this example is 40.6′, the tables inside the back cover are used. An extract from the appropriate table appears in Figure 5-10.

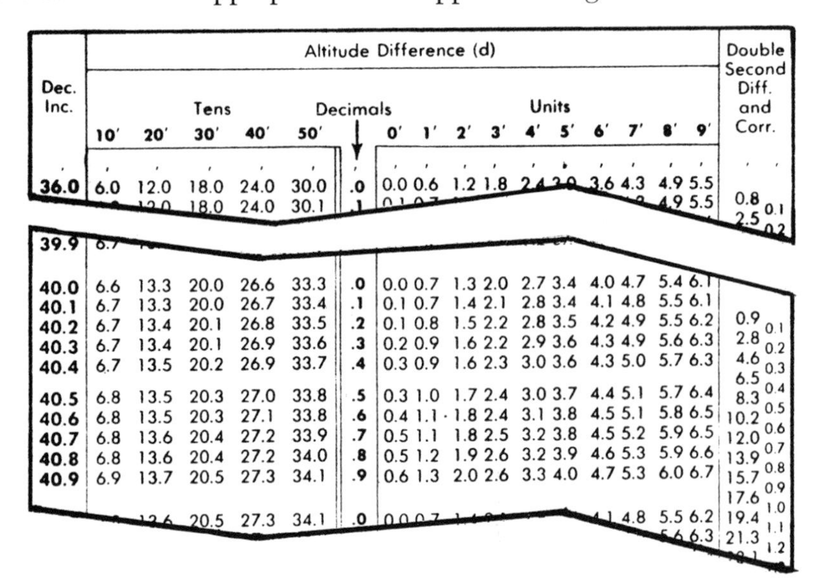

Dec. Inc.	Altitude Difference (d)															
	Tens					Decimals	Units									
	10′	20′	30′	40′	50′	↓	0′	1′	2′	3′	4′	5′	6′	7′	8′	9′
36.0	6.0	12.0	18.0	24.0	30.0	.0	0.0	0.6	1.2	1.8	2.4	3.0	3.6	4.3	4.9	5.5
		12.0	18.0	24.0	30.1	.1									4.9	5.5
39.9	6.7															
40.0	6.6	13.3	20.0	26.6	33.3	.0	0.0	0.7	1.3	2.0	2.7	3.4	4.0	4.7	5.4	6.1
40.1	6.7	13.3	20.0	26.7	33.4	.1	0.1	0.7	1.4	2.1	2.8	3.4	4.1	4.8	5.5	6.1
40.2	6.7	13.4	20.1	26.8	33.5	.2	0.1	0.8	1.5	2.2	2.8	3.5	4.2	4.9	5.5	6.2
40.3	6.7	13.4	20.1	26.9	33.6	.3	0.2	0.9	1.6	2.2	2.9	3.6	4.3	4.9	5.6	6.3
40.4	6.7	13.5	20.2	26.9	33.7	.4	0.3	0.9	1.6	2.3	3.0	3.6	4.3	5.0	5.7	6.3
40.5	6.8	13.5	20.3	27.0	33.8	.5	0.3	1.0	1.7	2.4	3.0	3.7	4.4	5.1	5.7	6.4
40.6	6.8	13.5	20.3	27.1	33.8	.6	0.4	1.1	1.8	2.4	3.1	3.8	4.5	5.1	5.8	6.5
40.7	6.8	13.6	20.4	27.2	33.9	.7	0.5	1.1	1.8	2.5	3.2	3.8	4.5	5.2	5.9	6.5
40.8	6.8	13.6	20.4	27.2	34.0	.8	0.5	1.2	1.9	2.6	3.2	3.9	4.6	5.3	5.9	6.6
40.9	6.9	13.7	20.5	27.3	34.1	.9	0.6	1.3	2.0	2.6	3.3	4.0	4.7	5.3	6.0	6.7
			20.5	27.3	34.1	.0							4.1	4.8	5.5	6.2
															5.6	6.3

Double Second Diff. and Corr.

Diff.	Corr.
0.8	
	0.1
2.5	
0.9	
	0.1
2.8	
	0.2
4.6	
	0.3
6.5	
	0.4
8.3	
	0.5
10.2	
	0.6
12.0	
	0.7
13.9	
	0.8
15.7	
	0.9
17.6	
	1.0
19.4	
	1.1
21.3	
	1.2

Figure 5-10. Extract from interpolation tables, Tables No. 229.

The interpolation table is entered first for the "tens" increment and then for the "units" increment of the interpolation correction. For the tens increment the table is entered directly across from the declination increment, 40.6′ in this case, and the increment beneath the appropriate number of tens of altitude difference is recorded. Here, no tens increment is extracted, since the altitude difference d is 08.7′.

To find the units increment, the appropriate units column (in this case the one headed 8′) is followed down the page in the group of tabulations most nearly opposite the declination increase until the appropriate decimal fraction is reached (.7 in this case). For this example, the units increment is −5.9; it is negative, since the altitude difference d is negative.

Had it been necessary to find a double second difference correction, the difference between the two d values directly above and directly below the d value corresponding to the integral portion of the actual declination is mentally computed. Using this "double second difference" as an entering argument, the right-hand side of the interpolation table is used to find the correction. There are several complete DS interpolation sections on each page; the section used is the one most nearly opposite the original declination increment. If a DS correction is necessary, it is entered on the form and always added to the total of the tens and units increments to form the total interpolation correction.

After extracting, recording, and summing the two increments to form the total interpolation correction, −5.9 in this case, the correction is added to the tabulated Hc recorded earlier to obtain the final computed altitude of 50° 45.2′. To find the intercept distance a, this value is compared with the observed altitude Ho 50° 39.4′ determined earlier. Since Ho is less than Hc in this case, the intercept distance, 5.8 miles, is labeled "Away" (A) from the direction of the GP of the body.

To find the true azimuth of the GP from the AP, the azimuth angle Z must be determined. To compute the value for Z, it is necessary to interpolate between the values of Z tabulated in *Tables No. 229* for the whole degrees of declination bracketing the true declination in the same latitude column used previously. Inasmuch as the difference between successive tabulated azimuth angles is normally small, interpolation is usually done mentally. The interpolation tables can be used for this interpolation, however, by substituting the difference between the two tabulated azimuth angles as a difference d. In this example, the azimuth angle tabulated for declination 52° is 47.4° and the angle for declination 53° is 45.8°. Hence, the value of the azimuth angle for a true declination 52° 40.6′ is 46.3° $\left(47.4° - 45.8° = (-)1.6°;\ \frac{40.6'}{60'} \times (-)1.6° = (-)1.1°;\ 47.4° \quad (-)1.1° = 46.3°\right)$. Since the assumed latitude lies in the southern hemisphere and the body lies west of the observer (LHA is less than 180°), the prefix S and the suffix W are applied: S 46.3° W.

Sight Reduction using Tables 229

C: 230°

S: 25 kts

Body		CANOPUS	
Ho (Obs Alt)		50-39.4	
Date		16 DEC	
DR Lat		34-19.0 S	
DR Long		163-05.7 E	
Obs. Time		04-11-33	
WE (S+, F-)		36 (s)	
ZT		04-12-09	
ZD (W+, E-)		-11	
GMT		17-12-09	
Date (GMT)		15 DEC	
Tab GHA	v	338-56.1	
GHA incr'mt.		3-02.7	
SHA or v Corr.		264-09.8	
GHA		606-08.6	
±360 if needed		246-08.6	
aλ (-W, +E)		162-51.4 E	
LHA		49	
Tab Dec	d		
d Corr (+ or -)			
True Dec		S 52-40.6	
a Lat (N or S)		34 S Same Cont.	
Dec Inc	(±)d	40.6	-8.7
Hc (Tab. Alt.)		50-51.1	
tens	DS Diff.	0	
units	DS Corr.	-5.9	+
Tot. Corr. (+ or -)		-5.9	
Hc (Comp. Alt.)		50-45.2	
Ho (Obs. Alt.)		50-39.4	
a (Intercept)		5.8 A	
Z		S 46.3 W	
Zn (°T)		226.3°T	

Figure 5-11. Complete solution for the star Canopus by Tables No. 229.

The final step in completing the sight reduction form is to convert the azimuth angle just computed to the true azimuth Zn of the body from the AP of the observer. One method of accomplishing this by sketching the angle was presented in Chapter 1. The easiest method, however, is simply to use the conversion formulas printed on each page of *Tables No. 229:*

N. Lat. L.H.A. greater than 180° . . . . . . Zn = Z
L.H.A. less than 180° . . . . Zn = 360° − Z

S. Lat. {L.H.A. greater than 180° . . . . . . Zn = 180° − Z
L.H.A. less than 180° . . . . Zn = 180° + Z

By either method the azimuth angle S 46°.3° W is converted to a true azimuth 226.3°T.

In Chapter 7 the plot of the resulting Canopus LOP will be combined with the LOPs resulting from the Venus and moon sight reductions partially completed earlier in this chapter to form a celestial fix.

Completing the Sight Reduction by *Tables No. 229*–Second Example

As the second example of the use of *Tables No. 229,* the sun sight reduction begun as the last example of the use of the *Nautical Almanac* will be completed here. The entering arguments for the tables for this example follow:

LHA	299
True Dec	S 23–18.9
a Lat (N or S)	31 N Same (Cont.)

Opening the tables to the LHA 299° page shown in Figure 5-9B, the tabulated Hc corresponding to declination 23° and latitude 31° is first recorded, along with the altitude difference d; they are 10° 26.7′ and −38.9, respectively. To eliminate the necessity to refer back to the LHA page, the azimuth angle Z values corresponding to declinations 23° and 24° are also recorded on a sheet of scratch paper; they are 125.0° and 125.8°, respectively.

Turning to the interpolation tables inside the front cover, the appropriate portion of which is shown in Figure 5-12, the correction to the tabulated Hc is first found. The tens increment is 9.5′, and the units increment is 2.7′; these values are boxed in the figure. No double second difference correction applies, so the total correction to the

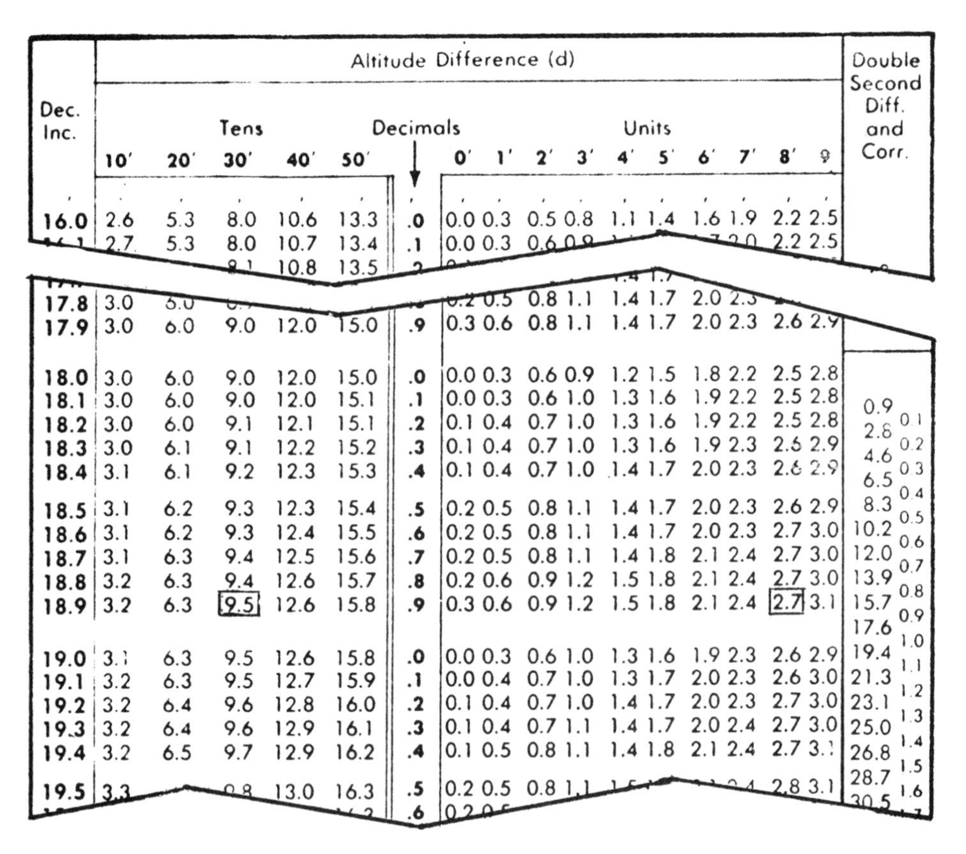

Dec. Inc.	Tens 10′	20′	30′	40′	50′	Decimals	Units 0′	1′	2′	3′	4′	5′	6′	7′	8′	9′
16.0	2.6	5.3	8.0	10.6	13.3	.0	0.0	0.3	0.5	0.8	1.1	1.4	1.6	1.9	2.2	2.5
16.1	2.7	5.3	8.0	10.7	13.4	.1	0.0	0.3	0.6	0.9				2.0	2.2	2.5
			8.1	10.8	13.5	.2										
											1.4	1.7				
17.8	3.0	6.0					0.2	0.5	0.8	1.1	1.4	1.7	2.0	2.3		
17.9	3.0	6.0	9.0	12.0	15.0	.9	0.3	0.6	0.8	1.1	1.4	1.7	2.0	2.3	2.6	2.9
18.0	3.0	6.0	9.0	12.0	15.0	.0	0.0	0.3	0.6	0.9	1.2	1.5	1.8	2.2	2.5	2.8
18.1	3.0	6.0	9.0	12.0	15.1	.1	0.0	0.3	0.6	1.0	1.3	1.6	1.9	2.2	2.5	2.8
18.2	3.0	6.0	9.1	12.1	15.1	.2	0.1	0.4	0.7	1.0	1.3	1.6	1.9	2.2	2.5	2.8
18.3	3.0	6.1	9.1	12.2	15.2	.3	0.1	0.4	0.7	1.0	1.3	1.6	1.9	2.3	2.6	2.9
18.4	3.1	6.1	9.2	12.3	15.3	.4	0.1	0.4	0.7	1.0	1.4	1.7	2.0	2.3	2.6	2.9
18.5	3.1	6.2	9.3	12.3	15.4	.5	0.2	0.5	0.8	1.1	1.4	1.7	2.0	2.3	2.6	2.9
18.6	3.1	6.2	9.3	12.4	15.5	.6	0.2	0.5	0.8	1.1	1.4	1.7	2.0	2.3	2.7	3.0
18.7	3.1	6.3	9.4	12.5	15.6	.7	0.2	0.5	0.8	1.1	1.4	1.8	2.1	2.4	2.7	3.0
18.8	3.2	6.3	9.4	12.6	15.7	.8	0.2	0.6	0.9	1.2	1.5	1.8	2.1	2.4	2.7	3.0
18.9	3.2	6.3	[9.5]	12.6	15.8	.9	0.3	0.6	0.9	1.2	1.5	1.8	2.1	2.4	[2.7]	3.1
19.0	3.1	6.3	9.5	12.6	15.8	.0	0.0	0.3	0.6	1.0	1.3	1.6	1.9	2.3	2.6	2.9
19.1	3.2	6.3	9.5	12.7	15.9	.1	0.0	0.4	0.7	1.0	1.3	1.7	2.0	2.3	2.6	3.0
19.2	3.2	6.4	9.6	12.8	16.0	.2	0.1	0.4	0.7	1.0	1.4	1.7	2.0	2.3	2.7	3.0
19.3	3.2	6.4	9.6	12.9	16.1	.3	0.1	0.4	0.7	1.1	1.4	1.7	2.0	2.4	2.7	3.0
19.4	3.2	6.5	9.7	12.9	16.2	.4	0.1	0.5	0.8	1.1	1.4	1.8	2.1	2.4	2.7	3.1
19.5	3.3			13.0	16.3	.5	0.2	0.5	0.8	1.1					2.8	3.1
						.6										

Double Second Diff.	and Corr.
0.9	
	0.1
2.8	
	0.2
4.6	
	0.3
6.5	
	0.4
8.3	
	0.5
10.2	
	0.6
12.0	
	0.7
13.9	
	0.8
15.7	
	0.9
17.6	
	1.0
19.4	
	1.1
21.3	
	1.2
23.1	
	1.3
25.0	
	1.4
26.8	
	1.5
28.7	
	1.6
30.5	

Figure 5-12. Extract from interpolation tables, Tables No. 229.

tabulated Hc is the sum of the tens and units increments, or 12.2′. Applying this correction yields the computed altitude 10° 14.5′.

Comparing this Hc with the Ho 10° 16.8′ results in an intercept distance of 2.3 miles; since Ho is greater than Hc, the intercept is labeled "Toward" (T) the GP of the body.

Referring to the azimuth angles noted earlier, the difference between the two is .8°. Entering the interpolation tables once again, using this figure as a difference d, a correction to the tabulated angle of .2° results. Hence, the azimuth angle is N 125.2° E (east, because the LHA is greater than 180°). Conversion of this azimuth angle yields a true azimuth of 125.2°T.

The completed sun sight reduction form appears in Figure 5-13 opposite. In Chapter 7, the resulting LOP will be combined with a second LOP obtained from a later sun observation to form a running fix.

Sight Reduction using Tables 229

C: 140 °

S: 20 kts.

M W E G g m

Body	SUN (LL)	
Corr. to Ha		
Ho (Obs Alt)	10-16.8	
Date	16 DEC	
DR Lat	31-08.1 N	
DR Long	64-48.0 W	
Obs. Time	08-12-06	
WE (S+, F-)	0	
ZT	08-12-06	
ZD (W+, E-)	+4	
GMT	12-12-06	
Date (GMT)	16 DEC	
Tab GHA / v	1-07.0	
GHA incr'mt.	3-01.5	
SHA or v Corr.		
GHA	4-08.5	
±360 if needed	364-08.5	
aλ (-W, +E)	65-08.5 W	
LHA	299	
Tab Dec / d	S 23-18.9	0.1
d Corr (+ or -)	0.0	
True Dec	S 23-18.9	
a Lat (N or S)	31 N Same (Cont.)	
Dec Inc / (±)d	18.9	-38.9
Hc (Tab. Alt.)	10-26.7	
tens / DS Diff.	-9.5	
units / DS Corr.	-2.7	+
Tot. Corr. (+ or -)	-12.2	
Hc (Comp. Alt.)	10-14.5	
Ho (Obs. Alt.)	10-16.8	
a (Intercept)	2.3 A/T	
Z	N 125.2 E	
Zn (°T)	125.2°T	

Figure 5-13. Complete solution for the sun (LL) by Tables No. 229.

Summary

In this chapter, the reduction of the observation of a celestial body by the use of the *Nautical Almanac* in conjunction with the *Sight Reduction Tables for Marine Navigation, No. 229,* was demonstrated by means of several examples. Use of these publications during sight reduction is rather tedious, but has the positive advantage that a high degree of accuracy is attainable in the resulting celestial LOP. In the following chapter, the use of the *Air Almanac* with the *Sight Reduction Tables for Air Navigation* is demonstrated for several of the same observations presented as examples in this chapter. The increased speed of computation using these latter publications will become obvious, but it should be borne in mind that the precision of the resulting Hc and Zn is less than that achieved by use of the *Nautical Almanac* and *Tables No. 229.*

6

The Complete Solution by the *Air Almanac* and *Tables No. 249*

In the last chapter the complete solution for a celestial line of position by the altitude-intercept method using the *Nautical Almanac* in conjunction with the *Sight Reduction Tables for Marine Navigation, No. 229,* was discussed. In this chapter an alternative method for the complete solution will be examined, using the *Air Almanac* with the *Sight Reduction Tables for Air Navigation, No. 249.* Although either of these two almanacs could be used with either of these two sight reduction tables, in practice they are usually employed as the titles of the chapters indicate; the *Nautical Almanac* and *Tables No. 229* are generally preferred in marine navigation in order to take advantage of the finer precision inherent in this combination. The greater ease of obtaining the complete solution by the *Air Almanac* and *Tables No. 249,* however, has resulted in their adoption by many marine navigators, particularly when operating in rough weather on the high seas, or in small vessels in which highly precise celestial observations are very difficult to make even under optimal weather conditions.

The *Air Almanac*

The contents of the *Air Almanac* are basically the same as the *Nautical Almanac,* the primary differences being in the arrangement and precision of the data. The *Air Almanac* is designed for speed and ease of computation, at the cost of some small degree of precision. As an example, when the *Air Almanac* is to be used to obtain sextant altitude corrections to obtain Ho, the hs measured by the sextant is normally rounded off to the nearest whole minute. All corrections to be applied to hs are tabulated to the nearest minute in various critical type tables, with the exception of the refraction correction; at very low altitudes the refraction correction is taken to the nearest two or five minutes. Figure 6-1A shows the dip table for the *Air Almanac,* which is located on the outside back cover, and Figure 6-1B depicts the refraction correction table on the inside back cover.

The sextant altitude hs corrected for IC is the entering argument for both tables. For surface navigation the left-hand column in the refraction table headed 0 (for 0 altitude) is used. The refraction correction R_o corresponding to the range of sextant altitude containing

CORRECTIONS TO BE APPLIED TO MARINE SEXTANT ALTITUDE

MARINE SEXTANT ERROR	CORRECTIONS
Sextant Number	In addition to sextant error and dip, corrections are to be applied for:
Index Error	Refraction Semi-diameter (for the Sun and Moon) Parallax (for the Moon) Dome refraction (if applicable)

CORRECTION FOR DIP OF THE HORIZON

To be subtracted from sextant altitude

Ht.	Dip	Ht.	Dip	Ht.	Dip	Ht.	Dip	Ht.	I
Ft.	′	Ft.	′	Ft.	′	Ft.	′	Ft.	
0		114		437		968		1 707	
	1		11		21		31		
2		137		481		1 033		1 792	
	2		12		22		32		
6		162		527		1 099		1 880	
	3		13		23		33		
12		189		575		1 168		1 970	
	4		14		24		34		
21		218		625		1 239		2 061	
	5		15		25		35		
31		250		677		1 311		2 155	
	6		16		26		36		
43		283		731		1 386		2 251	
	7		17		27		37		
58		318		787		1 463		2 349	
	8		18		28		38		
75		356		845		1 543		2 449	
	9		19		29		39		
93		395		906		1 624		2 551	
	10		20		30		40		
114		437		968		1 707		2 655	

Figure 6-1A. Dip correction table, Air Almanac.

the given hs is the value extracted; it is always subtracted from h
Under abnormal temperature conditions a multiplication factor *f*
obtained from the lower part of the table and applied to the bas
R_0 correction by means of the nomogram on the right side. (Note th
instruction in the lower right corner to use $R = R_0$ where the
correction is less than 10′.)

When a sun or moon observation is made, semidiameter (SD) an
parallax corrections are made using data on the daily pages. Th
semidiameter of the moon is always given to the nearest minute, whi
the semidiameter of the sun is given to the nearest tenth of a minut
The SDs are either added to or subtracted from hs, depending o
whether a lower or upper limb observation was made. The value o
parallax for the moon is found from a critical value table, using th
hs corrected for IC, dip, refraction, and SD as entering argument;
is always added to the corrected hs.

The yellow daily pages of the *Air Almanac* contain much of th
same data found on the corresponding white daily pages of the *Naut
cal Almanac*, the main difference being in the frequency of tabulation
Whereas in the *Nautical Almanac* tabulated values of GHA and decl
nation appear for every whole hour of GMT, in the *Air Almanac* th
tabulations are for every 10 minutes of GMT. The increased frequenc
of tabulations eliminates the necessity for any *v* or *d* corrections t
the GHA and declination. In fact, because the exact calculated valu
of the GHA of the sun correct to the nearest tenth is printed fo
every 10 minutes of GMT, it is possible to obtain a more precise valu
of GHA for the sun than when using the *Nautical Almanac*, where
the *v* correction for the sun is ignored. The GHA of Aries is also give
to the nearest tenth, but the GHAs and declinations of the navigationa

CORRECTIONS TO BE APPLIED TO SEXTANT ALTITUDE

REFRACTION

To be subtracted from sextant altitude (referred to as observed altitude in A.P. 3270)

R_0	0	5	10	15	20	25	30	35	40	45	50	55	R_0	0·9	1·0	1·1	1·2
	Height above sea level in units of 1 000 ft. — Sextant Altitude													$R = R_0 \times f$ — f			
′	° ′	° ′	° ′	° ′	° ′	° ′	° ′	° ′	° ′	° ′	° ′	° ′	′	′	′	′	′
	90	90	90	90	90	90	90	90	90	90	90	90					
0													0	0	0	0	0
	63	59	55	51	46	41	36	31	26	20	17	13					
1													1	1	1	1	1
	33	29	26	22	19	16	14	11	9	7	6	4					
2													2	2	2	2	2
	21	19	16	14	12	10	8	7	5	4	2 40	1 40					
3													3	3	3	3	4
	16	14	12	10	8	7	6	5	3 10	2 20	1 30	0 40					
4													4	4	4	4	5
	12	11	9	8	7	5	4 00	3 10	2 10	1 30	0 39	+0 05					
5													5	5	5	5	6
	10	9	7	5 50	4 50	3 50	3 10	2 20	1 30	0 49	+0 11	−0 19					
6													6	5	6	7	7
	8 10	6 50	5 50	4 50	4 00	3 00	2 20	1 50	1 10	0 24	−0 11	−0 38					
7													7	6	7	8	8
	6 50	5 50	5 00	4 00	3 10	2 30	1 50	1 20	0 38	+0 04	−0 28	−0 54					
8													8	7	8	9	10
	6 00	5 10	4 10	3 20	2 40	2 00	1 30	1 00	0 19	−0 13	−0 42	−1 08					
9													9	8	9	10	11
	5 20	4 30	3 40	2 50	2 10	1 40	1 10	0 35	+0 03	−0 27	−0 53	−1 18					
10													10	9	10	11	12
	4 30	3 40	2 50	2 20	1 40	1 10	0 37	+0 11	−0 16	−0 43	−1 08	−1 31					
12													12	11	12	13	14
	3 30	2 50	2 10	1 40	1 10	0 34	+0 09	−0 14	−0 37	−1 00	−1 23	−1 44					
14													14	13	14	15	17
	2 50	2 10	1 40	1 10	0 37	+0 10	−0 13	−0 34	−0 53	−1 14	−1 35	−1 56					
16													16	14	16	18	19
	2 20	1 40	1 20	0 43	+0 15	−0 08	−0 31	−0 52	−1 08	−1 27	−1 46	−2 05					
18													18	16	18	20	22
	1 50	1 20	0 49	+0 23	−0 02	−0 26	−0 46	−1 06	−1 22	−1 39	−1 57	−2 14					
20													20	18	20	22	24
	1 12	0 44	+0 19	−0 06	−0 28	−0 48	−1 09	−1 27	−1 42	−1 58	−2 14	−2 30					
25													25	22	25	28	30
	0 34	+0 10	−0 13	−0 36	−0 55	−1 14	−1 32	−1 51	−2 06	−2 21	−2 34	−2 49					
30													30	27	30	33	36
	+0 06	−0 16	−0 37	−0 59	−1 17	−1 33	−1 51	−2 07	−2 23	−2 37	−2 51	−3 04					
35													35	31	35	38	42
	−0 18	−0 37	−0 58	−1 16	−1 34	−1 49	−2 06	−2 22	−2 35	−2 49	−3 03	−3 16					
40													40	36	40	44	48
		−0 53	−1 14	−1 31	−1 47	−2 03	−2 18	−2 33	−2 47	−2 59	−3 13	−3 25					
45													45	40	45	50	54
		−1 10	−1 28	−1 44	−1 59	−2 15	−2 28	−2 43	−2 56	−3 08	−3 22	−3 33					
50													50	45	50	55	60
			−1 40	−1 53	−2 09	−2 24	−2 38	−2 52	−3 04	−3 17	−3 29	−3 41					
55													55	49	55	60	66
				−2 03	−2 18	−2 33	−2 46	−3 01	−3 12	−3 25	−3 37	−3 48					
60													60	54	60	66	72
							−2 53	−3 07	−3 19	−3 31	−3 42	−3 53					

f	0	5	10	15	20	25	30	35	40	45	50	55	f
	Temperature in °C.												
	+47	+36	+27	+18	+10	+ 3	− 5	−13	For these heights no temperature correction is necessary, so use $R = R_0$				
0·9													0·9
	+26	+16	+ 6	− 4	−13	−22	−31	−40					
1·0													1·0
	+ 5	− 5	−15	−25	−36	−46	−57	−68					
1·1													1·1
	−16	−25	−36	−46	−58	−71	−83	−95					
1·2													1·2
	−37	−45	−56	−67	−81	−95							

0·9	1·0	1·1	1·2
f			
Where R_0 is less than 10′ or the height greater than 35 000 ft. use $R = R_0$			

Choose the column appropriate to height, in units of 1 000 ft., and find the range of altitude in which the sextant altitude lies; the corresponding value of R_0 is the refraction, to be subtracted from sextant altitude, unless conditions are extreme. In that case find f from the lower table, with critical argument temperature. Use the table on the right to form the refraction, $R = R_0 \times f$.

Figure 6-1B. Refraction correction table, Air Almanac.

planets and the moon are given only to the nearest minute.

Each calendar day is covered by a single leaf in the *Air Almanac.* The front side of each leaf, or right-hand pages as viewed in the almanac, contain tabulations for every 10 minutes of GMT from 0 hours 0 minutes to 11 hours 50 minutes. The reverse sides—the left-hand pages—are for times from 12 hours 0 minutes to 23 hours 50 minutes. Because of the large number of daily pages necessary to cover a single year, the *Air Almanac* is issued by the U.S. Naval Observatory two times each year, with each volume covering a six-month period. The volumes are bound with color-coded red and blue plastic rings to assist in identification and to allow the pages to be removed for more convenient use.

A typical set of daily pages from the July-December volume of the *Air Almanac* for 15 December of the same year in which the examples in Chapter 5 are set is shown in Figures 6-2A and 6-2B. These may be compared with the corresponding daily pages from the *Nautical Almanac* shown in Figures 5-1A and 5-1B on pages 92 and 93.

The average value of SHA and declination for the four-month period covered for each of the 57 navigational stars is tabulated on the inside front cover and again on a fold-out page following the daily pages. Also appearing at the same locations is a table for interpolation of GHA for the sun, Aries, the planets, and the moon for time increments between the 10-minute tabulations in the daily pages. Both of these tables appear in Figure 6-3 on page 125.

Tables for finding the times of civil twilight, sunrise and sunset, and for interpolating the time of moonrise and moonset are included on white pages in the back of each volume, along with various other tables and sky diagrams. The tables are arranged in inverse order of use, so that the most commonly used data appears on the back cover and on the pages immediately adjacent to it.

Use of the *Air Almanac* during Sight Reduction

The *Air Almanac*, like the *Nautical Almanac*, is used during sight reduction to obtain various celestial coordinates from which certain of the entering arguments for the sight reduction tables to be used are derived. *Tables No. 249* are made up of three volumes. Volume I uses integral values for the local hour angle of Aries (LHA♈) and the assumed latitude of the observer as entering arguments, while the other two volumes use the same arguments as *Tables No. 229*—integral degrees of declination and LHA of the body, and the integral assumed latitude of the observer. When Volume I of *Tables No. 249* is to be used, the *Air Almanac* furnishes the GHA♈ from which LHA♈ is produced; for Volumes II and III, it provides the declination and GHA of the body, from which the LHA is derived.

There are a number of sight reduction forms that have been developed for use with the *Air Almanac* and *Tables No. 249*. Two forms developed by the U.S. Naval Academy for use with the *Air Almanac* and either Volume I or Volumes II or III of *Tables No. 249* will be used as the media on which the sample complete solutions demonstrated in the remainder of this chapter will be worked.

As was the case earlier when the role of the *Nautical Almanac* during sight reduction was examined, the use of the *Air Almanac* during sight reduction is best explained by means of examples. For purposes of comparison, the examples below are two of the same observations discussed in Chapter 5—the star Canopus and the upper limb of the moon. Because no v or d correction is necessary when

GREENWICH A. M. DECEMBER 15 (TUESDAY)

GMT	☉ SUN GHA	SUN Dec.	ARIES GHA ♈	VENUS −4.4 GHA	VENUS Dec.	JUPITER −1.3 GHA	JUPITER Dec.	SATURN 0.1 GHA	SATURN Dec.	◑ MOON GHA	MOON Dec.
h m	° ′	° ′	° ′	° ′	° ′	° ′	° ′	° ′	° ′	° ′	° ′
00 00	181 18.0	S23 14.3	83 14.2	221 29	S13 14	211 11	S17 59	38 17	N14 30	333 25	N25 19
10	183 47.9	14.3	85 44.6	223 59		213 41		40 47		335 50	18
20	186 17.9	14.3	88 15.0	226 29		216 12		43 18		338 15	17
30	188 47.8	· 14.3	90 45.5	228 59	· ·	218 42	· ·	45 48	· ·	340 40	· 16
40	191 17.8	14.4	93 15.9	231 29		221 12		48 19		343 04	15
50	193 47.7	14.4	95 46.3	234 00		223 43		50 49		345 29	13
01 00	196 17.7	S23 14.4	98 16.7	236 30	S13 14	226 13	S17 59	53 19	N14 30	347 54	N25 12
10	198 47.6	14.4	100 47.1	239 00		228 43		55 50		350 19	11
20	201 17.6	14.5	103 17.5	241 30		231 14		58 20		352 43	10
30	203 47.5	· 14.5	105 47.9	244 00	· ·	233 44	· ·	60 51	· ·	355 08	· 09
40	206 17.5	14.5	108 18.3	246 31		236 14		63 21		357 33	08
50	208 47.4	14.5	110 48.7	249 01		238 45		65 52		359 57	07
02 00	211 17.4	S23 14.5	113 19.1	251 31	S13 14	241 15	S17 59	68 22	N14 30	2 22	N25 06
10	213 47.4	14.6	115 49.6	254 01		243 45		70 52		4 47	05
20	216 17.3	14.6	118 20.0	256 31		246 16		73 23		7 12	03
30	218 47.3	· 14.6	120 50.4	259 02	· ·	248 46	· ·	75 53	· ·	9 36	· 02
40	221 17.2	14.6	123 20.8	261 32		251 16		78 24		12 01	01
50	223 47.2	14.7	125 51.2	264 02		253 47		80 54		14 26	25 00
03 00	226 17.1	S23 14.7	128 21.6	266 32	S13 15	256 17	S17 59	83 25	N14 30	16 51	N24 59
10	228 47.1	14.7	130 52.0	269 02		258 47		85 55		19 15	58
20	231 17.0	14.7	133 22.4	271 33		261 18		88 25		21 40	57
30	233 47.0	· 14.8	135 52.8	274 03	· ·	263 48	· ·	90 56	· ·	24 05	· 55
40	236 16.9	14.8	138 23.3	276 33		266 18		93 26		26 30	54
50	238 46.9	14.8	140 53.7	279 03		268 48		95 57		28 55	53
04 00	241 16.8	S23 14.8	143 24.1	281 33	S13 15	271 19	S17 59	98 27	N14 30	31 19	N24 52
10	243 46.8	14.8	145 54.5	284 04		273 49		100 58		33 44	51
20	246 16.7	14.9	148 24.9	286 34		276 19		103 28		36 09	50
30	248 46.7	· 14.9	150 55.3	289 04	· ·	278 50	· ·	105 59	· ·	38 34	· 49
40	251 16.6	14.9	153 25.7	291 34		281 20		108 29		40 59	47
50	253 46.6	14.9	155 56.1	294 04		283 50		110 59		43 23	46
05 00	256 16.5	S23 15.0	158 26.5	296 35	S13 15	286 21	S17 59	113 30	N14 30	45 48	N24 45
10	258 46.5	15.0	160 57.0	299 05		288 51		116 00		48 13	44
20	261 16.4	15.0	163 27.4	301 35		291 21		118 31		50 38	43
30	263 46.4	· 15.0	165 57.8	304 05	· ·	293 52	· ·	121 01	· ·	53 03	· 41
40	266 16.3	15.1	168 28.2	306 35		296 22		123 32		55 27	40
50	268 46.3	15.1	170 58.6	309 06		298 52		126 02		57 52	39
06 00	271 16.2	S23 15.1	173 29.0	311 36	S13 15	301 23	S17 59	128 32	N14 30	60 17	N24 38
10	273 46.2	15.1	175 59.4	314 06		303 53		131 03		62 42	37
20	276 16.1	15.1	178 29.8	316 36		306 23		133 33		65 07	36
30	278 46.1	· 15.2	181 00.2	319 06	· ·	308 54	· ·	136 04	· ·	67 31	· 34
40	281 16.0	15.2	183 30.6	321 36		311 24		138 34		69 56	33
50	283 46.0	15.2	186 01.1	324 07		313 54		141 05		72 21	32
07 00	286 15.9	S23 15.2	188 31.5	326 37	S13 15	316 25	S17 59	143 35	N14 30	74 46	N24 31
10	288 45.9	15.3	191 01.9	329 07		318 55		146 05		77 11	29
20	291 15.8	15.3	193 32.3	331 37		321 25		148 36		79 35	28
30	293 45.8	· 15.3	196 02.7	334 07	· ·	323 56	· ·	151 06	· ·	82 00	· 27
40	296 15.7	15.3	198 33.1	336 38		326 26		153 37		84 25	26
50	298 45.7	15.4	201 03.5	339 08		328 56		156 07		86 50	25
08 00	301 15.6	S23 15.4	203 33.9	341 38	S13 15	331 27	S18 00	158 38	N14 30	89 15	N24 23
10	303 45.6	15.4	206 04.3	344 08		333 57		161 08		91 40	22
20	306 15.5	15.4	208 34.8	346 38		336 27		163 38		94 05	21
30	308 45.5	· 15.4	211 05.2	349 09	· ·	338 58	· ·	166 09	· ·	96 29	· 20
40	311 15.4	15.5	213 35.6	351 39		341 28		168 39		98 54	18
50	313 45.4	15.5	216 06.0	354 09		343 58		171 10		101 19	17
09 00	316 15.3	S23 15.5	218 36.4	356 39	S13 15	346 29	S18 00	173 40	N14 29	103 44	N24 16
10	318 45.3	15.5	221 06.8	359 09		348 59		176 11		106 09	15
20	321 15.2	15.6	223 37.2	1 40		351 29		178 41		108 34	13
30	323 45.2	· 15.6	226 07.6	4 10	· ·	354 00	· ·	181 12	· ·	110 59	· 12
40	326 15.1	15.6	228 38.0	6 40		356 30		183 42		113 23	11
50	328 45.1	15.6	231 08.5	9 10		359 00		186 12		115 48	10
10 00	331 15.0	S23 15.6	233 38.9	11 40	S13 15	1 31	S18 00	188 43	N14 29	118 13	N24 08
10	333 45.0	15.7	236 09.3	14 11		4 01		191 13		120 38	07
20	336 14.9	15.7	238 39.7	16 41		6 31		193 44		123 03	06
30	338 44.9	· 15.7	241 10.1	19 11	· ·	9 01	· ·	196 14	· ·	125 28	· 04
40	341 14.8	15.7	243 40.5	21 41		11 32		198 45		127 53	03
50	343 44.8	15.8	246 10.9	24 11		14 02		201 15		130 18	02
11 00	346 14.7	S23 15.8	248 41.3	26 42	S13 15	16 32	S18 00	203 45	N14 29	132 42	N24 01
10	348 44.7	15.8	251 11.7	29 12		19 03		206 16		135 07	23 59
20	351 14.6	15.8	253 42.1	31 42		21 33		208 46		137 32	58
30	353 44.6	· 15.8	256 12.6	34 12	· ·	24 03	· ·	211 17	· ·	139 57	· 57
40	356 14.5	15.9	258 43.0	36 42		26 34		213 47		142 22	55
50	358 44.5	15.9	261 13.4	39 13		29 04		216 18		144 47	54

Lat.	Moon-rise	Diff.
N		
°	h m	m
72	□	*
70	□	*
68	□	*
66	15 14	*
64	16 14	*
62	16 47	53
60	17 11	47
58	17 31	44
56	17 47	41
54	18 01	39
52	18 13	38
50	18 23	37
45	18 45	34
40	19 03	32
35	19 18	31
30	19 31	30
20	19 52	28
10	20 11	26
0	20 29	25
10	20 46	23
20	21 05	22
30	21 26	20
35	21 39	18
40	21 53	17
45	22 10	15
50	22 31	13
52	22 41	12
54	22 52	10
56	23 04	08
58	23 19	06
60	23 36	03
S		

Moon's P. in A.

° Alt.	′ + Corr.	° Alt.	′ + Corr.
0		55	
	56		31
3		56	
	55		30
11		57	
	54		29
15		59	
	53		28
19		60	
	52		27
22		61	
	51		26
24		62	
	50		25
27		63	
	49		24
29		65	
	48		23
31		66	
	47		22
33		67	
	46		21
35		68	
	45		20
36		69	
	44		19
38		70	
	43		18
40		71	
	42		17
41		72	
	41		16
43		73	
	40		15
44		74	
	39		14
46		75	
	38		13
47		77	
	37		12
48		78	
	36		11
50		79	
	35		10
51		80	
	34		
52			
	33		
54			
	32		
55			
	31		
56			

Sun SD 16.3′

Moon SD 15′

Age 16d

Figure 6-2A. Air Almanac, *Greenwich* A.M., *15 December.*

GREENWICH P. M. DECEMBER 15 (TUESDAY)

GMT	☉ SUN GHA	SUN Dec.	ARIES GHA ♈	VENUS –4.4 GHA	VENUS Dec.	JUPITER –1.3 GHA	JUPITER Dec.	SATURN 0.1 GHA	SATURN Dec.	☽ MOON GHA	MOON Dec.
h m	° ′	° ′	° ′	° ′	° ′	° ′	° ′	° ′	° ′	° ′	° ′
12 00	1 14.4	S23 15.9	263 43.8	41 43	S13 16	31 34	S18 00	218 48	N14 29	147 12	N23 53
10	3 44.4	15.9	266 14.2	44 13		34 05		221 18		149 37	51
20	6 14.3	16.0	268 44.6	46 43		36 35		223 49		152 02	50
30	8 44.3	· 16.0	271 15.0	49 13	· ·	39 05	· ·	226 19	· ·	154 27	· 49
40	11 14.2	16.0	273 45.4	51 44		41 36		228 50		156 52	47
50	13 44.2	16.0	276 15.8	54 14		44 06		231 20		159 16	46
13 00	16 14.1	S23 16.0	278 46.3	56 44	S13 16	46 36	S18 00	233 51	N14 29	161 41	N23 45
10	18 44.1	16.1	281 16.7	59 14		49 07		236 21		164 06	44
20	21 14.0	16.1	283 47.1	61 44		51 37		238 52		166 31	42
30	23 44.0	· 16.1	286 17.5	64 15	· ·	54 07	· ·	241 22	· ·	168 56	· 41
40	26 13.9	16.1	288 47.9	66 45		56 38		243 52		171 21	40
50	28 43.9	16.2	291 18.3	69 15		59 08		246 23		173 46	38
14 00	31 13.8	S23 16.2	293 48.7	71 45	S13 16	61 38	S18 00	248 53	N14 29	176 11	N23 37
10	33 43.8	16.2	296 19.1	74 15		64 09		251 24		178 36	35
20	36 13.7	16.2	298 49.5	76 45		66 39		253 54		181 01	34
30	38 43.7	· 16.2	301 20.0	79 16	· ·	69 09	· ·	256 25	· ·	183 26	· 33
40	41 13.6	16.3	303 50.4	81 46		71 40		258 55		185 51	31
50	43 43.6	16.3	306 20.8	84 16		74 10		261 25		188 16	30
15 00	46 13.5	S23 16.3	308 51.2	86 46	S13 16	76 40	S18 00	263 56	N14 29	190 41	N23 29
10	48 43.5	16.3	311 21.6	89 16		79 11		266 26		193 06	27
20	51 13.4	16.3	313 52.0	91 47		81 41		268 57		195 31	26
30	53 43.4	· 16.4	316 22.4	94 17	· ·	84 11	· ·	271 27	· ·	197 55	· 25
40	56 13.3	16.4	318 52.8	96 47		86 41		273 58		200 20	23
50	58 43.3	16.4	321 23.2	99 17		89 12		276 28		202 45	22
16 00	61 13.2	S23 16.4	323 53.6	101 47	S13 16	91 42	S18 01	278 59	N14 29	205 10	N23 20
10	63 43.2	16.5	326 24.1	104 18		94 12		281 29		207 35	19
20	66 13.1	16.5	328 54.5	106 48		96 43		283 59		210 00	18
30	68 43.1	· 16.5	331 24.9	109 18	· ·	99 13	· ·	286 30	· ·	212 25	· 16
40	71 13.0	16.5	333 55.3	111 48		101 43		289 00		214 50	15
50	73 43.0	16.5	336 25.7	114 18		104 14		291 31		217 15	14
17 00	76 12.9	S23 16.6	338 56.1	116 49	S13 16	106 44	S18 01	294 01	N14 29	219 40	N23 12
10	78 42.9	16.6	341 26.5	119 19		109 14		296 32		222 05	11
20	81 12.8	16.6	343 56.9	121 49		111 45		299 02		224 30	09
30	83 42.8	· 16.6	346 27.3	124 19	· ·	114 15	· ·	301 32	· ·	226 55	· 08
40	86 12.7	16.6	348 57.8	126 49		116 45		304 03		229 20	06
50	88 42.7	16.7	351 28.2	129 20		119 16		306 33		231 45	05
18 00	91 12.6	S23 16.7	353 58.6	131 50	S13 16	121 46	S18 01	309 04	N14 29	234 10	N23 04
10	93 42.6	16.7	356 29.0	134 20		124 16		311 34		236 35	02
20	96 12.5	16.7	358 59.4	136 50		126 47		314 05		239 00	23 01
30	98 42.5	· 16.8	1 29.8	139 20	· ·	129 17	· ·	316 35	· ·	241 25	22 59
40	101 12.4	16.8	4 00.2	141 50		131 47		319 05		243 50	58
50	103 42.4	16.8	6 30.6	144 21		134 18		321 36		246 15	57
19 00	106 12.3	S23 16.8	9 01.0	146 51	S13 16	136 48	S18 01	324 06	N14 29	248 40	N22 55
10	108 42.3	16.8	11 31.5	149 21		139 18		326 37		251 05	54
20	111 12.2	16.9	14 01.9	151 51		141 49		329 07		253 30	52
30	113 42.2	· 16.9	16 32.3	154 21	· ·	144 19	· ·	331 38	· ·	255 55	· 51
40	116 12.1	16.9	19 02.7	156 52		146 49		334 08		258 20	49
50	118 42.1	16.9	21 33.1	159 22		149 19		336 38		260 45	48
20 00	121 12.0	S23 16.9	24 03.5	161 52	S13 17	151 50	S18 01	339 09	N14 29	263 10	N22 46
10	123 42.0	17.0	26 33.9	164 22		154 20		341 39		265 35	45
20	126 11.9	17.0	29 04.3	166 52		156 50		344 10		268 01	43
30	128 41.9	· 17.0	31 34.7	169 23	· ·	159 21	· ·	346 40	· ·	270 26	· 42
40	131 11.8	17.0	34 05.1	171 53		161 51		349 11		272 51	41
50	133 41.8	17.0	36 35.6	174 23		164 21		351 41		275 16	39
21 00	136 11.7	S23 17.1	39 06.0	176 53	S13 17	166 52	S18 01	354 12	N14 29	277 41	N22 38
10	138 41.7	17.1	41 36.4	179 23		169 22		356 42		280 06	36
20	141 11.6	17.1	44 06.8	181 54		171 52		359 12		282 31	35
30	143 41.6	· 17.1	46 37.2	184 24	· ·	174 23	· ·	1 43	· ·	284 56	· 33
40	146 11.5	17.2	49 07.6	186 54		176 53		4 13		287 21	32
50	148 41.5	17.2	51 38.0	189 24		179 23		6 44		289 46	30
22 00	151 11.4	S23 17.2	54 08.4	191 54	S13 17	181 54	S18 01	9 14	N14 29	292 11	N22 29
10	153 41.4	17.2	56 38.8	194 24		184 24		11 45		294 36	27
20	156 11.3	17.2	59 09.3	196 55		186 54		14 15		297 01	26
30	158 41.2	· 17.3	61 39.7	199 25	· ·	189 25	· ·	16 45	· ·	299 26	· 24
40	161 11.2	17.3	64 10.1	201 55		191 55		19 16		301 51	23
50	163 41.1	17.3	66 40.5	204 25		194 25		21 46		304 16	21
23 00	166 11.1	S23 17.3	69 10.9	206 55	S13 17	196 56	S18 01	24 17	N14 29	306 42	N22 20
10	168 41.0	17.3	71 41.3	209 26		199 26		26 47		309 07	18
20	171 11.0	17.4	74 11.7	211 56		201 56		29 18		311 32	17
30	173 40.9	· 17.4	76 42.1	214 26	· ·	204 27	· ·	31 48	· ·	313 57	· 15
40	176 10.9	17.4	79 12.5	216 56		206 57		34 18		316 22	14
50	178 40.8	17.4	81 43.0	219 26		209 27		36 49		318 47	12

Lat.	Moon-set	Diff.
N °	h m	m
72	· □	*
70	□	*
68	□	*
66	13 29	*
64	12 29	*
62	11 55	-01
60	11 30	+05
58	11 10	08
56	10 53	11
54	10 39	12
52	10 27	14
50	10 16	15
45	09 53	17
40	09 34	19
35	09 19	20
30	09 05	21
20	08 42	23
10	08 22	24
0	08 03	26
10	07 44	27
20	07 23	29
30	06 59	30
35	06 45	31
40	06 29	33
45	06 09	34
50	05 45	36
52	05 33	37
54	05 19	39
56	05 03	41
58	04 44	43
60	04 21	47
S		

Moon's P. in A.

° Alt.	′ + Corr.	° Alt.	′ + Corr.
0		56	
	55		30
10		57	
	54		29
14		59	
	53		28
18		60	
	52		27
21		61	
	51		26
24		62	
	50		25
26		63	
	49		24
28		64	
	48		23
30		66	
	47		22
32		67	
	46		21
34		68	
	45		20
36		69	
	44		19
38		70	
	43		18
39		71	
	42		17
41		72	
	41		16
42		73	
	40		15
44		74	
	39		14
45		75	
	38		13
47		76	
	37		12
48		78	
	36		11
50		79	
	35		10
51		80	
	34		
52			
	33		
54			
	32		
55			
	31		
56			
	30		
57			

Sun SD 16.3′

Moon SD 15′

Age 17d

Figure 6-2B. Air Almanac, *Greenwich* P.M., *15 December.*

STARS, SEPT.—DEC.

No.	Name	Mag.	S.H.A.	Dec.
			° ′	° ′
7*	*Acamar*	3·1	315 42	S.40 25
5*	*Achernar*	0·6	335 50	S.57 23
30*	*Acrux*	1·1	173 46	S.62 56
19	*Adhara* †	1·6	255 38	S.28 56
10*	*Aldebaran* †	1·1	291 26	N.16 27
32*	*Alioth*	1·7	166 49	N.56 07
34*	*Alkaid*	1·9	153 24	N.49 27
55	*Al Na'ir*	2·2	28 24	S.47 06
15	*Alnilam* †	1·8	276 19	S. 1 13
25*	*Alphard* †	2·2	218 28	S. 8 32
41*	*Alphecca* †	2·3	126 38	N.26 49
1*	*Alpheratz* †	2·2	358 17	N.28 56
51*	*Altair* †	0·9	62 40	N. 8 47
2	*Ankaa*	2·4	353 47	S.42 28
42*	*Antares* †	1·2	113 06	S.26 22
37*	*Arcturus* †	0·2	146 25	N.19 20
43	*Atria*	1·9	108 37	S.68 59
22	*Avior*	1·7	234 31	S.59 25
13	*Bellatrix* †	1·7	279 06	N. 6 20
16*	*Betelgeuse* †	0·1–1·2	271 36	N. 7 24
17*	*Canopus*	−0·9	264 10	S.52 40
12*	*Capella*	0·2	281 22	N.45 58
53*	*Deneb*	1·3	49 53	N.45 11
28*	*Denebola* †	2·2	183 06	N.14 44
4*	*Diphda* †	2·2	349 28	S.18 09
27*	*Dubhe*	2·0	194 31	N.61 54
14	*Elnath* †	1·8	278 53	N.28 35
47	*Eltanin*	2·4	91 01	N.51 30
54*	*Enif* †	2·5	34 19	N. 9 45
56*	*Fomalhaut* †	1·3	15 59	S.29 47
31	*Gacrux*	1·6	172 37	S.56 57
29*	*Gienah* †	2·8	176 26	S.17 23
35	*Hadar*	0·9	149 34	S.60 14
6*	*Hamal* †	2·2	328 37	N.23 20
48	*Kaus Aust.*	2·0	84 26	S.34 24
40*	*Kochab*	2·2	137 19	N.74 16
57	*Markab* †	2·6	14 10	N.15 03
8*	*Menkar* †	2·8	314 48	N. 3 59
36	*Menkent*	2·3	148 46	S.36 14
24*	*Miaplacidus*	1·8	221 47	S.69 35
9*	*Mirfak*	1·9	309 26	N.49 46
50*	*Nunki* †	2·1	76 38	S.26 20
52*	*Peacock*	2·1	54 09	S.56 50
21*	*Pollux* †	1·2	244 07	N.28 06
20*	*Procyon* †	0·5	245 33	N. 5 18
46*	*Rasalhague* †	2·1	96 36	N.12 35
26*	*Regulus* †	1·3	208 18	N.12 07
11*	*Rigel* †	0·3	281 43	S. 8 14
38*	*Rigil Kent.*	0·1	140 36	S.60 43
44	*Sabik* †	2·6	102 50	S.15 41
3*	*Schedar*	2·5	350 17	N.56 23
45*	*Shaula*	1·7	97 06	S.37 05
18*	*Sirius* †	−1·6	259 02	S.16 40
33*	*Spica* †	1·2	159 05	S.11 01
23*	*Suhail*	2·2	223 16	S.43 19
49*	*Vega*	0·1	81 01	N.38 45
39	*Zuben'ubi* †	2·9	137 41	S.15 55

INTERPOLATION OF G.H.A.

Increment to be added for intervals of G.M.T. to G.H.A. of: Sun, Aries (♈) and planets; Moon

SUN, etc.		MOON	SUN, etc.		MOON	SUN, etc.		MOON
m s	° ′	m s	m s	° ′	m s	m s	° ′	m s
00 00		00 00	03 17		03 25	06 37		06 52
	0 00			0 50			1 40	
01		00 02	21		03 29	41		06 56
	0 01			0 51			1 41	
05		00 06	25		03 33	45		07 00
	0 02			0 52			1 42	
09		00 10	29		03 37	49		07 04
	0 03			0 53			1 43	
13		00 14	33		03 41	53		07 08
	0 04			0 54			1 44	
17		00 18	37		03 45	06 57		07 13
	0 05			0 55			1 45	
21		00 22	41		03 49	07 01		07 17
	0 06			0 56			1 46	
25		00 26	45		03 54	05		07 21
	0 07			0 57			1 47	
29		00 31	49		03 58	09		07 25
	0 08			0 58			1 48	
33		00 35	53		04 02	13		07 29
	0 09			0 59			1 49	
37		00 39	03 57		04 06	17		07 33
	0 10			1 00			1 50	
41		00 43	04 01		04 10	21		07 37
	0 11			1 01			1 51	
45		00 47	05		04 14	25		07 42
	0 12			1 02			1 52	
49		00 51	09		04 19	29		07 46
	0 13			1 03			1 53	
53		00 55	13		04 23	33		07 50
	0 14			1 04			1 54	
00 57		01 00	17		04 27	37		07 54
	0 15			1 05			1 55	
01 01		01 04	21		04 31	41		07 58
	0 16			1 06			1 56	
05		01 08	25		04 35	45		08 02
	0 17			1 07			1 57	
09		01 12	29		04 39	49		08 06
	0 18			1 08			1 58	
13		01 16	33		04 43	53		08 11
	0 19			1 09			1 59	
17		01 20	37		04 48	07 57		08 15
	0 20			1 10			2 00	
21		01 24	41		04 52	08 01		08 19
	0 21			1 11			2 01	
25		01 29	45		04 56	05		08 23
	0 22			1 12			2 02	
29		01 33	49		05 00	09		08 27
	0 23			1 13			2 03	
33		01 37	53		05 04	13		08 31
	0 24			1 14			2 04	
37		01 41	04 57		05 08	17		08 35
	0 25			1 15			2 05	
41		01 45	05 01		05 12	21		08 40
	0 26			1 16			2 06	
45		01 49	05		05 17	25		08 44
	0 27			1 17			2 07	
49		01 53	09		05 21	29		08 48
	0 28			1 18			2 08	
53		01 58	13		05 25	33		08 52
	0 29			1 19			2 09	
01 57		02 02	17		05 29	37		08 56
	0 30			1 20			2 10	
02 01		02 06	21		05 33	41		09 00
	0 31			1 21			2 11	
05		02 10	25		05 37	45		09 04
	0 32			1 22			2 12	
09		02 14	29		05 41	49		09 09
	0 33			1 23			2 13	
13		02 18	33		05 46	53		09 13
	0 34			1 24			2 14	
17		02 22	37		05 50	08 57		09 17
	0 35			1 25			2 15	
21		02 27	41		05 54	09 01		09 21
	0 36			1 26			2 16	
25		02 31	45		05 58	05		09 25
	0 37			1 27			2 17	
29		02 35	49		06 02	09		09 29
	0 38			1 28			2 18	
33		02 39	53		06 06	13		09 33
	0 39			1 29			2 19	
37		02 43	05 57		06 10	17		09 38
	0 40			1 30			2 20	
41		02 47	06 01		06 15	21		09 42
	0 41			1 31			2 21	
45		02 51	05		06 19	25		09 46
	0 42			1 32			2 22	
49		02 56	09		06 23	29		09 50
	0 43			1 33			2 23	
53		03 00	13		06 27	33		09 54
	0 44			1 34			2 24	
02 57		03 04	17		06 31	37		09 58
	0 45			1 35			2 25	
03 01		03 08	21		06 35	41		10 00
	0 46			1 36			2 26	
05		03 12	25		06 39	45		
	0 47			1 37			2 27	
09		03 16	29		06 44	49		
	0 48			1 38			2 28	
13		03 20	33		06 48	53		
	0 49			1 39			2 29	
17		03 25	37		06 52	09 57		
	0 50			1 40			2 30	
03 21		03 29	06 41		06 56	10 00		

* Stars used in H.O. 249 (A.P. 3270) Vol. 1.

† Stars that may be used with Vols. 2 and 3.

Figure 6-3. Inside front cover, Air Almanac.

the *Air Almanac* is used, the procedures for finding the GHA and declination of a planet or the sun are virtually identical to the procedures for finding these same coordinates for a star or the moon. Separate examples of the complete solutions of the planets and the sun, therefore, will not be presented in this chapter.

Finding the Complete Solution of a Star by the *Air Almanac*

As the first example of the complete solution for a star by the use of the *Air Almanac,* let us consider once again the Canopus sight used as an example in Chapter 5. The star Canopus was observed at GMT 17-12-09 on 15 December and found to have a sextant altitude hs of 50° 46′, rounded off to the nearest whole minute. The DR position at the time of the observation was L 34° 19.0′ S, λ 163° 05.7′ E, and the index error was .3′ off the arc.

The first step in the complete solution using the *Air Almanac* is to find the observed altitude Ho by application of the corrections for IC, dip, and refraction to the hs. The index correction in this case is considered insignificant, the dip correction from the dip table in Figure 6-1A is 7′, and the refraction correction R_o from the refraction table of Figure 6-1B is −1′. Entering these values on the form and adding them algebraically yields the total correction to hs of −8′. Applying this correction to hs, an Ho of 50° 38′ results; note that this result differs by 1.4′ from the Ho as computed by the *Nautical Almanac* (see Figure 5-2). At this point, the Canopus form appears as shown in Figure 6-4A.

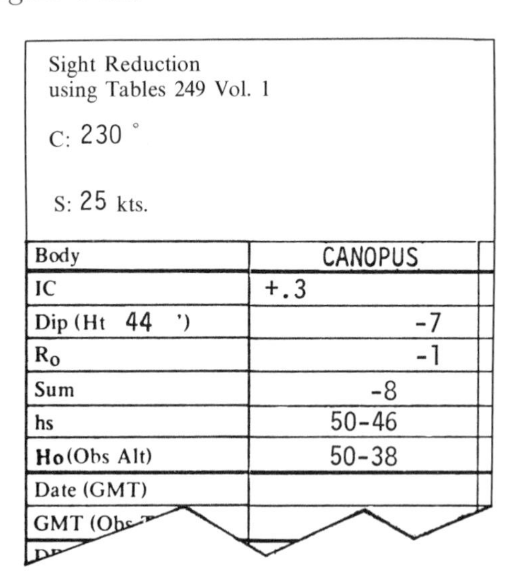

Sight Reduction
using Tables 249 Vol. 1

C: 230 °

S: 25 kts.

Body	CANOPUS
IC	+.3
Dip (Ht 44 ')	-7
Ro	-1
Sum	-8
hs	50-46
Ho (Obs Alt)	50-38
Date (GMT)	
GMT (Obs T	

Figure 6-4A. Canopus Air Almanac *form,* Ho *computed.*

Inasmuch as Volume I of *Tables No. 249* can be used to find the computed altitude Hc and the true azimuth Zn of Canopus in this case, the only coordinate of interest that must be obtained from the *Air Almanac* is the GHA♈ for the time of observation, for later conversion to an integral LHA♈, one of the two entering arguments for Volume I; the other argument is the star name. The tabulated value of GHA♈ for GMT 17-10-00 on 15 December from the daily page reproduced in Figure 6-2B is 341° 26.5′; this figure is entered on the form opposite "Tab. GHA♈." Next, the increment to GHA♈ corresponding to the remaining 2 minutes 9 seconds of GMT is found from the GHA interpolation table on the inside front cover, pictured in Figure 6-3. It is +0° 32′. Note that 02-09 is an exact tabulated argument; as is the case in all critical value tables of this type, the correction corresponding to the preceding interval is always extracted when entering arguments are identical to tabulated arguments. Adding the GHA♈ increment to the tabulated GHA♈ yields a total GHA♈ for GMT 17-12-09 of 341′ 58.5′.

If one of the 57 navigational stars other than the seven listed in Volume I of *Tables No. 249* had been observed, it would be necessary to determine the LHA and declination of the star for later use with either Volume II or III. In this case, the *Air Almanac*, like the *Nautical Almanac*, would be used to find the GHA and declination of the star. The GHA of the star is found by the formula

$$\text{GHA}☆ = \text{GHA}♈ + \text{SHA}☆$$

and the local hour angle by the relationship

$$\text{LHA}☆ = \text{GHA}☆ \begin{matrix} + a\lambda(E) \\ - a\lambda(W) \end{matrix}$$

The sidereal hour angle (SHA) and declination of the star is found from one of the two tables included in the *Air Almanac* for this purpose, located inside the front cover and following the daily pages.

Since an integral LHA♈ is required as an entering argument for Volume I of *Tables No. 249*, and the DR longitude is east, an assumed longitude ($a\lambda$) is chosen such that when it is added to the total GHA♈ a whole degree of LHA♈ will result. As before, the longitude selected should also lie within 30′ of the DR longitude. Thus, the appropriate assumed longitude for this example is 163° 01.5 E. Adding this value to GHA♈ and subtracting 360° results in a LHA♈ of 145°. The form now appears as shown in Figure 6-4B on the following page, ready for use with *Tables No. 249*.

Sight Reduction
using Tables 249 Vol. I

C: 230 °

S: 25 kts.

Body	CANOPUS
IC	+.3
Dip (Ht 44 ')	-7
R_o	-1
Sum	-8
hs	50-46
Ho (Obs Alt)	50-38
Date (GMT)	15 DEC
GMT (Obs Time)	17-12-09
DR Lat	34-19.0 S
DR Long	163-05.7 E
Tab GHA ♈	341-26.5
GHA ♈ incr'mt	0-32.0
GHA ♈	341-58.5
a λ (-W, +E)	163-01.5 E
LHA ♈	145
a Lat (N or S)	34 S
Hc (Comp...	

Figure 6-4B. Canopus Air Almanac *form ready for* Tables No. 249.

Determining the Ho, GHA, Declination, and LHA of the Moon

When the complete solution for any body other than one of the seven navigational stars featured in Volume I of *Tables No. 249* is to be found, the *Air Almanac* is used to obtain the GHA and declination of the body at the time of observation for later use with Volume II or III. In the following example, the upper limb of the moon was observed on the same occasion as the star Canopus of the example above. The GMT of the observation was 16-58-57 and its sextant altitude hs to the nearest minute was 28° 10′. The DR position at the time of the sight was L 34° 15.5′ S, λ 163° 11.7′ E.

As in the previous example, the first step in finding the complete solution is to obtain the observed altitude Ho. Since the moon was observed, additional corrections for semidiameter and parallax must be applied to hs, along with the usual IC, dip, and refraction corrections. The IC correction is 0, the dip correction from the dip table is −7′, and the refraction correction R_0 from the refraction table is

−2′. The semidiameter of the moon on this date is found from the daily page (Figure 6-2B) as 15′; since the upper limb of the moon was observed, the sign of this SD correction is negative. Finally, the parallax correction for the moon is found from the "P in A" table on the appropriate side of the daily page. The entering argument for this table is the sextant altitude hs corrected for IC, dip, refraction and semidiameter; in this case, a value of +49′ is extracted, corresponding to a corrected hs of 27° 46′. Applying this "P in A" correction results in an Ho of 28° 35′, a value that differs only .4′ from the corresponding Ho found by the use of the *Nautical Almanac.* The partially completed form appears in Figure 6-5A.

Sight Reduction
using Tables 249 Vol. II and III

C: 230 °

S: 25 kts.

Body	MOON (UL)
IC	+.3
Dip (Ht 44 ')	-7
Ro	-2
S.D.	-15
Sum	-24
hs	28-10
P in A (Moon)	+49
Ho(Obs Alt)	28-35
Date (GMT)	

Figure 6-5A. Moon Air Almanac *form,* Ho *computed.*

To find the GHA and declination of the moon, the values for these quantities tabulated for the 10 minutes of GMT immediately preceding the time of observation are extracted. The tabulated GHA of the moon for GMT 16-50-00 on 15 December is 217° 15′, and the tabulated declination is N23° 14′. A GHA increment of 2° 10′ is obtained from the GHA interpolation table for the moon for the additional 8 minutes and 57 seconds of time. This value is added to the tabulated GHA to obtain the total GHA for 16-58-57 of 219° 25′. No correction to the tabulated declination is necessary.

Finally, the LHA of the moon to the nearest whole degree is computed in the usual manner by applying a suitably chosen assumed longitude to the total GHA:

$$219° \; 25' + 163° \; 35' = 383°$$

Subtracting 360° yields an LHA of the moon at the time of observation of 23°. At this point the form is ready for use with *Tables No. 249*, and it appears as shown in Figure 6-5B below.

Determining the Ho, GHA, Declination, and LHA of the Sun and Planets

The procedure for determining the Ho, GHA, declination, and LHA of the sun is basically the same as in the example just discussed for the moon; the only difference is that no parallax correction to the sextant altitude hs is necessary for the sun. The procedure for the navigational planets is identical to the procedure for a star not listed

Sight Reduction
using Tables 249 Vols. II and III

C: 230 °

S: 25 kts.

Body	MOON(UL)
IC	+.3
Dip (Ht 44 ')	-7
Ro	-2
S.D.	-15
Sum	-24
hs	28-10
P in A (Moon)	+49
Ho (Obs Alt)	28-35
Date (GMT)	15 DEC
GMT (Obs Time)	16-58-57
DR Lat	34-15.5 S
DR Long	163-11.7 E
Tab GHA	217-15
GHA incr'mt	2-10
SHA (Star)	
GHA	219-25
±360 if needed	
a λ (-W, +E)	163-35 E
LHA	383 = 23
Tab Dec	N 23-14
a Lat (N or S)	34 S Same (Cont)
Dec Inc ( )d	

Figure 6-5B. Moon Air Almanac *form ready for* Tables No. 249.

in Volume I of *Tables No. 249.* The IC, dip, and refraction corrections are first found and applied to the hs to obtain Ho, then the daily pages and the GHA interpolation tables are used to obtain the values of GHA and declination. An integral value of LHA is determined by choosing a suitable assumed longitude and applying it to the total GHA in the normal manner.

Sight Reduction Tables for Air Navigation, No. 249

After either the LHA♈ or the LHA and declination of the body have been determined by means of the *Air Almanac,* the sight reduction form is ready for use with the *Sight Reduction Tables for Air Navigation, No. 249.* As the name indicates, these tables are designed primarily for the use of air navigators, but like the *Air Almanac,* they have found considerable favor with surface navigators in cases where their speed and ease of use offset the less precise computations they contain.

Tables No. 249 are published in three volumes. Like *Tables No. 229,* they are inspection tables designed for use with an assumed position, but they differ in the degree of precision. Whereas in *Tables No. 229* all computations are made to the nearest tenth of a minute, in *Tables No. 249* computed altitudes are stated only to the nearest whole minute. There are other differences as well. Volume I is specifically designed to reduce the time and effort required to obtain a celestial fix to a minimum, by listing calculated computed altitudes and true azimuths for a given set of seven navigational stars selected on a worldwide basis. Volumes II and III are similar to *Tables No. 229* in concept. They provided a computed Hc and azimuth angle Z for all bodies having a declination less than 30° north or south.

Volume I of *Tables No. 249* is arranged for entering with a whole degree of assumed latitude, a whole degree of local hour angle of Aries, and the appropriate star name. All integral degrees of latitude from 89° north to 89° south are included. For each degree of latitude and of LHA♈, seven selected stars are tabulated; they are selected chiefly for good distribution in azimuth, for their magnitude and altitude, and for continuity both in latitude and hour angle. Each combination of seven stars is tabulated for 15 degrees of LHA♈, and the three with the best distribution in azimuth and altitude for fixing purposes are marked with asterisks. First magnitude star names are capitalized, and second and third magnitude stars appear in lowercase letters. Inasmuch as the coordinates of the selected stars change slightly from year to year primarily as a result of precession and nutation of the earth's axis, Volume I is recomputed and reissued once every five years. The year for which Volume I is issued is called an *epoch* year (not necessarily the year for which tabulations in Volume I are computed). The tabulations in the volume for *Epoch 1975,* for example, were calculated for the year 1974, and the *Epoch 1980* volume calculations were done for 1980. This procedure causes no difficulties, however, because a precession and nutation table is included in the back of each issue for adjusting LOP and fix positions obtained using the tabulated figures for any year in the period three years before and five years after the epoch year.

In practice, when Volume I is to be used for sight reduction purposes, the navigator first determines approximate values of LHA♈ and the assumed latitude based upon his projected DR position for the probable time of observation. He then enters Volume I, using these coordinates as arguments to find which seven stars are tabulated for that approximate position and time. Thus, he can determine beforehand exactly which stars he should observe to obtain a fix using Volume I of *Tables No. 249* and, of the seven, which three will provide the best fix. Furthermore, because the calculated Hc and true azimuth Zn are given for each star, he knows precisely where to look in the heavens for each of the stars he will observe. Hence, Volume I can be considered to be a kind of *starfinder* for selected stars, in addition to its prime function as a sight reduction table.

For every degree of latitude between 89° north and 89° south, there are two pages in Volume I. Left-hand pages contain tabulated Hc's and Zn's for seven selected stars for LHAs between 0° and 179°, while right-hand pages contain tabulations for LHAs between 180° and 359°. The appropriate Volume I page for latitude 34° south and LHAs from 0°-179° for the epoch of the example problems of this chapter is reproduced in Figure 6-6A.

As mentioned previously, Volumes II and III of *Tables No. 249* are similar in format to *Tables No. 229,* with the main differences being that the tabulated Hc and Z values are precise only to the nearest whole minute, and the range of declination covered is from N 30° to S 30°. Hence, complete data for the reduction of sights of the sun, moon, navigational planets, and most but not all navigational stars are provided. Volume II contains tabulations for assumed latitudes from 0° to 39°, and Volume III is for assumed latitudes from 40° to 89°. The entering arguments are a whole degree of assumed latitude, a whole degree of declination of same or contrary name to the latitude, and a whole degree of LHA. For each set of entering arguments, the tables contain a computed altitude under the heading "Hc," a computed altitude difference d in minutes between the tabulated Hc and the Hc for a declination one degree higher, and an azimuth angle under the heading "Z." A typical page from Volume II appears in Figure 6-6B.

LAT 34°S

LHA ♈	Hc	Zn	Hc	Zn	Hc	Zn	Hc	Zn	Hc	Zn	Hc	Zn	Hc	Zn
	*Alpheratz		Hamal		*RIGEL		CANOPUS		ACHERNAR		*Peacock		Enif	
0	27 03	002	25 23	032	14 17	090	23 11	139	61 29	152	47 11	221	35 28	317
1	27 04	001	25 49	031	15 07	090	23 44	139	61 52	153	46 38	221	34 54	316
2	27 04	000	26 14	030	15 57	089	24 17	138	62 14	154	46 06	221	34 19	315
3	27 04	359	26 39	029	16 47	089	24 50	138	62 35	155	45 33	221	33 44	314
4	27 02	358	27 03	028	17 36	088	25 23	138	62 56	156	45 01	221	33 08	313
5	27 00	357	27 26	027	18 26	088	25 57	138	63 16	157	44 28	221	32 31	312
6	26 57	356	27 48	026	19 16	087	26 30	137	63 35	158	43 55	221	31 54	311
7	26 53	355	28 10	025	20 05	086	27 04	137	63 54	159	43 22	221	31 16	310
8	26 48	354	28 31	024	20 55	086	27 38	137	64 11	160	42 50	221	30 38	310
9	26 42	353	28 51	024	21 45	085	28 12	137	64 28	161	42 17	221	29 59	309
10	26 35	352	29 11	023	22 34	085	28 46	136	64 44	162	41 44	221	29 20	308
11	26 28	351	29 30	022	23 24	084	29 21	136	64 58	163	41 11	221	28 41	307
12	26 20	350	29 47	021	24 13	083	29 55	136	65 12	164	40 38	221	28 01	306
13	26 10	349	30 05	020	25 03	083	30 30	136	65 25	165	40 05	221	27 21	305
14	26 01	348	30 21	019	25 52	082	31 05	136	65 37	167	39 33	221	26 40	305
	*Hamal		ALDEBARAN		RIGEL		*CANOPUS		ACHERNAR		*FOMALHAUT		Alpheratz	
15	30 36	017	18 18	054	26 41	082	31 40	135	65 48	168	63 27	271	25 50	347
16	30 51	016	18 58	054	27 30	081	32 15	135	65 58	169	62 37	270	25 38	346
17	31 04	015	19 38	053	28 19	080	32 50	135	66 07	170	61 47	269	25 26	345
18	31 17	014	20 18	052	29 08	080	33 25	135	66 15	172	60 58	269	25 13	344
19	31 29	013	20 57	051	29 57	079	34 00	135	66 21	173	60 08	268	24 59	343
20	31 40	012	21 35	051	30 46	078	34 36	134	66 27	174	59 18	268	24 44	342
21	31 50	011	22 14	050	31 35	078	35 11	134	66 31	176	58 28	267	24 29	341
22	31 59	010	22 52	049	32 23	077	35 47	134	66 34	177	57 39	267	24 13	341
23	32 08	009	23 29	048	33 12	076	36 23	134	66 36	178	56 49	266	23 56	340
24	32 15	008	24 06	047	34 00	076	36 59	134	66 37	180	55 59	266	23 38	339
25	32 22	007	24 42	047	34 48	075	37 35	134	66 36	181	55 10	265	23 20	338
26	32 27	006	25 18	046	35 36	074	38 11	134	66 35	183	54 20	265	23 01	337
27	32 32	005	25 54	045	36 24	074	38 47	134	66 32	184	53 31	265	22 41	336
28	32 35	004	26 28	044	37 12	073	39 23	133	66 28	185	52 41	264	22 20	335
29	32 38	003	27 03	043	37 59	072	39 59	133	66 23	187	51 52	264	21 59	334
	Hamal		ALDEBARAN		*SIRIUS		CANOPUS		*ACHERNAR		FOMALHAUT		*Alpheratz	
30	32 40	001	27 37	042	24 48	094	40 35	133	66 17	188	51 02	263	21 37	333
31	32 41	000	28 10	042	25 38	093	41 11	133	66 09	189	50 13	263	21 15	333
32	32 41	359	28 43	041	26 28	093	41 48	133	66 01	190	49 24	262	20 52	332
33	32 39	358	29 15	040	27 17	092	42 24	133	65 51	192	48 34	262	20 28	331
34	32 37	357	29 46	039	28 07	092	43 00	133	65 41	193	47 45	261	20 03	330
35	32 34	356	30 17	038	28 57	091	43 37	133	65 29	194	46 56	261	19 38	329
36	32 31	355	30 47	037	29 47	091	44 13	133	65 16	195	46 07	261	19 13	329
37	32 26	354	31 17	036	30 36	090	44 49	133	65 03	197	45 18	260	18 46	328
38	32 20	353	31 46	035	31 26	090	45 26	133	64 48	198	44 29	260	18 19	327
39	32 13	352	32 14	034	32 16	089	46 02	133	64 33	199	43 40	259	17 52	326
40	32 06	351	32 41	033	33 06	088	46 38	133	64 16	200	42 51	259	17 24	325
41	31 57	350	33 08	032	33 55	088	47 15	133	63 59	201	42 02	259	16 55	325
42	31 48	349	33 34	031	34 45	087	47 51	133	63 41	202	41 14	258	16 26	324
43	31 37	347	33 59	030	35 35	087	48 27	133	63 22	203	40 25	258	15 57	323
44	31 26	346	34 24	029	36 24	086	49 04	133	63 02	204	39 36	257	15 27	322
	*ALDEBARAN		BETELGEUSE		SIRIUS		*CANOPUS		ACHERNAR		*FOMALHAUT		Hamal	
45	34 47	028	31 42	053	37 14	086	49 40	133	62 42	205	38 48	257	31 14	345
46	35 10	027	32 21	052	38 03	085	50 16	134	62 21	206	37 59	257	31 01	344
47	35 32	026	33 01	051	38 53	084	50 52	134	61 59	206	37 11	256	30 47	343
48	35 53	025	33 39	051	39 42	084	51 28	134	61 36	207	36 23	256	30 32	342
49	36 13	023	34 17	050	40 32	083	52 03	134	61 13	208	35 35	255	30 17	341
50	36 33	022	34 55	049	41 21	082	52 39	134	60 49	209	34 47	255	30 00	340
51	36 51	021	35 32	048	42 10	082	53 15	134	60 25	210	33 59	254	29 43	339
52	37 09	020	36 09	047	43 00	081	53 50	135	60 00	210	33 11	254	29 25	338
53	37 25	019	36 44	046	43 49	080	54 26	135	59 35	211	32 23	254	29 06	337
54	37 41	018	37 20	045	44 38	080	55 01	135	59 09	212	31 35	253	28 46	336
55	37 55	017	37 55	044	45 27	079	55 36	135	58 43	212	30 48	253	28 26	335
56	38 09	015	38 29	043	46 15	078	56 10	136	58 16	213	30 00	252	28 05	334
57	38 22	014	39 02	042	47 04	077	56 45	136	57 49	213	29 13	252	27 43	333
58	38 33	013	39 35	041	47 52	077	57 19	136	57 22	214	28 26	252	27 20	332
59	38 44	012	40 07	040	48 41	076	57 51	137	56 54	214	27 39	251	26 57	331
	ALDEBARAN		*BETELGEUSE		SIRIUS		*CANOPUS		ACHERNAR		*FOMALHAUT		Hamal	
60	38 54	011	40 38	038	49 29	075	58 27	137	56 26	215	26 52	251	26 32	331
61	39 02	009	41 09	037	50 17	074	59 01	138	55 57	215	26 05	250	26 08	330
62	39 10	008	41 38	036	51 05	073	59 34	138	55 28	216	25 18	250	25 42	329
63	39 16	007	42 07	035	51 52	073	60 07	139	54 59	216	24 31	249	25 16	328
64	39 22	006	42 35	034	52 40	072	60 40	139	54 30	216	23 45	249	24 49	327
65	39 26	004	43 03	033	53 27	071	61 12	140	54 00	217	22 58	249	24 22	326
66	39 29	003	43 29	031	54 14	070	61 44	140	53 30	217	22 12	248	23 54	325
67	39 32	002	43 54	030	55 00	069	62 16	141	53 00	218	21 26	248	23 25	324
68	39 33	001	44 19	029	55 46	068	62 47	142	52 29	218	20 40	247	22 56	323
69	39 33	359	44 42	028	56 32	067	63 17	143	51 59	218	19 54	247	22 26	323
70	39 32	358	45 05	026	57 18	066	63 47	143	51 28	218	19 09	246	21 55	322
71	39 30	357	45 26	025	58 03	065	64 17	144	50 57	219	18 23	246	21 24	321
72	39 27	356	45 47	024	58 48	064	64 45	145	50 26	219	17 38	246	20 53	320
73	39 22	354	46 06	022	59 32	062	65 14	146	49 55	219	16 53	245	20 21	319
74	39 17	353	46 25	021	60 16	061	65 41	147	49 23	219	16 07	245	19 48	319
	BETELGEUSE		*PROCYON		Suhail		*ACRUX		ACHERNAR		*Diphda		ALDEBARAN	
75	46 42	020	35 52	051	42 01	120	21 11	153	48 52	219	30 54	268	39 11	352
76	46 58	018	36 31	050	42 44	120	21 34	153	48 20	220	30 05	268	39 03	351
77	47 13	017	37 09	049	43 27	120	21 57	152	47 48	220	29 15	267	38 55	350
78	47 27	015	37 46	048	44 10	120	22 20	152	47 16	220	28 25	266	38 45	348
79	47 39	014	38 23	047	44 53	120	22 43	152	46 44	220	27 36	266	38 35	347
80	47 50	012	38 59	046	45 36	120	23 07	152	46 12	220	26 46	265	38 23	346
81	48 01	011	39 35	045	46 20	119	23 30	151	45 40	220	25 56	265	38 10	345
82	48 09	010	40 10	044	47 03	119	23 54	151	45 08	220	25 07	264	37 57	344
83	48 17	008	40 45	043	47 47	119	24 18	151	44 36	220	24 17	264	37 42	342
84	48 23	007	41 18	042	48 30	119	24 43	151	44 04	220	23 28	263	37 27	341
85	48 28	005	41 51	041	49 13	119	25 07	150	43 31	220	22 39	263	37 10	340
86	48 32	004	42 24	040	49 57	119	25 32	150	42 59	221	21 49	262	36 53	339
87	48 35	002	42 55	039	50 41	119	25 56	150	42 27	221	21 00	262	36 35	338
88	48 36	001	43 26	038	51 24	119	26 21	150	41 55	221	20 11	261	36 15	337
89	48 36	359	43 56	036	52 08	119	26 46	150	41 22	221	19 22	261	35 55	336

LHA ♈	Hc	Zn	Hc	Zn	Hc	Zn	Hc	Zn	Hc	Zn	Hc	Zn	Hc	Zn
	PROCYON		REGULUS		*Suhail		ACRUX		*ACHERNAR		RIGEL		*BETELGEUSE	
90	44 25	035	15 29	063	52 51	119	27 12	149	40 50	221	62 04	335	48 34	358
91	44 53	034	16 13	063	53 35	119	27 37	149	40 18	221	61 42	333	48 31	356
92	45 21	033	16 58	062	54 19	119	28 03	149	39 45	220	61 18	331	48 27	355
93	45 47	031	17 41	061	55 02	119	28 28	149	39 13	220	60 53	329	48 22	353
94	46 13	030	18 25	061	55 46	119	28 54	149	38 41	220	60 27	327	48 15	352
95	46 37	029	19 08	060	56 30	119	29 20	149	38 08	220	59 59	325	48 08	350
96	47 01	027	19 51	059	57 13	119	29 46	148	37 36	220	59 30	324	47 58	349
97	47 23	026	20 33	058	57 57	119	30 12	148	37 04	220	59 00	322	47 48	347
98	47 44	025	21 16	058	58 40	119	30 38	148	36 32	220	58 29	320	47 36	346
99	48 05	023	21 57	057	59 24	119	31 05	148	36 00	220	57 56	319	47 24	344
100	48 24	022	22 39	056	60 07	119	31 31	148	35 28	220	57 23	317	47 10	343
101	48 42	021	23 20	055	60 51	119	31 58	148	34 56	220	56 49	316	46 54	342
102	48 59	019	24 01	055	61 34	119	32 24	148	34 24	220	56 14	314	46 38	340
103	49 14	018	24 41	054	62 17	120	32 51	147	33 53	220	55 37	313	46 21	339
104	49 29	016	25 21	053	63 01	120	33 18	147	33 21	219	55 01	311	46 02	337
	PROCYON		REGULUS		*Gienah		ACRUX		*ACHERNAR		RIGEL		*BETELGEUSE	
105	49 42	015	26 01	052	18 54	099	33 45	147	32 50	219	54 23	310	45 42	336
106	49 54	013	26 40	052	19 43	098	34 12	147	32 18	219	53 45	309	45 22	335
107	50 04	012	27 19	051	20 32	098	34 39	147	31 47	219	53 06	308	45 00	333
108	50 14	010	27 57	050	21 22	097	35 06	147	31 16	219	52 26	306	44 37	332
109	50 22	008	28 35	049	22 11	097	35 33	147	30 44	219	51 45	305	44 13	331
110	50 28	007	29 12	048	23 00	096	36 00	147	30 13	218	51 05	304	43 49	330
111	50 34	005	29 49	047	23 50	096	36 27	147	29 42	218	50 23	303	43 23	328
112	50 38	004	30 25	046	24 39	095	36 54	147	29 12	218	49 41	302	42 56	327
113	50 40	002	31 01	045	25 29	094	37 22	147	28 41	218	48 59	301	42 29	326
114	50 42	001	31 36	045	26 19	094	37 49	147	28 11	218	48 16	300	42 01	325
115	50 42	359	32 11	044	27 08	093	38 16	147	27 40	217	47 32	299	41 32	324
116	50 40	358	32 45	043	27 58	093	38 43	147	27 10	217	46 48	298	41 02	322
117	50 37	356	33 18	042	28 48	092	39 11	147	26 40	217	46 04	297	40 31	321
118	50 33	354	33 51	041	29 37	092	39 38	147	26 10	217	45 20	296	40 00	320
119	50 28	353	34 23	040	30 27	091	40 05	147	25 40	217	44 35	295	39 27	319
	REGULUS		*SPICA		ACRUX		*CANOPUS		RIGEL		BETELGEUSE		*PROCYON	
120	34 55	039	13 37	094	40 33	147	64 35	215	43 49	294	38 55	318	50 21	351
121	35 26	038	14 27	094	41 00	147	64 06	216	43 04	293	38 21	317	50 13	350
122	35 56	037	15 16	093	41 27	147	63 36	217	42 18	292	37 47	316	50 03	348
123	36 25	036	16 06	093	41 54	147	63 06	218	41 32	292	37 12	315	49 52	347
124	36 54	035	16 56	092	42 21	147	62 36	218	40 45	291	36 36	314	49 40	345
125	37 22	034	17 45	091	42 48	147	62 04	219	39 59	290	36 00	313	49 27	344
126	37 49	032	18 35	091	43 15	147	61 33	220	39 12	289	35 24	312	49 12	342
127	38 15	031	19 25	090	43 42	147	61 01	220	38 25	288	34 46	311	48 56	341
128	38 40	030	20 15	090	44 09	147	60 28	221	37 38	288	34 09	310	48 39	339
129	39 05	029	21 04	089	44 36	147	59 56	221	36 50	287	33 30	309	48 21	338
130	39 29	028	21 54	089	45 03	148	59 22	222	36 02	286	32 52	308	48 02	336
131	39 52	027	22 44	088	45 29	148	58 49	222	35 14	285	32 12	307	47 42	335
132	40 14	026	23 33	088	45 56	148	58 15	223	34 26	285	31 33	307	47 20	334
133	40 35	024	24 23	087	46 22	148	57 41	223	33 38	284	30 53	306	46 57	332
134	40 55	023	25 13	086	46 48	148	57 07	224	32 50	283	30 12	305	46 34	331
	*REGULUS		SPICA		*ACRUX		CANOPUS		*RIGEL		BETELGEUSE		PROCYON	
135	41 14	022	26 02	086	47 14	148	56 33	224	32 01	283	29 31	304	46 09	330
136	41 32	021	26 52	085	47 40	149	55 58	224	31 13	282	28 50	303	45 44	328
137	41 49	019	27 42	085	48 06	149	55 23	225	30 24	281	28 08	302	45 17	327
138	42 05	018	28 31	084	48 32	149	54 48	225	29 35	281	27 26	302	44 50	326
139	42 20	017	29 21	083	48 57	149	54 13	225	28 46	280	26 43	301	44 21	325
140	42 34	016	30 10	083	49 22	150	53 38	225	27 57	279	26 00	300	43 52	323
141	42 47	014	30 59	082	49 47	150	53 02	226	27 08	279	25 17	299	43 22	322
142	42 58	013	31 48	082	50 12	150	52 26	226	26 19	278	24 34	299	42 51	321
143	43 09	012	32 38	081	50 36	151	51 51	226	25 30	277	23 50	298	42 19	320
144	43 18	010	33 27	080	51 01	151	51 15	226	24 40	277	23 06	297	41 47	319
145	43 27	009	34 16	080	51 25	151	50 39	226	23 51	276	22 21	296	41 14	318
146	43 34	008	35 05	079	51 49	152	50 03	226	23 01	276	21 37	296	40 40	317
147	43 40	006	35 53	078	52 12	152	49 27	227	22 12	275	20 52	295	40 05	316
148	43 45	005	36 42	078	52 35	152	48 51	227	21 22	274	20 07	294	39 30	314
149	43 49	004	37 30	077	52 58	153	48 14	227	20 33	274	19 21	294	38 55	313
	REGULUS		*Denebola		SPICA		*ACRUX		CANOPUS		*SIRIUS		PROCYON	
150	43 51	002	34 57	032	38 19	076	53 21	153	47 38	227	42 55	279	38 18	312
151	43 53	001	35 23	031	39 07	075	53 43	154	47 02	227	42 06	278	37 41	311
152	43 53	000	35 48	030	39 55	075	54 05	154	46 25	227	41 17	278	37 04	310
153	43 52	358	36 13	029	40 43	074	54 26	155	45 49	227	40 28	277	36 25	310
154	43 50	357	36 37	028	41 31	073	54 48	155	45 13	227	39 38	276	35 47	309
155	43 47	356	37 00	027	42 18	072	55 08	156	44 36	227	38 49	276	35 08	308
156	43 42	354	37 22	026	43 05	072	55 29	156	44 00	227	37 59	275	34 28	307
157	43 37	353	37 43	025	43 53	071	55 48	157	43 24	227	37 10	274	33 48	306
158	43 30	351	38 03	023	44 39	070	56 08	157	42 47	227	36 20	274	33 07	305
159	43 22	350	38 22	022	45 26	069	56 27	158	42 11	227	35 30	273	32 26	304
160	43 13	349	38 41	021	46 12	068	56 45	158	41 35	227	34 41	273	31 45	303
161	43 03	348	38 58	020	46 58	067	57 03	159	40 58	227	33 51	272	31 03	302
162	42 51	346	39 14	019	47 44	066	57 21	160	40 22	227	33 01	271	30 21	302
163	42 39	345	39 30	017	48 30	065	57 38	160	39 46	227	32 12	271	29 39	301
164	42 25	344	39 44	016	49 15	065	57 54	161	39 10	227	31 22	270	28 56	300
	ARCTURUS		*ANTARES		ACRUX		*CANOPUS		SIRIUS		PROCYON		*REGULUS	
165	19 25	049	20 41	109	58 10	162	38 34	226	30 32	270	28 12	299	42 11	342
166	20 02	048	21 28	108	58 25	163	37 58	226	29 42	269	27 29	298	41 55	341
167	20 39	047	22 15	108	58 40	163	37 22	226	28 53	269	26 45	298	41 38	340
168	21 15	046	23 03	107	58 54	164	36 46	226	28 03	268	26 01	297	41 21	339
169	21 51	046	23 50	107	59 07	165	36 10	226	27 13	268	25 16	296	41 02	337
170	22 26	045	24 38	106	59 20	166	35 34	226	26 23	267	24 31	295	40 42	336
171	23 01	044	25 26	106	59 32	166	34 59	226	25 34	267	23 46	295	40 22	335
172	23 35	043	26 14	105	59 43	167	34 23	226	24 44	266	23 01	294	40 00	334
173	24 09	042	27 02	105	59 54	168	33 48	225	23 55	265	22 16	293	39 38	333
174	24 42	041	27 50	104	60 04	169	33 12	225	23 05	265	21 30	293	39 14	331
175	25 14	041	28 38	104	60 13	170	32 37	225	22 15	264	20 44	292	38 50	330
176	25 46	040	29 26	104	60 22	171	32 02	225	21 26	264	19 58	291	38 25	329
177	26 18	039	30 15	103	60 30	171	31 27	225	20 37	263	19 11	291	37 59	328
178	26 49	038	31 03	103	60 37	172	30 52	224	19 47	263	18 24	290	37 32	327
179	27 19	037	31 52	102	60 43	173	30 17	224	18 58	262	17 38	289	37 05	326

Figure 6-6A. Left-hand page, Lat 34° S, LHAs 0°–179°, Volume 1 Tables No. 249.

N. Lat. {LHA greater than 180°....... Zn=Z; LHA less than 180°........... Zn=360−Z}

DECLINATION (15°-29°) CONTRARY NAME TO LATITUDE

LHA	15° Hc	d	Z	16° Hc	d	Z	17° Hc	d	Z	18° Hc	d	Z	19° Hc	d	Z
°	° ′	′	°	° ′	′	°	° ′	′	°	° ′	′	°	° ′	′	°
69	08 11	38	114	07 33	37	115	06 56	38	116	06 18	38	117	05 40	37	118
68	08 56	38	115	08 18	38	116	07 40	38	117	07 02	38	117	06 24	38	118
67	09 41	38	116	09 03	38	116	08 25	38	117	07 47	39	118	07 08	38	119
66	10 26	39	116	09 47	38	117	09 09	39	118	08 30	38	119	07 52	39	119
65	11 10	−38	117	10 32	−39	118	09 53	−39	118	09 14	−39	119	08 35	−39	120
64	11 54	39	118	11 15	39	118	10 36	39	119	09 57	39	120	09 18	39	121
63	12 38	39	118	11 59	39	119	11 20	40	120	10 40	39	120	10 01	40	121
62	13 22	39	119	12 43	40	120	12 03	40	120	11 23	40	121	10 43	40	122
61	14 06	40	119	13 26	40	120	12 46	41	121	12 05	40	122	11 25	40	122
60	14 49	−40	120	14 09	−41	121	13 28	−40	122	12 48	−41	122	12 07	−41	123
59	15 32	41	121	14 51	41	122	14 10	41	122	13 29	41	123	12 48	41	124
58	16 14	41	121	15 33	41	122	14 52	41	123	14 11	41	124	13 30	42	124
57	16 57	42	122	16 15	41	123	15 34	42	124	14 52	41	124	14 11	42	125
56	17 39	42	123	16 57	42	124	16 15	42	124	15 33	42	125	14 51	42	126
55	18 20	−42	124	17 38	−42	124	16 56	−42	125	16 14	−43	126	15 31	−42	127
54	19 01	42	124	18 19	42	125	17 37	43	126	16 54	43	127	16 11	43	127
53	19 42	42	125	19 00	43	126	18 17	43	127	17 34	44	127	16 50	43	128
52	20 23	43	126	19 40	43	126	18 57	44	127	18 13	44	128	17 29	43	129
51	21 03	43	127	20 20	44	127	19 36	44	128	18 52	44	129	18 08	44	129
50	21 43	−44	127	20 59	−44	128	20 15	−44	129	19 31	−45	129	18 46	−44	130
49	22 22	44	128	21 38	44	129	20 54	45	129	20 09	45	130	19 24	45	131
48	23 01	44	129	22 17	45	130	21 32	45	130	20 47	45	131	20 02	46	132
47	23 40	45	130	22 55	45	130	22 10	46	131	21 24	45	132	20 39	46	132
46	24 18	45	130	23 33	46	131	22 47	46	132	22 01	46	132	21 15	46	133
45	24 56	−46	131	24 10	−46	132	23 24	−46	133	22 38	−47	133	21 51	−46	134
44	25 33	46	132	24 47	47	133	24 00	46	133	23 14	47	134	22 27	47	135
43	26 10	47	133	25 23	47	134	24 36	47	134	23 49	47	135	23 02	48	136
42	26 46	47	134	25 59	47	134	25 12	48	135	24 24	47	136	23 37	48	136
41	27 22	48	135	26 34	47	135	25 47	48	136	24 59	48	137	24 11	49	137
40	27 57	−48	135	27 09	−48	136	26 21	−48	137	25 33	−49	137	24 44	−49	138
39	28 32	49	136	27 43	48	137	26 55	49	138	26 06	49	138	25 17	49	139
38	29 06	49	137	28 17	49	138	27 28	49	138	26 39	49	139	25 50	50	140
37	29 39	49	138	28 50	49	139	28 01	50	139	27 11	49	140	26 22	50	141
36	30 12	49	139	29 23	50	140	28 33	50	140	27 43	50	141	26 53	51	142
35	30 45	−50	140	29 55	−50	141	29 05	−51	141	28 14	−51	142	27 23	−50	142
34	31 17	51	141	30 26	51	141	29 35	50	142	28 45	51	143	27 54	52	143
33	31 48	51	142	30 57	51	142	30 06	52	143	29 14	51	144	28 23	52	144
32	32 18	51	143	31 27	52	143	30 35	51	144	29 44	52	145	28 52	52	145
31	32 48	52	144	31 56	52	144	31 04	52	145	30 12	52	146	29 20	52	146
30	33 17	−52	145	32 25	−53	145	31 32	−52	146	30 40	−53	146	29 47	−52	147
29	33 45	52	146	32 53	53	146	32 00	53	147	31 07	53	147	30 14	53	148
28	34 13	53	147	33 20	53	147	32 27	53	148	31 34	54	148	30 40	53	149
27	34 40	53	148	33 47	54	148	32 53	54	149	31 59	54	149	31 05	54	150
26	35 06	54	149	34 12	54	149	33 18	54	150	32 24	54	150	31 30	54	151
25	35 31	−54	150	34 37	−54	150	33 43	−55	151	32 48	−54	151	31 54	−55	152
24	35 56	55	151	35 01	54	152	34 07	55	152	33 12	55	153	32 17	55	153
23	36 20	55	152	35 25	55	153	34 30	56	153	33 34	55	154	32 39	55	154
22	36 43	56	153	35 47	55	154	34 52	56	154	33 56	56	155	33 00	55	155
21	37 05	56	154	36 09	56	155	35 13	56	155	34 17	56	156	33 21	56	156
20	37 26	−56	155	36 30	−57	156	35 33	−56	156	34 37	−56	157	33 41	−57	157
19	37 46	56	157	36 50	57	157	35 53	57	157	34 56	56	158	34 00	57	158
18	38 05	56	158	37 09	57	158	36 12	57	159	35 15	57	159	34 18	57	159
17	38 24	57	159	37 27	58	159	36 29	57	160	35 32	57	160	34 35	57	160
16	38 41	57	160	37 44	58	160	36 46	57	161	35 49	58	161	34 51	58	162
15	38 58	−58	161	38 00	−58	162	37 02	−58	162	36 04	−58	162	35 06	−57	163
14	39 13	58	162	38 15	58	163	37 17	58	163	36 19	58	163	35 21	58	164
13	39 28	59	164	38 29	58	164	37 31	58	164	36 33	59	165	35 34	58	165
12	39 41	58	165	38 43	59	165	37 44	58	165	36 46	59	166	35 47	59	166
11	39 54	59	166	38 55	59	166	37 56	59	167	36 57	59	167	35 58	58	167
10	40 05	−59	167	39 06	−59	168	38 07	−59	168	37 08	−59	168	36 09	−59	168
9	40 15	59	169	39 16	59	169	38 17	59	169	37 18	59	169	36 19	60	169
8	40 25	60	170	39 25	59	170	38 26	59	170	37 27	60	170	36 27	59	171
7	40 33	60	171	39 33	59	171	38 34	60	171	37 34	59	172	36 35	60	172
6	40 40	59	172	39 41	60	173	38 41	60	173	37 41	59	173	36 42	60	173
5	40 46	−60	174	39 46	−59	174	38 47	−60	174	37 47	−60	174	36 47	−60	174
4	40 51	60	175	39 51	59	175	38 52	60	175	37 52	60	175	36 52	60	175
3	40 55	60	176	39 55	60	176	38 55	60	176	37 55	60	176	36 55	59	177
2	40 58	60	177	39 58	60	178	38 58	60	178	37 58	60	178	36 58	60	178
1	40 59	59	179	40 00	60	179	39 00	60	179	38 00	60	179	37 00	60	179
0	41 00	−60	180	40 00	−60	180	39 00	−60	180	38 00	−60	180	37 00	−60	180
	15°			16°			17°			18°			19°		

LHA	20° Hc	d	Z	21° Hc	d	Z	22° Hc	d	Z	23° Hc	d	Z	24° Hc	d	Z
°	° ′	′	°	° ′	′	°	° ′	′	°	° ′	′	°	° ′	′	°
69	05 03	38	118	04 25	38	119	03 47	38	120	03 09	38	121	02 31	38	121
68	05 46	38	119	05 08	38	120	04 30	38	120	03 52	38	121	03 14	39	122
67	06 30	39	120	05 51	38	120	05 13	39	121	04 34	38	122	03 56	39	123
66	07 13	39	120	06 34	39	121	05 55	39	122	05 16	39	122	04 37	39	123
65	07 56	−39	121	07 17	−39	122	06 38	−40	122	05 58	−39	123	05 19	−39	124
64	08 39	40	121	07 59	39	122	07 20	40	123	06 40	40	124	06 00	40	124
63	09 21	40	122	08 41	40	123	08 01	40	124	07 21	40	124	06 41	40	125
62	10 03	40	123	09 23	40	123	08 43	41	124	08 02	40	125	07 22	41	126
61	10 45	41	123	10 04	40	124	09 24	41	125	08 43	41	126	08 02	41	126
60	11 26	−41	124	10 45	−41	125	10 04	−41	125	09 23	−41	126	08 42	−41	127
59	12 07	41	125	11 26	41	125	10 45	42	126	10 03	41	127	09 22	42	128
58	12 48	41	125	12 07	42	126	11 25	42	127	10 43	42	127	10 01	42	128
57	13 29	42	126	12 47	43	127	12 04	42	127	11 22	42	128	10 40	43	129
56	14 09	43	127	13 26	42	127	12 44	43	128	12 01	43	129	11 18	42	129
55	14 49	−43	127	14 06	−43	128	13 23	−43	129	12 40	−43	129	11 57	−44	130
54	15 28	43	128	14 45	43	129	14 02	44	129	13 18	43	130	12 35	44	131
53	16 07	44	129	15 23	43	129	14 40	44	130	13 56	44	131	13 12	44	132
52	16 46	44	129	16 02	44	130	15 18	45	131	14 33	44	132	13 49	44	132
51	17 24	44	130	16 40	45	131	15 55	44	132	15 11	45	132	14 26	45	133
50	18 02	−45	131	17 17	−45	132	16 32	−45	132	15 47	−45	133	15 02	−45	134
49	18 39	45	132	17 54	45	132	17 09	46	133	16 23	45	134	15 38	46	134
48	19 16	45	132	18 31	46	133	17 45	46	134	16 59	46	134	16 13	46	135
47	19 53	46	133	19 07	46	134	18 21	46	134	17 35	47	135	16 48	46	136
46	20 29	46	134	19 43	47	135	18 56	47	135	18 09	46	136	17 23	47	137
45	21 05	−47	135	20 18	−47	135	19 31	−47	136	18 44	−47	137	17 57	−48	137
44	21 40	47	135	20 53	48	136	20 05	47	137	19 18	48	137	18 30	47	138
43	22 14	47	136	21 27	48	137	20 39	48	138	19 51	48	138	19 03	48	139
42	22 49	48	137	22 01	48	138	21 13	49	138	20 24	48	139	19 36	49	140
41	23 22	48	138	22 34	49	139	21 45	48	139	20 57	49	140	20 08	49	140
40	23 55	−48	139	23 07	−49	139	22 18	−49	140	21 29	−50	141	20 39	−49	141
39	24 28	49	140	23 39	50	140	22 49	49	141	22 00	50	141	21 10	50	142
38	25 00	50	140	24 10	49	141	23 21	50	142	22 31	50	142	21 41	51	143
37	25 32	50	141	24 42	51	142	23 51	50	142	23 01	51	143	22 10	50	144
36	26 02	50	142	25 12	51	143	24 21	50	143	23 31	51	144	22 40	51	144
35	26 33	−51	143	25 42	−51	144	24 51	−51	144	24 00	−52	145	23 08	−51	145
34	27 02	51	144	26 11	51	144	25 20	52	145	24 28	52	146	23 36	51	146
33	27 31	51	145	26 40	52	145	25 48	52	146	24 56	52	146	24 04	52	147
32	28 00	52	146	27 08	53	146	26 15	52	147	25 23	52	147	24 31	53	148
31	28 28	53	147	27 35	53	147	26 42	52	148	25 50	53	148	24 57	53	149
30	28 55	−53	148	28 02	−53	148	27 09	−54	149	26 15	−53	149	25 22	−53	150
29	29 21	53	149	28 28	54	149	27 34	53	150	26 41	54	150	25 47	54	151
28	29 47	54	150	28 53	54	150	27 59	54	151	27 05	54	151	26 11	54	151
27	30 11	54	150	29 17	54	151	28 23	54	151	27 29	54	152	26 35	55	152
26	30 36	55	151	29 41	54	152	28 47	55	152	27 52	55	153	26 57	55	153
25	30 59	−55	152	30 04	−55	153	29 09	−55	153	28 14	−55	154	27 19	−55	154
24	31 22	55	153	30 27	56	154	29 31	55	154	28 36	56	155	27 40	55	155
23	31 44	56	154	30 48	56	155	29 52	55	155	28 57	56	156	28 01	56	156
22	32 05	56	156	31 09	56	156	30 13	56	156	29 17	56	157	28 21	56	157
21	32 25	56	157	31 29	57	157	30 32	56	157	29 36	56	158	28 40	57	158
20	32 44	−56	158	31 48	−57	158	30 51	−56	158	29 55	−57	159	28 58	−57	159
19	33 03	57	159	32 06	57	159	31 09	57	159	30 12	57	160	29 15	57	160
18	33 21	57	160	32 24	58	160	31 26	57	160	30 29	57	161	29 32	58	161
17	33 38	58	161	32 40	57	161	31 43	58	161	30 45	58	162	29 47	57	162
16	33 53	57	162	32 56	58	162	31 58	58	163	31 00	58	163	30 02	58	163
15	34 09	−58	163	33 11	−58	163	32 13	−59	164	31 14	−58	164	30 16	−58	164
14	34 23	59	164	33 24	58	164	32 26	58	165	31 28	58	165	30 30	59	165
13	34 36	59	165	33 37	58	165	32 39	59	166	31 40	58	166	30 42	59	166
12	34 48	58	166	33 50	59	167	32 51	59	167	31 52	59	167	30 53	58	167
11	35 00	59	167	34 01	59	168	33 02	59	168	32 03	59	168	31 04	59	168
10	35 10	−59	169	34 11	−59	169	33 12	−59	169	32 13	−59	169	31 14	−60	169
9	35 19	59	170	34 20	59	170	33 21	59	170	32 22	60	170	31 22	59	170
8	35 28	59	171	34 29	60	171	33 29	59	171	32 30	59	171	31 31	60	171
7	35 35	59	172	34 36	60	172	33 36	59	172	32 37	60	172	31 37	59	173
6	35 42	60	173	34 42	59	173	33 43	60	173	32 43	60	173	31 43	59	174
5	35 47	−59	174	34 48	−60	174	33 48	−60	174	32 48	−60	175	31 48	−59	175
4	35 52	60	175	34 52	60	175	33 52	60	176	32 52	59	176	31 53	60	176
3	35 56	60	177	34 56	60	177	33 56	60	177	32 56	60	177	31 56	60	177
2	35 58	60	178	34 58	60	178	33 58	60	178	32 58	60	178	31 58	60	178
1	36 00	60	179	35 00	60	179	34 00	60	179	33 00	60	179	32 00	60	179
0	36 00	−60	180	35 00	−60	180	34 00	−60	180	33 00	−60	180	32 00	−60	180
	20°			21°			22°			23°			24°		

LHA	25° Hc	d	Z	26° Hc	d	Z	27° Hc	d	Z	28° Hc	d	Z	29° Hc	d	Z	LHA
°	° ′	′	°	° ′	′	°	° ′	′	°	° ′	′	°	° ′	′	°	
69	01 53	38	122	01 15	38	123	00 37	38	124	−0 01	38	124	−0 39	38	125	291
68	02 35	38	123	01 57	38	123	01 19	39	124	00 40	38	125	00 02	39	126	292
67	03 17	39	123	02 38	38	124	02 00	39	125	01 21	39	126	00 42	39	126	293
66	03 58	39	124	03 19	39	125	02 40	39	125	02 01	39	126	01 22	39	127	294
65	04 40	−40	125	04 00	−39	125	03 21	−40	126	02 41	−40	127	02 01	−39	128	295
64	05 20	39	125	04 41	40	126	04 01	40	127	03 21	40	127	02 41	40	128	296
63	06 01	40	126	05 21	41	127	04 40	40	127	04 00	40	128	03 20	41	129	297
62	06 41	40	126	06 01	41	127	05 20	41	128	04 39	41	129	03 58	41	129	298
61	07 21	41	127	06 40	41	128	05 59	41	128	05 18	41	129	04 37	42	130	299
60	08 01	−42	128	07 19	−41	128	06 38	−42	129	05 56	−41	130	05 15	−42	131	300
59	08 40	42	128	07 58	42	129	07 16	42	130	06 34	41	130	05 53	42	131	301
58	09 19	42	129	08 37	43	130	07 54	42	130	07 12	42	131	06 30	43	132	302
57	09 57	42	130	09 15	43	130	08 32	42	131	07 50	43	132	07 07	43	132	303
56	10 36	43	130	09 53	43	131	09 10	43	132	08 27	44	132	07 43	43	133	304
55	11 13	−43	131	10 30	−43	132	09 47	−44	132	09 03	−43	133	08 20	−44	134	305
54	11 51	44	132	11 07	44	132	10 23	44	133	09 39	44	134	08 55	44	134	306
53	12 28	44	132	11 44	44	133	11 00	45	134	10 15	44	134	09 31	45	135	307
52	13 05	45	133	12 20	45	134	11 35	44	134	10 51	45	135	10 06	45	135	308
51	13 41	45	134	12 56	45	134	12 11	45	135	11 26	46	136	10 40	45	136	309
50	14 17	−46	134	13 31	−45	135	12 46	−46	136	12 00	−45	136	11 15	−46	137	310
49	14 52	46	135	14 06	45	136	13 21	46	136	12 35	47	137	11 48	46	138	311
48	15 27	46	136	14 41	46	136	13 55	47	137	13 08	46	138	12 22	47	138	312
47	16 02	47	136	15 15	47	137	14 28	46	138	13 42	47	138	12 55	47	139	313
46	16 36	47	137	15 49	47	138	15 02	48	138	14 14	47	139	13 27	47	140	314
45	17 09	−47	138	16 22	−48	139	15 34	−47	139	14 47	−48	140	13 59	−48	140	315
44	17 43	48	139	16 55	48	139	16 07	48	140	15 19	49	141	14 30	48	141	316
43	18 15	48	139	17 27	49	140	16 38	48	141	15 50	49	141	15 01	48	142	317
42	18 47	48	140	17 59	49	141	17 10	49	141	16 21	49	142	15 32	49	143	318
41	19 19	49	141	18 30	49	142	17 41	50	142	16 51	49	143	16 02	50	143	319
40	19 50	−50	142	19 00	−49	142	18 11	−50	143	17 21	−50	144	16 31	−50	144	320
39	20 20	49	143	19 31	50	143	18 41	51	144	17 50	50	144	17 00	50	145	321
38	20 50	50	143	20 00	50	144	19 10	51	145	18 19	50	145	17 29	51	146	322
37	21 20	51	144	20 29	51	145	19 38	51	145	18 47	51	146	17 56	51	146	323
36	21 49	51	145	20 58	52	146	20 06	51	146	19 15	51	147	18 24	52	147	324
35	22 17	−52	146	21 25	−51	146	20 34	−52	147	19 42	−52	148	18 50	−52	148	325
34	22 45	52	147	21 53	52	147	21 01	53	148	20 08	52	148	19 16	52	149	326
33	23 12	53	148	22 19	52	148	21 27	53	149	20 34	52	149	19 42	53	150	327
32	23 38	53	148	22 45	53	149	21 52	52	149	21 00	53	150	20 07	53	150	328
31	24 04	53	149	23 11	54	150	22 17	53	150	21 24	53	151	20 31	54	151	329
30	24 29	−54	150	23 35	−53	151	22 42	−54	151	21 48	−54	152	20 54	−53	152	330
29	24 53	54	151	23 59	53	152	23 06	54	152	22 12	55	153	21 17	54	153	331
28	25 17	54	152	24 23	54	152	23 29	55	153	22 34	54	153	21 40	55	154	332
27	25 40	54	153	24 46	55	153	23 51	55	154	22 56	55	154	22 01	55	155	333
26	26 02	54	154	25 08	55	154	24 13	56	155	23 17	55	155	22 22	55	156	334
25	26 24	−55	155	25 29	−56	155	24 33	−55	156	23 38	−55	156	22 43	−56	156	335
24	26 45	56	156	25 49	55	156	24 54	56	157	23 58	56	157	23 02	56	157	336
23	27 05	56	157	26 09	56	157	25 13	56	157	24 17	56	158	23 21	56	158	337
22	27 25	57	158	26 28	56	158	25 32	56	158	24 36	57	159	23 39	56	159	338
21	27 43	56	159	26 47	57	159	25 50	57	159	24 53	56	160	23 57	57	160	339
20	28 01	−57	159	27 04	−57	160	26 07	−57	160	25 10	−57	161	24 13	−57	161	340
19	28 18	57	160	27 21	57	161	26 24	57	161	25 27	58	161	24 29	57	162	341
18	28 34	57	161	27 37	57	162	26 40	58	162	25 42	58	162	24 44	57	163	342
17	28 50	58	162	27 52	58	163	26 54	57	163	25 57	58	163	24 59	58	164	343
16	29 04	57	163	28 07	58	164	27 09	58	164	26 11	58	164	25 13	59	165	344
15	29 18	−58	164	28 20	−58	165	27 22	−58	165	26 24	−59	165	25 25	−58	166	345
14	29 31	58	165	28 33	59	166	27 34	58	166	26 36	59	166	25 37	58	166	346
13	29 43	58	166	28 45	59	167	27 46	59	167	26 47	58	167	25 49	59	167	347
12	29 55	59	167	28 56	59	168	27 57	59	168	26 58	59	168	25 59	59	168	348
11	30 05	59	169	29 06	59	169	28 07	59	169	27 08	59	169	26 09	59	169	349
10	30 14	−59	170	29 15	−59	170	28 16	−59	170	27 17	−59	170	26 18	−60	170	350
9	30 23	59	171	29 24	60	171	28 24	59	171	27 25	59	171	26 26	60	171	351
8	30 31	60	172	29 31	59	172	28 32	60	172	27 32	59	172	26 33	60	172	352
7	30 38	60	173	29 38	60	173	28 38	59	173	27 39	60	173	26 39	59	173	353
6	30 44	60	174	29 44	60	174	28 44	60	174	27 44	59	174	26 45	60	174	354
5	30 49	−60	175	29 49	−60	175	28 49	−60	175	27 49	−60	175	26 49	−59	175	355
4	30 53	60	176	29 53	60	176	28 53	60	176	27 53	60	176	26 53	60	176	356
3	30 56	60	177	29 56	60	177	28 56	60	177	27 56	60	177	26 56	60	177	357
2	30 58	60	178	29 58	60	178	28 58	60	178	27 58	60	178	26 58	60	178	358
1	31 00	60	179	30 00	60	179	29 00	60	179	28 00	60	179	27 00	60	179	359
0	31 00	−60	180	30 00	−60	180	29 00	−60	180	28 00	−60	180	27 00	−60	180	360
	25°			26°			27°			28°			29°			

S. Lat. {LHA greater than 180°....... Zn=180−Z; LHA less than 180°........... Zn=180+Z}

DECLINATION (15°-29°) CONTRARY NAME TO LATITUDE

LAT 34°

Figure 6-6B. Lat 34° page, Volume II, Tables No. 249.

The tables are entered in the column containing tabulations for the whole degree of declination less than the true declination. The tabulated Hc in this column directly across from the appropriate LHA is recorded, as well as the d value and the azimuth angle Z. The Hc thus extracted must then be adjusted to the proper value corresponding to the true declination. To do this, a "declination correction" is found mathematically by multiplying the tabulated d value by the additional minutes of the true declination divided by 60. For the sake of convenience, a table containing precomputed products for this multiplication is included in the back of Volumes II and III as Table 4; a portion of this table appears in Figure 6-6C. The table uses as entering arguments the d value and minutes of the true declination. After the correction is found, it is applied to the tabulated altitude according to the sign of the d value in the main table.

TABLE 4.—Correction to Tabulated Altitud[e]

d / ′	1 2 3	4 5 6	7 8 9	10 11 12	13 14 15	16 17 18	19 20 21	22 23 24	25 26 27	28 29 30	31 32 33
0	0 0 0	0 0 0	0 0 0	0 0 0	0 0 0	0 0 0	0 0 0	0 0 0	0 0 0	0 0 0	0 0 [illegible]
1	0 0 0	0 0 0	0 0 0	0 0 0	0 0 0	0 0 0	0 0 0	0 0 0	0 0 0	0 0 0	1 1 [illegible]
2	0 0 0	0 0 0	0 0 0	0 0 0	0 0 0	1 1 1	1 1 1	1 1 1	1 1 1	1 1 1	1 1 [illegible]
3	0 0 0	0 0 0	0 0 0	0 1 1	1 1 1	1 1 1	1 1 1	1 1 1	1 1 1	1 1 2	2 2 2
4	0 0 0	0 0 0	0 1 1	1 1 1	1 1 1	1 1 1	1 1 1	1 2 2	2 2 2	2 2 2	2 2 [illegible]
5	0 0 0	0 0 0	1 1 1	1 1 1	1 1 1	1 1 2	2 2 2	2 2 2	2 2 2	2 2 2	3 3 [illegible]
6	0 0 0	0 0 1	1 1 1	1 1 1	1 1 2	2 2 2	2 2 2	2 2 2	2 3 3	3 3 3	3 [illegible]
7	0 0 0	0 1 1	1 1 1	1 1 1	2 2 2	2 2 2	2 2 2	3 3 3	3 3 3	3 3 4	4 [illegible]
8	0 0 0	1 1 1	1 1 1	1 1 2	2 2 2	2 2 2	3 3 3	3 3 3	3 3 4	4 4 4	4 [illegible]
9	0 0 0	1 1 1	1 1 1	2 2 2	2 2 2	2 3 3	3 3 3	3 3 4	4 4 4	4 4 4	5 [illegible]
10	0 0 0	1 1 1	1 1 2	2 2 2	2 2 2	3 3 3	3 3 4	4 4 4	4 4 4	5 5 5	5 [illegible]
11	0 0 1	1 1 1	1 1 2	2 2 2	2 3 3	3 3 3	3 4 4	4 4 4	5 5 5	5 5 6	6 [illegible]
12	0 0 1	1 1 1	1 2 2	2 2 2	3 3 3	3 3 4	4 4 4	4 5 5	5 5 5	6 6 6	[illegible]
13	0 0 1	1 1 1	2 2 2	2 2 3	3 3 3	3 4 4	4 4 5	5 5 5	5 6 6	6 6 6	[illegible]
14	0 0 1	1 1 1	2 2 2	2 3 3	3 3 4	4 4 4	4 5 5	5 5 6	6 6 6	7 7 7	[illegible]
15	0 0 1	1 1 2	2 2 2	2 3 3	3 4 4	4 4 4	5 5 5	6 6 6	6 6 7	7 7 8	[illegible]
16	0 1 1	1 1 2	2 2 2	3 3 3	3 4 4	4 5 5	5 5 6	6 6 6	7 7 7	7 8 8	[illegible]
17	0 1 1	1 1 2	2 2 3	3 3 3	4 4 4	5 5 5	5 6 6	6 7 7	7 7 8	8 8 8	[illegible]
18	0 1 1	1 2 2	2 2 3	3 3 4	4 4 4	5 5 5	6 6 6	7 7 7	8 8 8	8 9 9	[illegible]
19	0 1 1	1 2 2	2 3 3	3 3 4	4 4 5	5 5 6	6 6 7	7 7 8	8 8 9	9 9 1[illegible]	[illegible]
20	0 1 1	1 2 2	2 3 3	3 4 4	4 5 5	5 6 6	6 7 7	7 8 8	8 9 9	9 10 1[illegible]	[illegible]
21	0 1 1	1 2 2	2 3 3	4 4 4	5 5 5	6 6 6	7 7 7	8 8 8	9 9 9	10 10 [illegible]	[illegible]
22	0 1 1	1 2 2	3 3 3	4 4 4	5 5 6	6 6 7	7 7 8	8 8 9	9 10 10	10 11 [illegible]	[illegible]
23	0 1 1	2 2 2	3 3 3	4 4 5	5 5 6	6 7 7	7 8 8	8 9 9	10 10 10	11 11 1[illegible]	[illegible]
24	0 1 1	2 2 2	3 3 4	4 4 5	5 6 6	6 7 7	8 8 8	9 9 10	10 10 11	11 12 1[illegible]	[illegible]
25	0 1 1	2 2 2	3 3 4	4 5 5	5 6 6	7 7 8	8 8 9	9 10 10	10 11 11	12 12 12	[illegible]
[illegible]	0 1 1	2 2 3	3 3 4	4 5 5	6 6 6	7 7 8	8 9 9	10 10 10	11 11 12	12 13 13	[illegible]
43	[illegible] 1	2 2 3	3 4 4	4 5 5	6 6 7	7 8 8	9 9 9	10 10 11	11 12 12	13 13 14	[illegible]
44	1 1 2	[illegible]	3 4 4	5 5 6	6 7 7	7 8 8	9 9 10	10 11 11	12 12 13	13 14 14	[illegible]
45	1 2 2	3 4 4	[illegible]	[illegible]	10 10 11	12 13 14	[illegible]	11 11 12	12 13 13	14 [illegible] 22	2[illegible]
46	1 2 2	3 4 5	5 6 7	8 8 9	10 11 12	12 13 14	15 15 16	17 1[illegible]	[illegible] 21	21 22 23	24
47	1 2 2	3 4 5	5 6 7	8 9 9	10 11 12	13 13 14	15 16 16	17 18 19	20 20 21	22 23 24	24
48	1 2 2	3 4 5	6 6 7	8 9 10	10 11 12	13 14 14	15 16 17	18 18 19	20 21 22	22 23 24	25
49	1 2 2	3 4 5	6 7 7	8 9 10	11 11 12	13 14 15	16 16 17	18 19 20	20 21 22	23 24 24	25
50	1 2 2	3 4 5	6 7 8	8 9 10	11 12 12	13 14 15	16 17 18	18 19 20	21 22 22	23 24 25	26
51	1 2 3	3 4 5	6 7 8	8 9 10	11 12 13	14 14 15	16 17 18	19 20 20	21 22 23	24 25 26	26
52	1 2 3	3 4 5	6 7 8	9 10 10	11 12 13	14 15 16	16 17 18	19 20 21	22 23 23	24 25 26	27
53	1 2 3	4 4 5	6 7 8	9 10 11	11 12 13	14 15 16	17 18 19	19 20 21	22 23 24	25 26 26	27
54	1 2 3	4 4 5	6 7 8	9 10 11	12 13 14	14 15 16	17 18 19	20 21 22	22 23 24	25 26 27	28 2[illegible]
55	1 2 3	4 5 6	6 7 8	9 10 11	12 13 14	15 16 16	17 18 19	20 21 22	23 24 25	26 27 28	28 2[illegible]
56	1 2 3	4 5 6	7 7 8	9 10 11	12 13 14	15 16 17	18 19 20	21 21 22	23 24 25	26 27 28	29 3[illegible]
57	1 2 3	4 5 6	7 8 9	10 10 11	12 13 14	15 16 17	18 19 20	21 22 23	24 25 26	27 28 28	29 3[illegible]
58	1 2 3	4 5 6	7 8 9	10 11 12	13 14 14	15 16 17	18 19 20	21 22 23	24 25 26	27 28 29	30 3[illegible]
59	1 2 3	4 5 6	7 8 9	10 11 12	13 14 15	16 17 18	19 20 21	22 23 24	25 26 27	28 29 30	30 3[illegible]

[illegible]	[illegible]8	49 50 51	52 53 54	55 56 57	58 59 60	d / ′
	0	0 0 0	0 0 0	0 0 0	0 0 0	0
	1	1 1 1	1 1 1	1 1 1	1 1 1	1
	2	2 2 2	2 2 2	2 2 2	2 2 2	2
	2	2 2 3	3 3 3	3 3 3	3 3 3	3
	3 3	3 3 3	3 4 4	4 4 4	4 4 4	4
	4 4	4 4 4	4 4 4	5 5 5	5 5 5	5
	5 5	5 5 5	5 5 5	6 6 6	6 6 6	6
	5 6	6 6 6	6 6 6	6 7 7	7 7 7	7
	6 6	7 7 7	7 7 7	7 7 8	8 8 8	8
	7 7	7 8 8	8 8 8	8 8 9	9 9 9	9
	8 8 8	8 8 8	9 9 9	9 9 10	10 10 10	10
	8 9 9	9 9 9	10 10 10	10 10 10	11 11 11	11
	9 9 10	10 10 10	10 11 11	11 11 11	12 12 12	12
	10 10 10	11 11 11	11 11 12	12 12 12	13 13 13	13
	11 11 11	11 12 12	12 12 13	13 13 13	14 14 14	14
	12 12 12	12 12 13	13 13 14	14 14 14	14 15 15	15
	12 13 13	13 13 14	14 14 14	15 15 15	15 16 16	16
3	13 13 14	14 14 14	15 15 15	16 16 16	16 17 17	17
4	14 14 14	15 15 15	16 16 16	16 17 17	17 18 18	18
4	15 15 15	16 16 16	16 17 17	17 18 18	18 19 19	19
15	15 16 16	16 17 17	17 18 18	18 19 19	19 20 20	20
16	16 16 17	17 18 18	18 19 19	19 20 20	20 21 21	21
16	17 17 18	18 18 19	19 19 20	20 21 21	21 22 22	22
17	18 18 18	19 19 20	20 20 21	21 21 22	22 23 23	23
18	18 19 19	20 20 20	21 21 22	22 22 23	23 24 24	24
[illegible]	19 20 20	20 21 21	22 22 22	23 23 24	24 25 25	25
	20 20 21	21 22 22	23 23 23	24 24 25	25 26 26	26
	21 21 22	22 22 23	23 24 24	25 25 26	26 27 27	27
	[illegible]1 22 22	23 23 24	24 25 25	26 [illegible]	[illegible] 28 28	28
	34 35 3[illegible]	24 24 25	25 2[illegible]	[illegible] 42 43	44 44 4[illegible]	[illegible]
	[illegible]5 36 37	38 38 3[illegible]	[illegible] 41 41	42 43 44	44 45 46	46
	[illegible]6 37 38	38 39 40	41 42 42	43 44 45	45 46 47	47
	[illegible]7 38 38	39 40 41	42 42 43	44 45 46	46 47 48	48
	[illegible]8 38 39	40 41 42	42 43 44	45 46 47	47 48 49	49
	39 40	41 42 42	43 44 45	46 47 48	48 49 50	50
	40 41	42 42 43	44 45 46	47 48 48	49 50 51	51
	41 42	42 43 44	45 46 47	48 49 49	50 51 52	52
	42 42	43 44 45	46 47 48	49 49 50	51 52 53	53
	42 43	44 45 46	47 48 49	50 50 51	52 53 54	54
	[illegible]3 44	45 46 47	48 49 50	50 51 52	53 54 55	55
	[illegible]4 45	46 47 48	49 49 50	51 52 53	54 55 56	56
	[illegible]5 46	47 48 48	49 50 51	52 53 54	55 56 57	57
	[illegible]6 46	47 48 49	50 51 52	53 54 55	56 57 58	58
	[illegible]5 47	48 49 50	51 52 53	54 55 56	57 58 59	59

Figure 6-6C. Portion of Table 4, Volume II, Tables No. 249.

After the Hc corresponding to the true declination is obtained, the azimuth angle Z is then converted to true azimuth Zn by the rules given on each page, which are identical to the rules given in *Tables No. 229* for the same purpose.

In the following two sections of this chapter, the examples begun earlier in the chapter will be completed using Volumes I and II of *Tables No. 249* as appropriate. The convenience afforded by Volume I of *Tables No. 249* in comparison to *Tables No. 229* or Volumes II and III of *Tables No. 249* will soon become apparent to the inexperienced navigator in the course of reading over these two examples.

Completing the Sight Reduction by *Tables No. 249*—First Example

As the first example of the completion of the sight reduction process by *Tables No. 249*, let us return to the Canopus form shown in Figure 6-4B. Since the star Canopus is one of the seven stars tabulated for 34° south latitude (the assumed latitude) and LHA♈ of 145°, Volume I can be used for direct extraction of the Hc and Zn. From Figure 6-6A, the tabulated Hc for Canopus is 50° 39′, and the tabulated Zn is 226°T. It remains only to compare this Hc with the Ho to find the intercept distance a. Since the Ho is 50° 38′ and the Hc is 50° 39′, the intercept distance is 1 mile away (A) from the assumed position at L 34° S, λ 163° 01.5′ E. The completed form appears in Figure 6-7 opposite.

Note that when Volume I of *Tables No. 249* is used, no interpolations of any kind are required. After the LHA♈ and assumed latitude have been determined, a simple table look-up is all that is necessary to find Hc and Zn. The LHA♈ can be found through the use of either the *Nautical* or *Air Almanac*, and the assumed latitude is simply the whole degree of latitude nearest the DR position of the observer at the time of the sight.

Completing the Sight Reduction by *Tables No. 249*—Second Example

As a second and perhaps more complicated example of the use of *Tables No. 249* in sight reduction, let us complete the solution for the moon sight of Figure 6-5B. From the form, the LHA as determined by the *Air Almanac* was 23°, and the declination is N 23° 14′. The remaining entering argument for Volume II of *Tables No. 249*, the assumed latitude, is 34° south.

The assumed latitude is the primary entering argument for both Volumes II and III of *Tables No. 249*. Once the proper latitude pages have been located, the page of interest is selected by referring to the declination. In this case, the latitude 34° page that contains tabulations for declinations of 15°–29° contrary to the name of the latitude and LHAs from 0° to 69° is the page chosen; this page appears in Figure 6-6B. The tabulated values of Hc and Z selected are those contained in the 23° declination column directly opposite the vertical 23° LHA argument. Hc is 28° 57′, and Z is 156°. To find the Hc

Sight Reduction
using Tables 249 Vol. 1

C: 230 °

S: 25 kts.

Body	CANOPUS
IC	+.3
Dip (Ht 44 ')	-7
R_o	-1
Sum	-8
hs	50-46
Ho (Obs Alt)	50-38
Date (GMT)	15 DEC
GMT (Obs Time)	17-12-09
DR Lat	34-19.0 S
DR Long	163-05.7 E
Tab GHA ♈	341-26.5
GHA ♈ incr'mt	0-32.0
GHA ♈	341-58.5
a λ (-W, +E)	163-01.5 E
LHA ♈	145
a Lat (N or S)	34 S
Hc (Comp Alt)	50-39
Ho (Obs Alt)	50-38
a (Intercept)	1 Ⓐ
Zn (°T)	226°T
P and N Corr'n	N/A

Figure 6-7. Completed Canopus sight form by Tables No. 249.

corresponding with the true declination 23° 14′, the difference d between the aforementioned Hc and the tabulation for the next higher degree of declination is recorded for use with Table 4. In this case the d value is −56′. Using this value as a vertical argument and the 14′ additional declination as a horizontal argument, the declination correction as found from Table 4 of Figure 6-6C is −13′. Thus, the computed Hc corresponding with true declination 23° 14′ is given by the following expression:

$$28^\circ\ 57' - 13' = 28^\circ\ 44'$$

No interpolation is necessary for the azimuth angle Z.

To complete the solution, Z is converted to Zn by the rule for south latitudes printed on each page:

S. Lat. $\begin{cases} \text{LHA greater than } 180° \ldots\ldots\ldots & Zn = 180 - Z \\ \text{LHA less than } 180° \ldots\ldots\ldots\ldots & Zn = 180 + Z \end{cases}$

Thus, Zn = 180° + 156° = 336°T. Comparison of Hc with the Ho determined previously, 28° 35′, yields an intercept distance a of 9 miles "away" from the GP of the moon—away, since Hc is greater than Ho. The completed moon sight reduction form appears in Figure 6-8.

Sight Reduction
using Tables 249 Vols. II and III

C: 230 °

S: 25 kts.

Body	MOON (UL)	
IC	+.3	
Dip (Ht 44 ')	-7	
R_o	-2	
S.D.	-15	
Sum	-24	
hs	28-10	
P in A (Moon)	+49	
Ho (Obs Alt)	28-35	
Date (GMT)	15 DEC	
GMT (Obs Time)	16-58-57	
DR Lat	34-15.5 S	
DR Long	163-11.7 E	
Tab GHA	217-15	
GHA incr'mt	2-10	
SHA (Star)		
GHA	219-25	
±360 if needed		
a λ (-W, +E)	163-35 E	
LHA	383 = 23	
Tab Dec	N 23-14	
a Lat (N or S)	34 S Same (Cont)	
Dec Inc | (±)d	14 | -56	
H_c (Tab Alt)	28-57	
Dec Corr'n	-13	
Hc (Comp Alt)	28-44	
Ho (Obs Alt)	28-35	
a (Intercept)	9 (A)	
Z	S 156 W	
Zn (°T)	336°T	

Figure 6-8. Completed moon sight form by Tables No. 249.

Plotting the Celestial LOP Determined by The *Air Almanac* and *Tables No. 249*

In surface navigation the celestial LOP as determined by the use of the *Air Almanac* and *Tables No. 249* is plotted just as is done for computations based on the *Nautical Almanac* and *Tables No. 229.* The assumed position corresponding to the assumed latitude and longitude chosen for the sight solution is first plotted, then the intercept distance a is laid down along the true azimuth line and the LOP is plotted relative to it. Several LOPs may be combined to form a fix or running fix, or an estimated position may be located on a single LOP, by methods to be discussed in the following chapter. Figure 6-9 pictures

Sight Reduction
using Tables 249 Vols. II and III

C: 230 °

S: 25 kts.

Body	MOON(UL)	VENUS
IC	+.3	+.3
Dip (Ht 44 ')	-7	-7
Ro	-2	-3
S.D.	-15	
Sum	-24	-10
hs	28-10	16-47
P in A (Moon)	+49	
Ho (Obs Alt)	28-35	16-37
Date (GMT)	15 DEC	15 DEC
GMT (Obs Time)	16-58-57	17-04-12
DR Lat	34-15.5 S	34-17.0 S
DR Long	163-11.7 E	163-09.1 E
Tab GHA	217-15	116-49
GHA incr'mt	2-10	1-03
SHA (Star)		
GHA	219-25	117-52
±360 if needed		
a λ (-W, +E)	163-35 E	163-08 E
LHA	383 = 23	281
Tab Dec	N 23-14	S 13-16
a Lat (N or S)	34 S Same (Cont)	34 S (Same) Cont
Dec Inc (±)d	14 -56	16 +32
Hc (Tab Alt)	28-57	16-15
Dec Corr'n	-13	+9
Hc (Comp Alt)	28-44	16-24
Ho (Obs Alt)	28-35	16-37
a (Intercept)	9 (A)	13 (T)
Z	S 156 W	S 84.7 E
Zn (°T)	336°T	095.3°T

Sight Reduction
using Tables 249 Vol. I

C: 230 °

S: 25 kts.

Body	CANOPUS
IC	+.3
Dip (Ht 44 ')	-7
Ro	-1
Sum	-8
hs	50-46
Ho (Obs Alt)	50-38
Date (GMT)	15 DEC
GMT (Obs Time)	17-12-09
DR Lat	34-19.0 S
DR Long	163-05.7 E
Tab GHA ♈	341-26.5
GHA ♈ incr'mt	0-32.0
GHA ♈	341-58.5
a λ (-W, +E)	163-01.5 E
LHA ♈	145
a Lat (N or S)	34 S
Hc (Comp Alt)	50-39
Ho (Obs Alt)	50-38
a (Intercept)	1 (A)
Zn (°T)	226°T
P and N Corr'n	N/A

Figure 6-9. Complete solutions for the moon, Venus, and Canopus by the Air Almanac *and* Tables No. 249.

the completed sight forms for Canopus and the moon discussed in the preceding sections, as well as a complete solution by the *Air Almanac* and *Tables No. 249* for the Venus sight of Chapter 5.

When Volume I of *Tables No. 249* is used for sight reduction in a year other than the designated Epoch year, it will frequently be necessary to correct the resulting LOP for the effects of precession and nutation of the earth. These corrections are tabulated in Table 5, located in the back of Volume I, for each year in which they apply during the period three years before and five years after the Epoch year. A portion of Table 5 containing corrections for 1982 from the Epoch 1980 edition of *Tables No. 249* is reproduced in Figure 6-10. Entering arguments for the table are the nearest tabulated values of LHA♈ and latitude. No interpolation is required. If, for instance, the year in which the observations in this chapter were made were 1982, and the Epoch 1980 edition of Volume I had been used to obtain Hc and Zn, the Canopus LOP would have to be shifted a distance of 1 mile in the direction 120°T.

TABLE 5—CORRECTION FOR PRECESSION AND NUTATION

L.H.A.	North latitudes								South latitudes							L.H.A.
♈	N. 89°	N. 80°	N. 70°	N. 60°	N. 50°	N. 40°	N. 20°	0°	S. 20°	S. 40°	S. 50°	S. 60°	S. 70°	S. 80°	S. 89°	♈
								1982								
°	′ °	′ °	′ °	′ °	′ °	′ °	′ °	′ °	′ °	′ °	′ °	′ °	′ °	′ °	′ °	°
0	1 010	1 030	1 040	1 050	1 060	1 060	2 070	2 070	2 070	1 060	1 060	1 050	1 040	1 020	1 000	0
30	1 040	1 050	1 060	1 060	2 070	2 070	2 070	2 070	2 070	1 060	1 050	1 040	1 020	1 350	1 330	30
60	1 060	1 070	1 080	1 080	2 080	2 080	2 080	2 080	1 080	1 070	1 060	0 —	0 —	0 —	1 300	60
90	1 090	1 090	1 090	1 090	2 090	2 090	2 090	2 090	1 090	1 090	1 100	0 —	0 —	0 —	1 270	90
120	1 120	1 110	1 110	1 110	2 110	2 100	2 100	2 100	1 110	1 110	1 120	1 140	0 —	1 220	1 240	120
150	1 150	1 140	1 130	1 120	1 120	2 110	2 110	2 110	2 110	(1 120)	1 130	1 140	1 160	1 180	1 210	150
180	1 180	1 160	1 140	1 130	1 130	1 120	2 120	2 110	2 120	1 120	1 120	1 130	1 140	1 160	1 170	180
210	1 210	1 190	1 160	1 140	1 130	1 120	2 110	2 110	2 110	2 110	2 110	1 120	1 120	1 130	1 150	210
240	1 240	0 —	0 —	0 —	1 120	1 110	1 100	2 100	2 100	2 100	2 100	1 100	1 110	1 110	1 120	240
270	1 270	0 —	0 —	0 —	1 090	1 090	1 090	2 090	2 090	2 090	2 090	1 090	1 090	1 090	1 090	270
300	1 310	1 320	0 —	1 040	1 060	1 070	1 070	2 080	2 080	2 080	2 080	1 070	1 070	1 070	1 060	300
330	1 340	1 000	1 020	1 040	1 050	1 060	2 070	2 070	2 070	2 070	1 060	1 060	1 050	1 040	1 030	330
360	1 010	1 030	1 040	1 050	1 060	1 060	2 070	2 070	2 070	1 060	1 060	1 050	1 040	1 020	1 000	360

Figure 6-10. 1982 corrections for precession and nutation, Table 5, Epoch 1980 edition of Volume I Tables No. 249.

If Volume I is exclusively used for all LOPs of a celestial fix, any required precession and nutation correction should be applied to the fix position itself rather than the individual LOPs.

The Use of *Tables No. 249* as a Starfinder

As was mentioned earlier in this chapter, Volume I of *Tables No. 249* may be used as a *starfinder* for precalculation of the approximate sextant altitudes and true azimuths of the optimum seven navigational stars for any particular location and time. To use Volume I in this manner, an LHA♈ and assumed latitude are chosen based on the projected DR track and predicted time of observation. The tables are then opened to the appropriate latitude page, and the names and coordinates of the stars tabulated for the precomputed LHA♈ are extracted and recorded. A sample list of seven stars preselected for observation when the assumed latitude is 34° south and the LHA♈ is 145° (see Figure 6-6A, page 133) appears below.

	Regulus	Spica	Acrux	Canopus	Rigel	Betelgeuse	Procyon
Hc's	43.5°	34.5°	51.5°	50.5°	24.0°	22.5°	41°
Zn's	009°	080°	151°	226°	276°	296°	318°

Chapter 8 discusses a more sophisticated starfinder called the Rude Starfinder, by means of which the coordinates of all possible navigational stars and planets observable from a given projected position may be determined. Volume I of *Tables No. 249*, however, is very convenient to use in this mode if a simple 3 or 4-star fix is all that is desired.

Summary

This chapter has examined the altitude-intercept method of the complete solution for one or more celestial LOPs using the *Air Almanac* in conjunction with *Sight Reduction Tables No. 249*. The relative ease of the sight solution using Volume I of *Tables No. 249* in comparison to Volumes II and III and *Tables No. 229* was stressed, and the secondary use of Volume I as a starfinder was discussed.

It should be emphasized that the convenience of the *Air Almanac* and *Tables No. 249* is gained at the expense of some precision in the resulting celestial LOP. The precision lost is roughly comparable to the observation errors experienced on a medium-size ship in rough weather. Until he has gained some proficiency with the sextant, the inexperienced navigator should take advantage of the greater precision afforded by the *Nautical Almanac* used in combination with *Tables No. 229* to offset any possible observation errors. After he has gained experience, the navigator can then switch to the less precise though more rapid *Air Almanac* and *Tables No. 249* in situations that do not demand high precision.

7

The Celestial Fix and Running Fix

The preceding chapters have dealt with the solution of the navigational triangle to obtain a celestial LOP by the altitude-intercept method, by use of either the *Nautical Almanac* or *Air Almanac* in conjunction with either the *Sight Reduction Tables for Marine Navigation, No. 229,* or *Air Navigation, No. 249.* In this chapter, the combination of several such LOPs to achieve the ultimate objective of celestial navigation—the celestial fix or running fix—will be addressed.

In celestial navigation, as in piloting, it should be emphasized that neatness and proper labeling standards are essential to accuracy in the navigation plot. It has been shown that the assumed position is based on the DR position at the time of the observation. If the DR position is greatly in error because of inaccuracies in the DR plot, the assumed position chosen may be so far distant from the ship's actual position as to cause the intercept distance to be excessive. This in turn can lead to an error of sizable proportions in the plotted celestial LOP, unless the assumed position is relocated to reduce the intercept distance. In circumstances requiring the ship's position to be speedily determined, the delay thus caused could be unacceptable; hence the importance of an accurate DR plot for celestial navigation.

The Celestial Fix

In Chapter 2 the altitude-intercept method of plotting a single celestial LOP given the assumed position, intercept distance, and true azimuth was demonstrated (see Figure 2-5). It was stated that two or preferably three or more simultaneous celestial LOPs could be crossed to determine the celestial fix position of the observer at the time of the observation.

In practice, sextant observations of the altitudes of celestial bodies are normally not made simultaneously, but rather over an interval of several minutes. Occasionally, as much as a half hour may elapse between the first and last observations during a round of sextant observations, especially when unfavorable weather conditions prevail.

Because the fix is formed by the intersection of two or more *simultaneous* LOPs, it is usually necessary to adjust the positions of all but one of the celestial LOPs resulting from a round of observations, to account for the moving ship's progress during the interval in which the observations were made.

The technique of adjusting the positions of the various LOPs to form a *celestial fix* is similar to that used during piloting to plot a running fix, with one important difference. Whereas in piloting an earlier LOP is always advanced to the time of a later LOP to form the running fix, in celestial navigation LOPs can either be *advanced* or *retired* up to 30 minutes to form the celestial fix. If two celestial LOPs were plotted 20 minutes apart, for example, the first LOP could be advanced 20 minutes to be crossed with the second LOP, or the second LOP could be retired 20 minutes to be crossed with the first LOP. Since the accuracy of a celestial fix is usually considered acceptable if it is within 1 to 2 miles from the true position, advancing or retiring a celestial LOP for up to 30 minutes in this manner does not unduly affect the accuracy of the resultant fix. Advancement of a celestial LOP for periods in excess of a half hour results in a *celestial running fix*, which will be discussed later in this chapter.

Although strictly speaking only two intersecting simultaneous celestial LOPs are required for a celestial fix, in the normal practice of celestial navigation as in piloting at least three or more LOPs having a good spread in azimuth are usually desired. The additional LOPs act as a check on the accuracy of the resulting position.

Plotting the Celestial Fix

The choice of whether to advance or retire celestial LOPs to form the celestial fix mainly depends on the preference of the individual navigator. It should be noted, however, that advancing the LOPs results in the advantage of a later time of fix than would be the case were subsequent LOPs retired to the time of an earlier LOP. For this reason, and also because it has been found that most students grasp the concept of advancing an LOP more readily than retiring an LOP, the former technique will be the one presented herein.

While it would be possible to first plot all the celestial LOPs obtained during a round of observations and then advance certain of them to form the celestial fix, in practice only the APs of the LOPs to be advanced are plotted and advanced. This enhances the neatness of the plot by eliminating unnecessary lines. After the APs have been advanced as necessary, the corresponding LOPs are then plotted relative to them in the usual manner.

As an example of the plot of a celestial fix, consider the thr observations used as examples in Chapters 5 and 6. The comple solutions by *Tables No. 229* for each of these bodies—the star Canopu the planet Venus, and the moon—are shown on a single sight reducti form as they would appear in practice in Figure 7-1 opposite.

Since the moon and Venus observations were made first, at zo times 0359 and 0404 respectively (rounded to the nearest minute), the APs will be advanced to the time of the Canopus observation, 041 For purposes of plotting the celestial fix, the times of observation a always rounded to the nearest minute in this manner, with no lo of accuracy resulting. The assumed position for the moon, because was the first body observed, is designated AP_1; it must be advance 13 minutes from 0359 to 0412. The assumed position for Venus, AP must be advanced 8 minutes from 0404 to 0412, while the assume position for Canopus, AP_3, need not be advanced at all. The mo convenient method of advancing AP_1 and AP_2 is to use the D plot constructed earlier to find the DR positions of the ship at th various times of observation, for use during the sight reduction proces It will be assumed here that such a plot has been constructed on properly labeled SAPS-35, oriented for southern latitudes as shown Figure 7-2 on page 146.

The first step in the plot of the celestial fix on the SAPS-35 is plot the initial positions of all APs based on the assumed latitude ar longitude entered on the sight reduction form. Next, the distance ar direction between the 0359 and 0412 DR positions are picked off ar used to advance AP_1. Similarly, the 0404 and 0412 DR positions a used as a reference for advancing AP_2. The advanced APs are show in Figure 7-2.

After the appropriate APs have been advanced, the associate celestial LOPs are then plotted in the normal manner by laying o the intercept distances along the true azimuths either toward or awa from the APs as required. Each LOP is labeled with the name of th body upon which it is based, and a one-eighth-inch diameter circ is drawn over the intersection of the three LOPs to indicate the f position. Since all LOPs can now be considered simultaneous, on the fix symbol is labeled with the time of the fix. To complete th plot, a new DR course and speed line is originated from the fix an properly labeled with the ordered course and speed. The complete plot is shown in Figure 7-3 on page 147.

After the navigator has become proficient in the techniques o celestial navigation, he will often plot only a single DR positio corresponding to the time of one of the observations obtained durin a given round of sights, and he will use this DR position to determin his assumed position for all sights reduced on that occasion. In th

Sight Reduction
using Tables 229

C: 230°

S: 25 kts

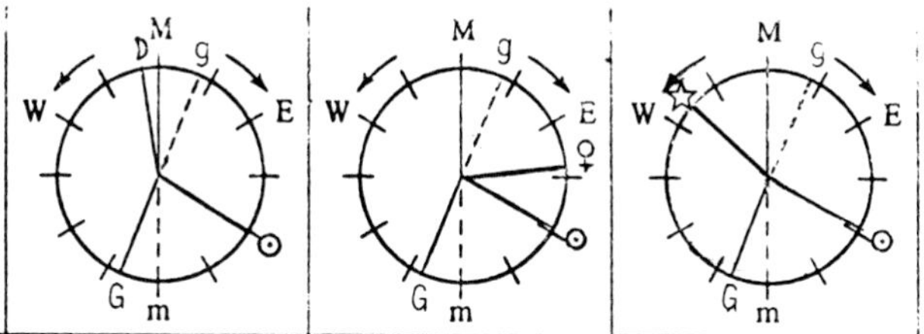

Body		MOON (UL)		VENUS		CANOPUS	
IC		+.3	-	+.3	-	+.3	-
Dip (Ht 44 ')		-6.4		-6.4		-6.4	
Sum		-6.1		-6.1		-6.1	
hs		28-09.6		16-47.4		50-46.3	
ha		28-03.5		16-41.3		50-40.2	
Alt. Corr		59.7		-3.2		-.8	
Add'l.				+.5			
H.P. (55.4)		2.2	-30				
Corr. to ha		31.9		-2.7		-.8	
Ho (Obs Alt)		28-35.4		16-38.6		50-39.4	
Date		16 DEC		16 DEC		16 DEC	
DR Lat		34-15.5 S		34-17.0 S		34-19.0 S	
DR Long		163-11.7 E		163-09.1 E		163-05.7 E	
Obs. Time		03-58-21		04-03-36		04-11-33	
WE (S+, F-)		36 (s)		36 (s)		36 (s)	
ZT		03-58-57		04-04-12		04-12-09	
ZD (W+, E-)		-11		-11		-11	
GMT		16-58-57		17-04-12		17-12-09	
Date (GMT)		15 DEC		15 DEC		15 DEC	
Tab GHA	v	205-10.3	10.9	116-48.5	1.1	338-56.1	
GHA incr'mt.		14-04.0		1-03.0		3-02.7	
SHA or v Corr.		10.6		0.1		264-09.8	
GHA		219-24.9		117-51.6		606-08.6	
±360 if needed						246-08.6	
aλ (-W, +E)		163-35.1 E		163-08.4 E		162-51.4 E	
LHA		23		281		49	
Tab Dec	d	N 23-21.1	-8.3	S 13-16.1	0.1		
d Corr (+ or -)		-8.1		0.0			
True Dec		N 23-13.0		S 13-16.1		S 52-40.6	
a Lat (N or S)		34 S Same (Cont.)		34 S (Same) Cont.		34 S (Same) Cont.	
Dec Inc	(±)d	13.0	-55.8	16.1	31.8	40.6	-8.7
Hc (Tab. Alt.)		28-56.7		16-15.3		50-51.1	
tens	DS Diff.	-10.8		8.0		0	
units	DS Corr.	-1.3	+	.5	+	-5.9	+
Tot. Corr. (+ or -)		-12.1		8.5		-5.9	
Hc (Comp. Alt.)		28-44.6		16-23.8		50-45.2	
Ho (Obs. Alt.)		28-35.4		16-38.6		50-39.4	
a (Intercept)		9.2 A		14.8 T		5.8 A	
Z		S 155.8 W		S 084.8 E		S 46.3 W	
Zn (°T)		335.8°T		095.2°T		226.3°T	

Figure 7-1. Complete solutions for the moon, Venus, and Canopus by Tables No. 229.

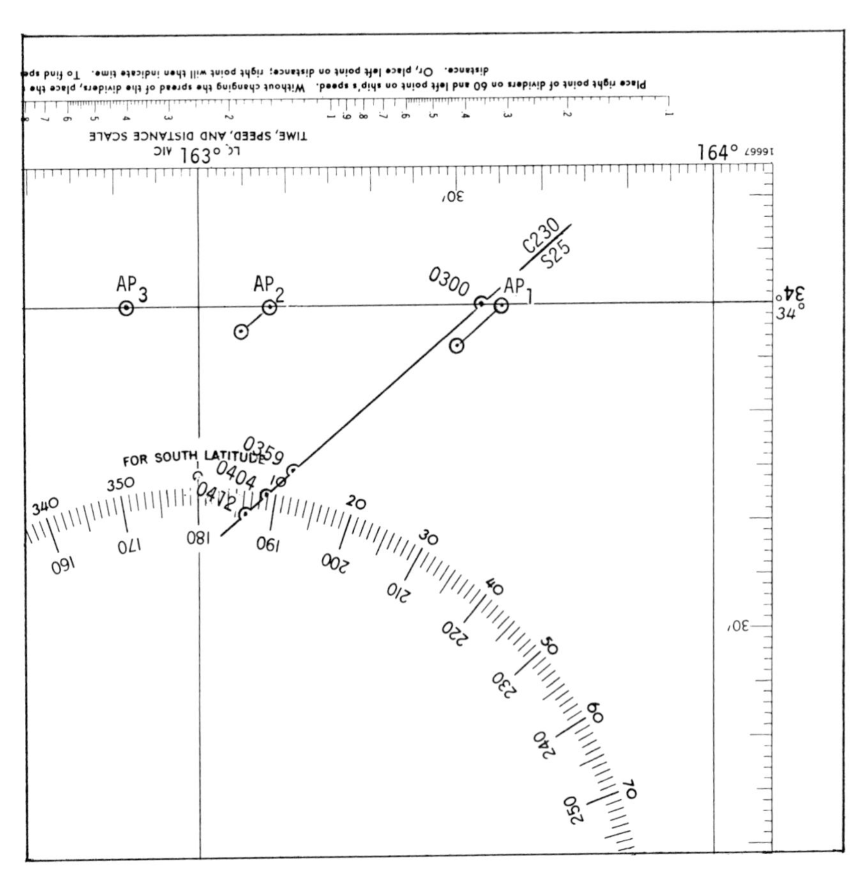

Figure 7-2. Advancing AP_1 and AP_2 using the DR plot.

situation the DR positions corresponding to the time of each sight are not plotted to use as a basis for advancing or retiring the appropriate APs. Instead, the navigator simply makes use of the ship's deck logbook to obtain the ship's ordered course and speed during the observation period, and he uses these as a basis for adjusting the appropriate APs.

If a triangle results rather than a point fix when three celestial LOPs are plotted, or if a polygon is formed by four or more LOPs, the fix position can be assumed to lie approximately at the center of the triangle or polygon if the LOPs have been obtained from celestial bodies well distributed in azimuth. On the other hand, if for some reason such a triangle or polygon is formed by LOPs obtained from bodies all within 180° of azimuth, as might occur when observations are made under cloudy or overcast skies, the fix position may lie *outside* the area enclosed by the LOPs.

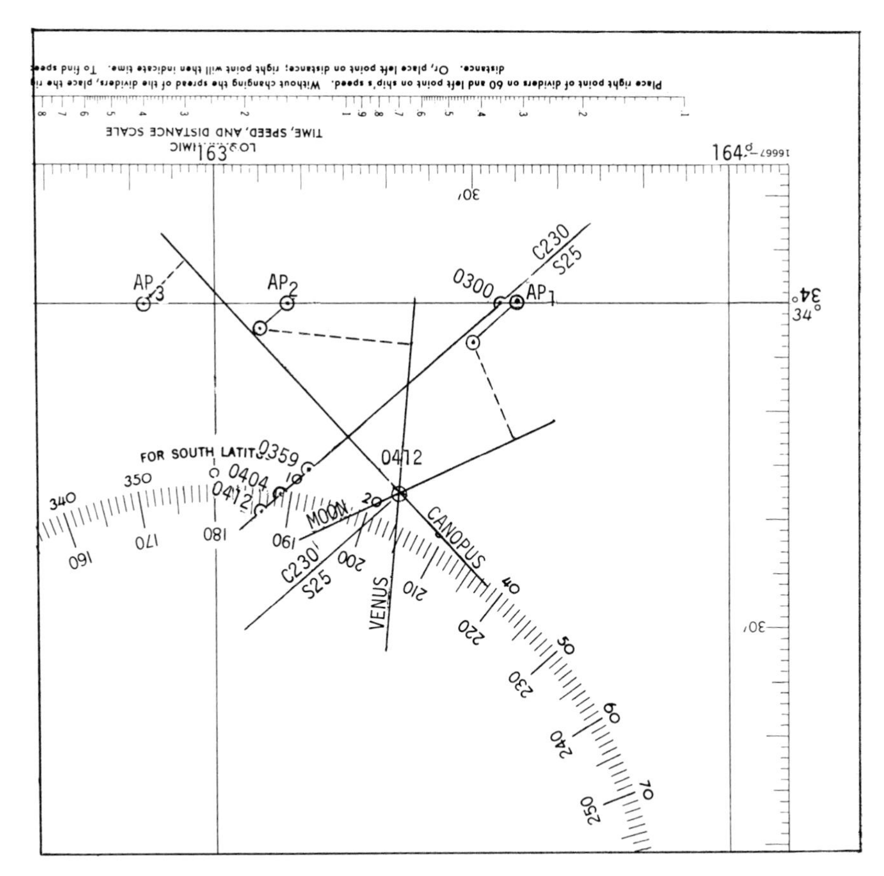

Figure 7-3. The completed celestial fix.

In situations in which three or more observations are made of bodies within 180° of azimuth, it is wise to use so-called *LOP bisectors* to determine the fix. In this technique, each angle formed by a pair of LOPs is bisected, with the bisector drawn from the vertex of the angle in the direction of the mean of the azimuths of the two bodies. As an example, consider Figure 7-4 on the following page, in which LOPs corresponding to three celestial bodies having true azimuths of 224°, 260°, and 340° have been plotted. The bisectors of each of the pairs of lines are computed below:

$$\frac{(224^\circ + 260^\circ)}{2} = 242^\circ$$

$$\frac{(260^\circ + 340^\circ)}{2} = 300^\circ$$

$$\frac{(224^\circ + 340^\circ)}{2} = 282^\circ$$

They appear as blue lines in Figure 7-4. The most probable position for the fix lies at the intersection of the three bisectors rather than within the rather large triangle formed by the LOPs themselves.

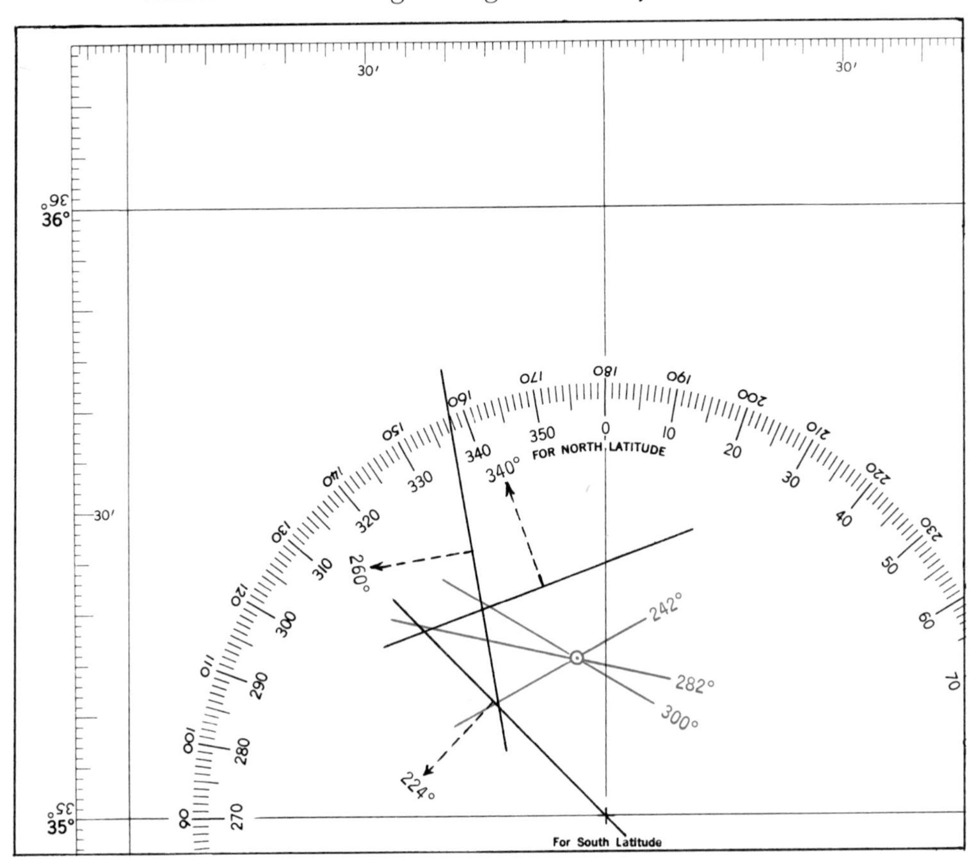

Figure 7-4. Finding an exterior fix position using bisectors.

If a triangle similar to that appearing in Figure 7-4 results because of a constant observation error, as for example an error caused by a hazy horizon, a point fix may be obtained by adding or subtracting a uniform correction determined by trial and error to all intercept distances.

The Estimated Position by Celestial Navigation

If an appreciable time has elapsed since the determination of the last fix of the ship's position at sea, the cumulative error inherent in the DR plot may build up to the point where the ship's actual position is a good distance away from her DR position for a given time. Under these circumstances, if a single celestial LOP can be obtained and plotted, an *estimated position* (*EP*) can be determined based on the

LOP. If the LOP is accurate, the ship must have been located somewhere on it at the moment of observation. In the absence of a second intersecting LOP, an estimate of the ship's position on the line can be made by dropping a perpendicular construction line to the single LOP from the ship's DR position at the time the LOP was obtained. The estimated position thus formed represents the closest point on the LOP to the ship's DR position for that time. Such an estimated position based on the sun line completed as an example in Chapter 5 (see Figure 5-13) is illustrated in Figure 7-5. It is labeled with a one-eighth-inch square and the time as shown.

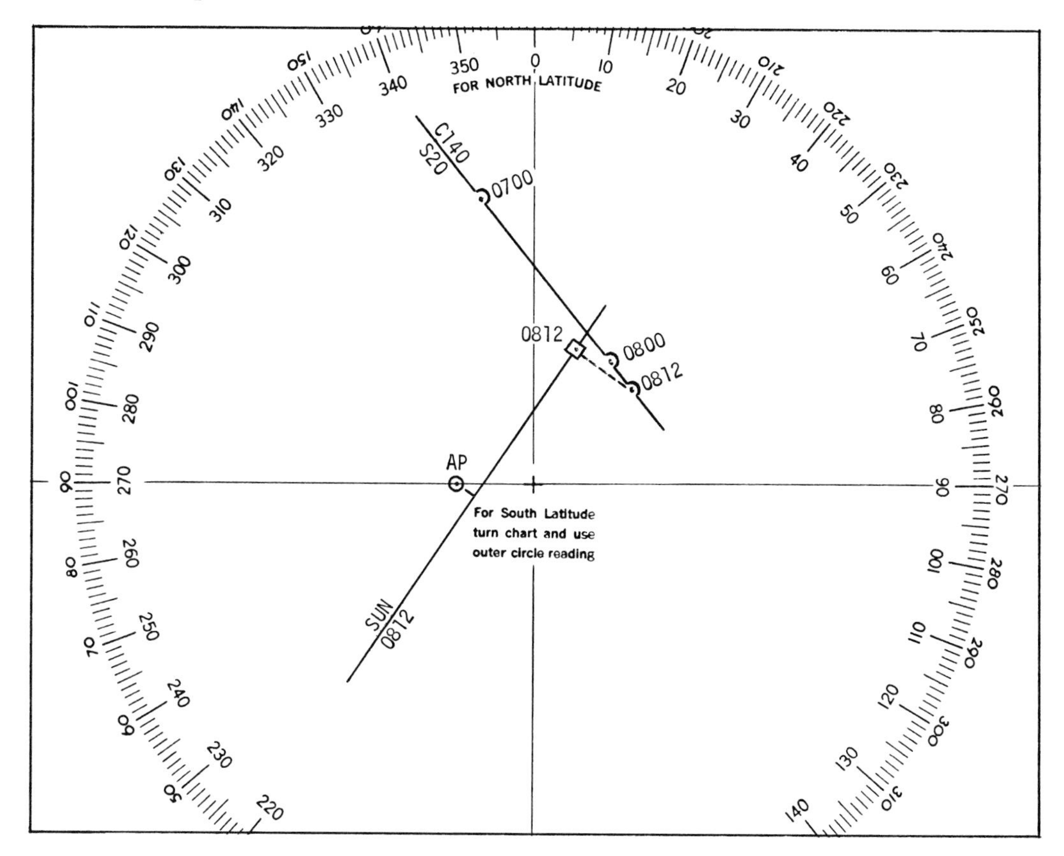

Figure 7-5. An estimated position based on a single celestial LOP.

After the EP has been plotted, if the ship is well out to sea and the navigator places a high degree of confidence in the single LOP, he might elect to project a DR track from it, so as to obtain more meaningful estimates of the ship's position until either a second LOP is obtained to form a running fix, or two or more simultaneous LOPs are obtained to fix the ship's position. The standard procedure, of course, would be to continue the original DR until either a fix or running fix was obtained.

The Celestial Running Fix

The procedure for plotting a running fix involving a celestial line of position at sea is virtually identical to the procedure used during piloting. A celestial LOP may be advanced for any reasonable time interval to be crossed with a subsequent LOP derived from any source, or any earlier LOP may be advanced and crossed with a later celestial LOP to determine a running fix. No arbitrary time limit exists for the advancement of an LOP at sea, because of the less stringent positioning requirements in this environment as opposed to piloting waters. The only criterion that should be applied is the angle of intersection of the advanced and subsequent LOPs.

If a celestial LOP is to be advanced to form a running fix, it is the usual practice to advance the line itself rather than its AP, as is done in plotting a celestial fix. As in piloting, the distance and direction through which the LOP is advanced is based on the DR plot. It is the normal procedure at sea to obtain running fixes using sun LOPs every few hours if possible during the course of the day. The noon position at sea, for example, is normally obtained by advancing a morning sun line to cross the LAN latitude line; a mid-afternoon running fix is normally plotted by advancing the LAN line to cross an afternoon sun line.

As an example of a celestial running fix produced from two celestial LOPs, consider the two sun sight solutions shown in Figure 7-6. The 0812 zone time sight was presented as an example earlier, and the 1100 zone time sight was obtained later in the morning of the same day.

The plot of the first sun line, and an EP based on it, is shown in Figure 7-5. The DR track was continued from the 0812 DR to the time of the second LOP, 1100. After the 1100 sun line is plotted, the SAPS-31 appears as shown in Figure 7-7A on page 152; note the course and speed change at 0900.

To plot the celestial running fix, the 0812 sun line is advanced to the time of the 1100 sun line by reference to the DR plot. For clarity in this example, a construction line has been drawn between the 0812 and 1100 DR positions, showing the distance and direction through which the 0812 line was advanced. The advanced line is so labeled, and the running fix position is symbolized by a one-eighth-inch circle placed over the intersection of the two sun lines. After the symbol is labeled with the time, a new DR course and speed line is originated from the running fix position and labeled. The completed plot appears in Figure 7-7B on page 153.

Sight Reduction using Tables 229 C: S:	M W E G m	M W E G m
Body	SUN (LL)	SUN (LL)
IC	+ -.3	+ -.3
Dip (Ht 44 ')	-6.4	-6.4
Sum	-6.7	-6.7
hs	10-12.4	35-13.1
ha	10-05.7	35-06.4
Alt. Corr	+11.1	+14.9
Add'l.		
H.P. ( )		
Corr. to ha	+11.1	+14.9
Ho (Obs Alt)	10-16.8	35-21.3
Date	16 DEC	16 DEC
DR Lat	31-08.1 N	30-25.0 N
DR Long	64-48.0 W	64-35.8 W
Obs. Time	08-12-06	11-00-00
WE (S+, F-)	0	0
ZT	08-12-06	11-00-00
ZD (W+, E-)	+4	+4
GMT	12-12-06	15-00-00
Date (GMT)	16 DEC	16 DEC
Tab GHA / v	1-07.0	46-06.1
GHA incr'mt.	3-01.5	
SHA or v Corr.		
GHA	4-08.5	46-06.1
±360 if needed	364-08.5	406-06.1
aλ (-W, +E)	65-08.5 W	64-06.1
LHA	299	342
Tab Dec / d	S 23-18.9 / 0.1	S 23-19.2 / 0.1
d Corr (+ or -)	0.0	0.0
True Dec	S 23-18.9	S 23-19.2
a Lat (N or S)	31 N Same (Cont.)	30 N Same (Cont.)
Dec Inc / (±)d	18.9 / -38.9	19.2 / -56.8
Hc (Tab. Alt.)	10-26.7	34-15.0
tens / DS Diff.	-9.5	-16.0
units / DS Corr.	-2.7 / +	-2.2 / +
Tot. Corr. (+ or -)	-12.2	-18.2
Hc (Comp. Alt.)	10-14.5	34-56.8
Ho (Obs. Alt.)	10-16.8	35-21.3
a (Intercept)	2.3 T	34.5 A
Z	N 125.2 E	N 160 E
Zn (°T)	125.2°T	160°T

Figure 7-6. Complete solutions for two sun lines by Tables No. 229.

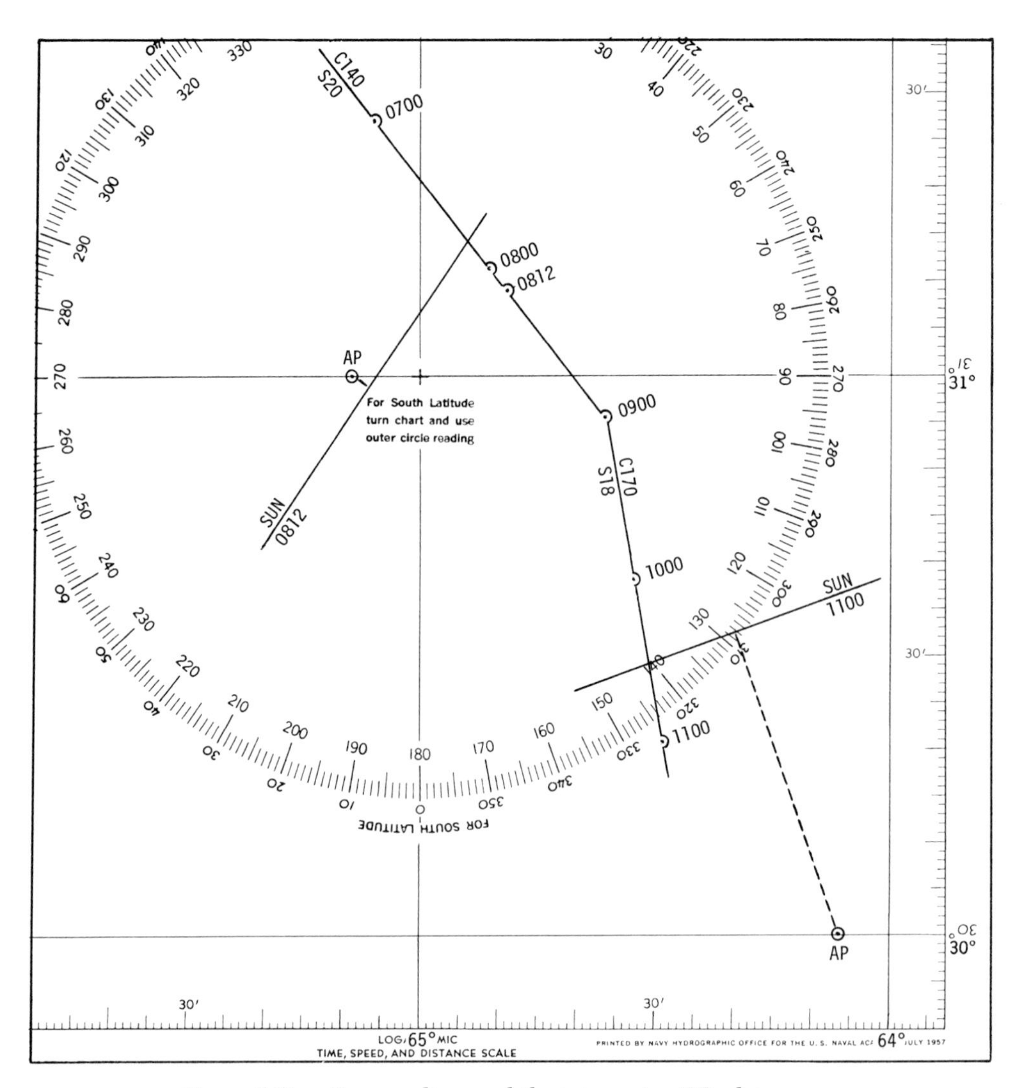

Figure 7-7A. Two sun lines and the intervening DR plot.

As an alternative to continuing the DR plot from the 0812 DR position, the navigator might have chosen to originate a new DR plot from the 0812 EP, especially if some time had elapsed since the last good fix. In this event, the 0812 sun line is advanced to the time of the subsequent 1100 sun line by reference to the distance between the 0812 EP and the 1100 DR. The resulting running fix is shown plotted in Figure 7-7C (page 154); the fix position so obtained is identical to that produced by the conventional method described above.

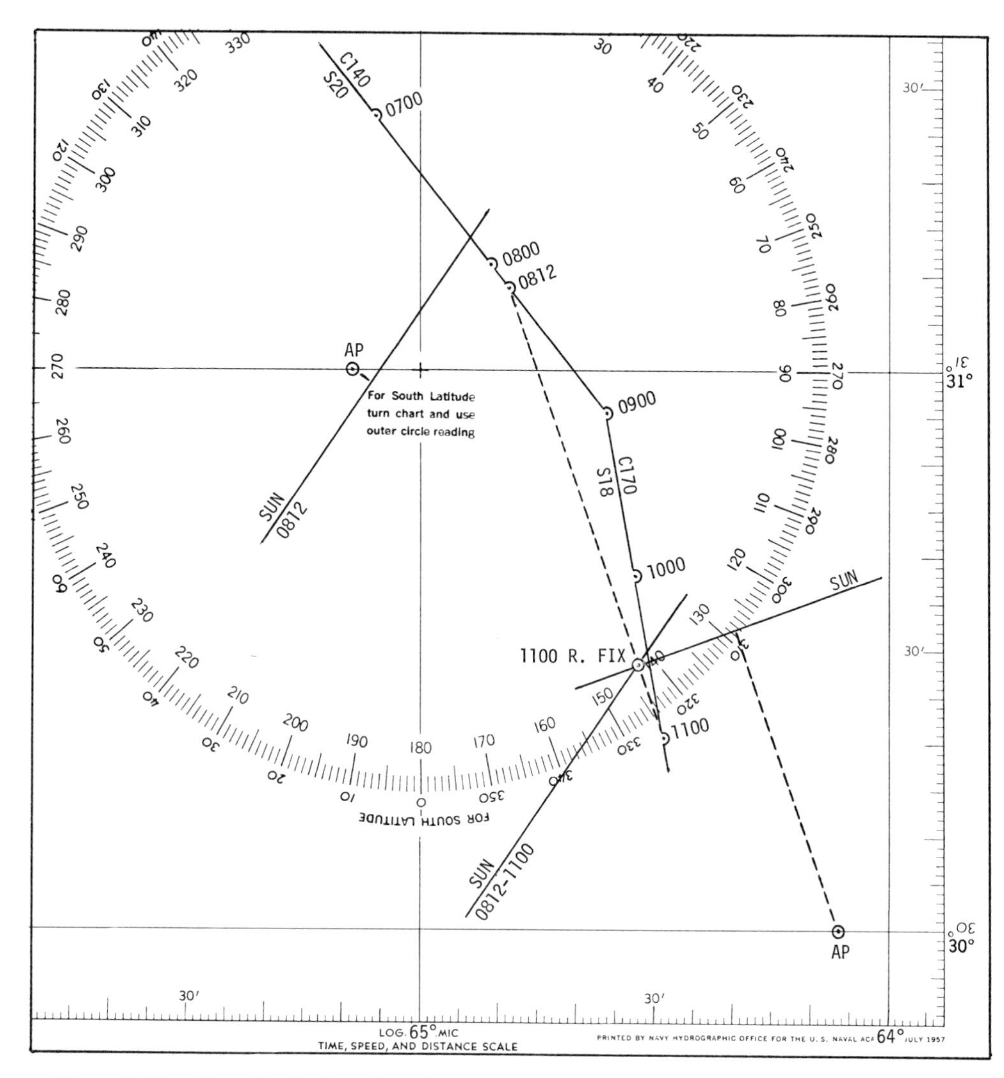

Figure 7-7B. Completed plot of a celestial running fix.

Summary

In this chapter, the procedures for plotting the celestial fix and running fix have been presented, together with a discussion of the method used to plot an estimated position at sea based on an observation of a single celestial body. Such a single LOP can be combined with an earlier or later LOP obtained from any other source to form a running fix, with the only restrictions being that the earlier LOP must not be advanced over an excessive time interval, and the two LOPs should form an acceptable angle with one another. The naviga-

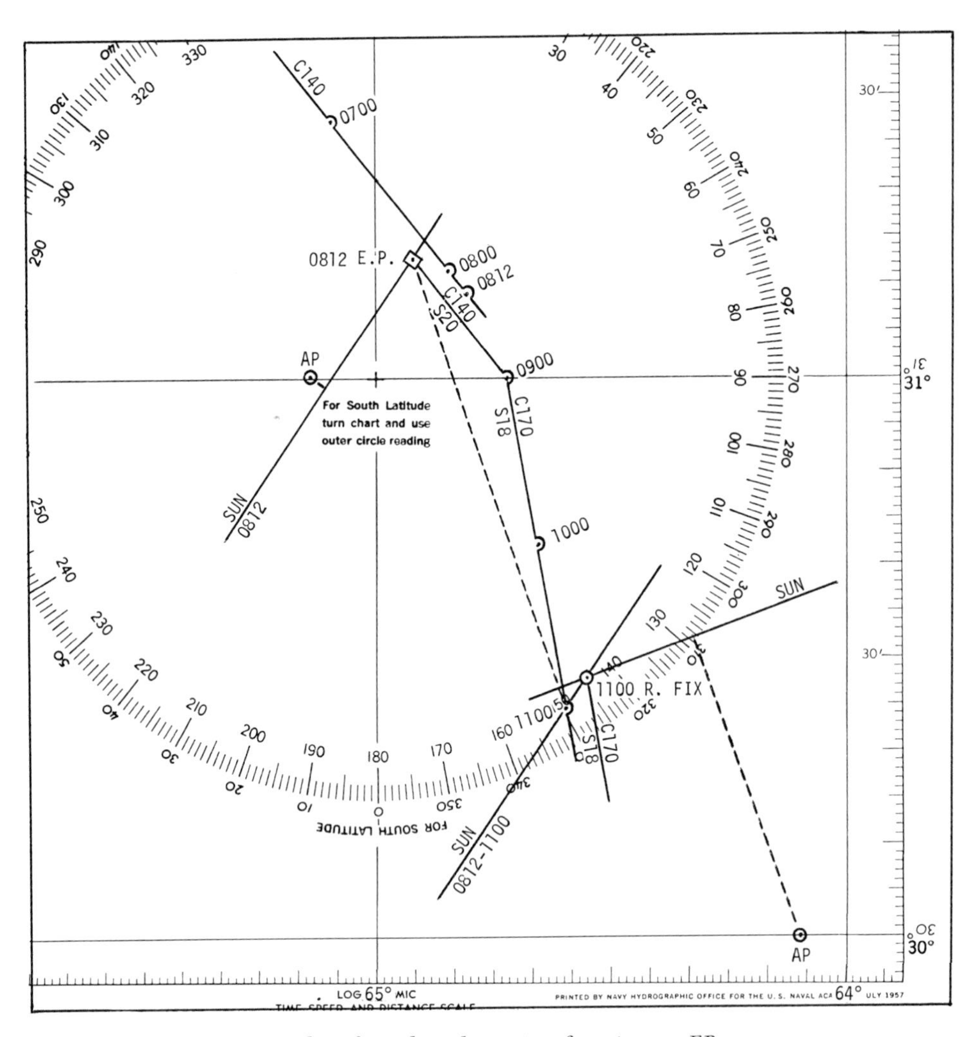

Figure 7-7C. Plot of a celestial running fix using an EP.

tor should never hesitate to combine a celestial LOP with an LOP determined from another source to obtain positioning information.

This chapter concludes the presentation of the determination of the celestial fix by the altitude-intercept method. The succeeding chapters on celestial navigation will each deal with special case solutions of the navigational or the celestial triangle for some specific purpose auxiliary to finding the ship's position. The interplay of celestial position-finding with electronically determined fixes will be presented in Chapter 16, which describes a typical day's work in navigation at sea.

8

The Rude Starfinder

To solve the navigational triangle for a computed altitude and true azimuth, the navigator must either know beforehand or be able to determine afterwards the name of the celestial body he has observed, so that he can obtain its GHA and declination from either the *Nautical Almanac* or the *Air Almanac*. Several aids are available to the navigator to assist him in identifying and locating celestial bodies, among which are star charts, sky diagrams, and starfinders. *Star charts* are representations resembling photographs of the night sky at certain times of the year; *sky diagrams* are drawings of the heavens as they would be seen from certain locations at various times; and *starfinders* are devices intended to furnish the approximate altitude and true azimuth of celestial bodies either before or after navigational observations. There are several different manuals, textbooks, and almanacs available that contain excellent star charts; among these are *Dutton's Navigation and Piloting*, the *American Practical Navigator* (Bowditch), and the *Nautical Almanac*. Each *Air Almanac* contains a set of sky diagrams for various latitudes for the fifteenth day of each of the six months it covers. The use of Volume I of *No. 249* as a starfinder has already been discussed in Chapter 6.

In this chapter, perhaps the most commonly used of all mechanical starfinders, the *Rude Starfinder*, will be examined. This starfinder was originally patented by Captain G. T. Rude, USC&GS, some years ago and was subsequently sold to the old U.S. Navy Hydrographic Office, which improved the device and made it available under the designation 2102-D. Most practicing navigators, however, continue to refer to it as the Rude Starfinder. Currently, it is issued on request to Navy and other government users by the Ship's Parts Control Center, Mechanicsburg, Pennsylvania, and it is produced for sale to commercial and private users by a number of marine supply firms.

Description of the Rude Starfinder

The Rude Starfinder, 2102-D, is designed to permit the determination of the approximate apparent altitude and azimuth of any of the 57 selected navigational stars tabulated in the *Nautical* and *Air*

Almanacs that appear above the observer's celestial horizon at any given place and time. With some minor manipulations it can also be set up to obtain the positions of the navigational planets, any unlisted stars of interest, and even the sun or moon if desired. The device can also be used to identify an unknown body, given its altitude and true azimuth. The accuracy of the starfinder is generally considered to be about ±3 to 5 degrees in both altitude and true azimuth determinations.

Essentially, the Rude Starfinder consists of the *star base*, an opaque white plastic circular base plate fitted with a peg in the center, and ten circular transparent *templates*.

On one side of the star base the north celestial pole appears at the center, and on the other side, the south celestial pole. All of the 57 selected stars are shown on each side at their positions relative to the appropriate pole in a type of projection called an azimuthal equidistant projection. In this projection, the positions of the stars relative to one another are distorted, but their true declinations and azimuths relative to the pole are correct; hence, the pattern of the stars on the star base does not correspond to their apparent positions as seen in the sky. Each star on the base is labeled, and its magnitude is indicated by its symbol—a large heavy ring indicates first magnitude, an intermediate-sized ring second magnitude, and a small thin ring third magnitude. The celestial equator appears as a solid circle about four inches in diameter on each side of the star base, and the periphery of each side is graduated to a half-degree of LHA♈.

There are 10 templates included for use with the star base. Nine of these are printed with blue ink and are designed for apparent altitude and azimuth determinations, while the tenth, printed in red ink, is intended for the plotting of bodies other than the 57 selected stars on the base plate. There is one blue template for every 10° of latitude between 5° and 85°; one side of each template is for use in north latitudes, the other for south latitudes. Each of these "latitude" templates is printed with a set of oval blue altitude curves at 5° intervals, with the outermost curve representing the observer's celestial horizon, and a second set of radial azimuth curves, also at 5° intervals. The red template is printed with a set of concentric declination circles, one for each 10° of declination, and a set of radial meridian angle lines. In use, the appropriate template is snapped in place on the star base like a record on a phonograph turntable.

Figure 8-1 shows the south celestial pole side of the star base with the red template affixed and the nine blue templates arrayed below.

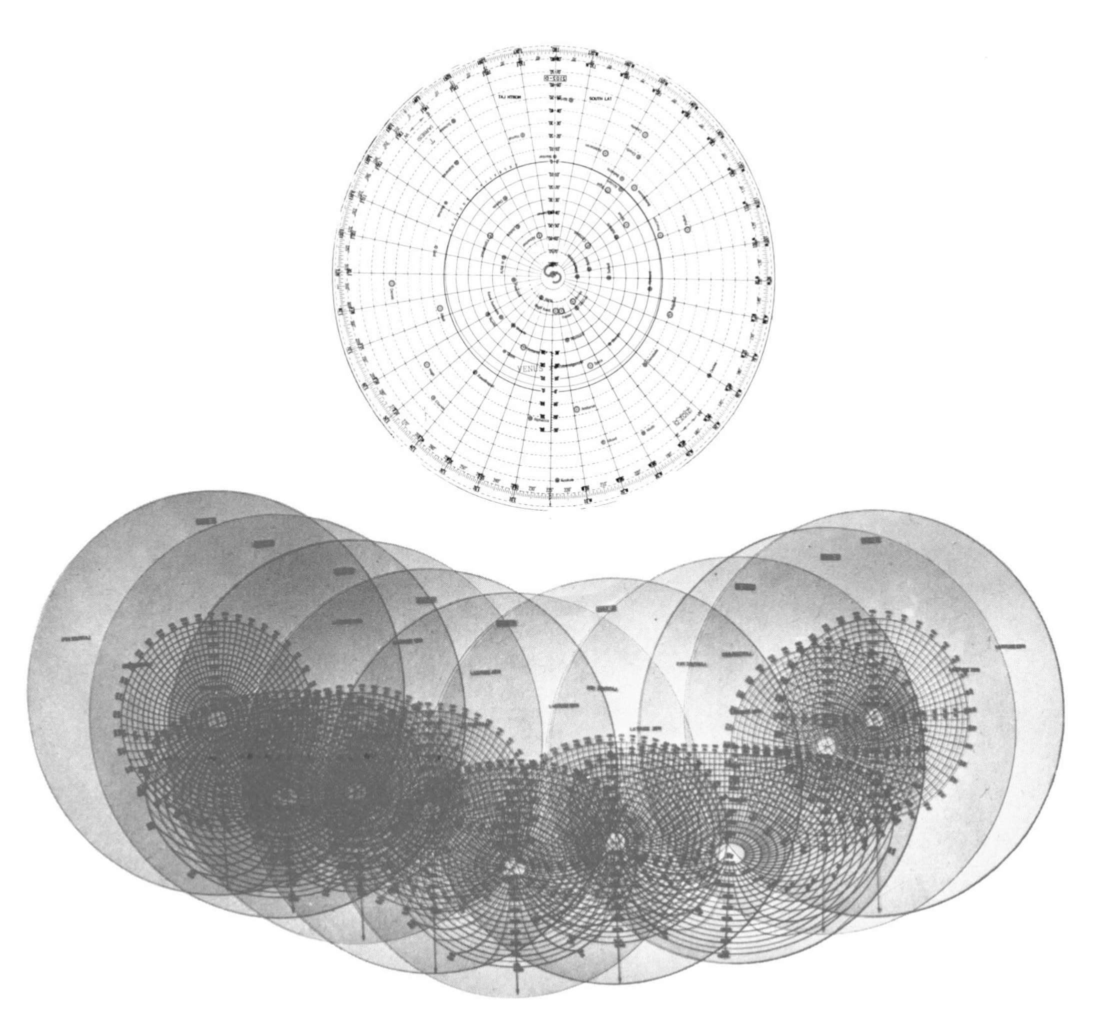

Figure 8-1. The Rude Starfinder, 2102-D.

Use of the Rude Starfinder for Observations of the 57 Selected Stars

Although it can be used for other bodies, the Rude Starfinder is most conveniently suited for determining which of the 57 selected navigational stars will be favorably situated for observation at a given DR position and time. To use the device for this purpose, the LHA♈ must first be determined for the midtime of the period during which the observations are scheduled to be made. In practice, the beginning of civil twilight is generally used for morning sights, and the ending of civil twilight is used for evening sights; both the definition and technique of prediction of times for these phenomena will be discussed in Chapter 11. After the LHA♈ for the appropriate time and projected DR position has been computed, the blue template for the latitude closest to the DR latitude is selected and placed over the proper side of the star base plate. The name of the side of the star base used, i.e., N or S, and the name of the template latitude should always be identical. Should the DR position lie exactly between two template latitudes, such as L. 10° 00.0′, either template may be selected.

After the appropriate template has been placed on the proper side of the base plate, the template is rotated until the index arrow (the 0°/180° azimuth line) points to the proper value of LHA♈ on the rim of the star base. The apparent altitude ha and true azimuth Zn of all stars of interest located within the perimeter of the blue grid (the outer curve of which represents the observer's celestial horizon) are then recorded to the nearest whole degree. Normally the stars are recorded in the order of increasing true azimuth.

As an example, suppose that the navigator wished to use the Rude Starfinder to record the name, apparent altitude, and true azimuth of all first magnitude stars with apparent altitudes greater than 10°, suitable for observation from the same position for which the use of Volume I of *Tables No. 249* as a starfinder was demonstrated in Chapter 6. The assumed latitude for this position was 34° south, and the LHA♈ was 145°.

Since the latitude is 34° south, the south pole side of the star base is used together with the blue template for latitude 35°S. After aligning the 0°/180° azimuth arrow with a LHA♈ of 145°, the device should appear as in Figure 8-2.

The following is a list of the ha's and Zn's of all first magnitude stars as read from the Starfinder in Figure 8-2 to the nearest whole degree:

Star	ha	Zn
*Regulus	42°	008°
*Spica	34°	079°
Hadar	42°	142°
Rigel Kent	38°	143°
*Acrux	53°	150°
Achernar	15°	208°
*Canopus	52°	227°
*Rigel	24°	277°
Sirius	47°	283°
*Betelgeuse	22°	297°
*Procyon	40°	318°
Pollux	21°	332°

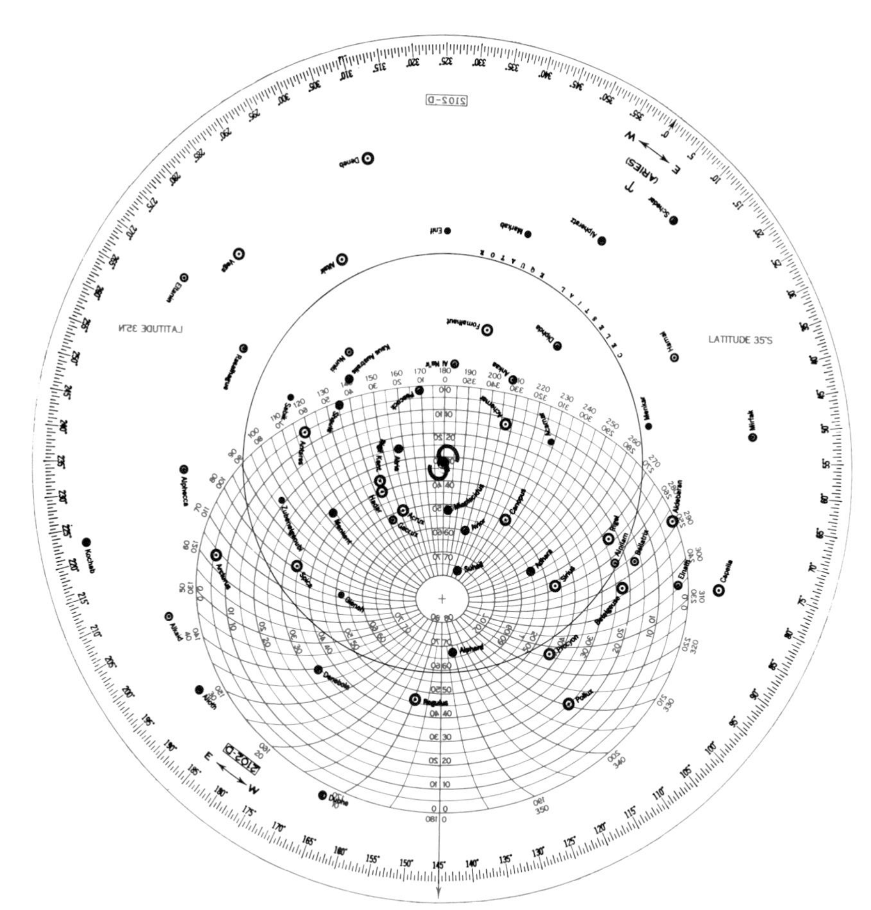

Figure 8-2. Rude Starfinder set up for LHA♈ 145°, aL. 34° S.

Those stars preceded by an asterisk in the preceding list appear also on the list obtained by the use of Volume I of *Tables No. 249* shown on page 137. As can be seen by a quick comparison of the two lists, the data agree quite closely.

In practice, the navigator should always list more stars than he actually expects to observe, as some may be obscured by clouds. Moreover, the stars listed should not be confined to those of the first magnitude; all the stars shown on the Starfinder are readily visible in clear weather. The most convenient band of apparent altitude for observation purposes lies roughly between 15° and 60°, but it is sometimes preferable to list stars lower or higher than those limits, rather than to have poor distribution in azimuth.

Use of the Starfinder for the Navigational Planets

The Rude Starfinder, 2102-D, may be used in a similar manner to find the apparent altitude and true azimuth of the navigational planets that will appear above the celestial horizon at a given position and time. Inasmuch as the planets are continually changing position relative to the "fixed" stars, their positions cannot be permanently printed on the star base along with the 57 selected stars; instead, they must be periodically plotted on the star base by means of the red template. In practice most navigators plot the positions of the planets on the star base about every 10 days to two weeks.

The first step in the plotting of a planet on the star base is to determine the value of an expression equivalent to the right ascension of the planet, expressed in terms of angular measurement. It might be recalled from Chapter 1 that right ascension is defined as the angular distance of the hour circle of a celestial body measured eastward along the celestial equator from the hour circle of Aries; this measurement expressed in time units is employed in the science of astronomy to locate bodies on the celestial sphere, rather than sidereal hour angle (SHA). The angular equivalent of right ascension is given by the expression 360° − SHA. If the *Nautical Almanac* is available, the average SHA for each navigational planet for each three-day period is tabulated at the bottom of the left-hand daily pages (see Figure 5-1A). If only the *Air Almanac* is on hand, the expression 360° − SHA may be found by subtracting the GHA of the planet from the GHA♈ for the time of the observation.

After the value of the right ascension angle has been obtained, the next step is to place the red template on the proper side of the star base; the north sides of the base plate and red template are used for north latitudes, and the south sides for south latitudes. On the red template, a radial line is printed to represent every 10° of meridian angle, and a concentric circle is printed for every 10° of declination, with the median circle being the celestial equator. When in place on the base plate, this median circle should be concurrent with the celestial equator circle on the base plate. The solid circles within the celestial equator circle then represent declinations of the same name as the base plate, while the dashed circles outside the equator represent declinations of contrary name.

The final step is to align the arrow at the end of the 0° meridian angle radial on the template with the calculated value of the right ascension angle on the base plate. After the template has been properly aligned, the body of interest is then plotted on the base plate with a pencil, using the declination scale printed for this purpose alongside the open slot on the 0° meridian angle radial. The declination of the body is obtained from an almanac. If it is of the *same* name as the center of the star base plate, the body is plotted on the side of the celestial equator circle *toward* the center; if the declination is of *contrary* name, the body is plotted on the side of the equator *away from* the center of the base plate.

As an example, let us plot the position of Venus on the south side of the star base plate for the time of the observation used as an example in the three previous chapters. The SHA of Venus for the time of the observation on 15 December can be found from the daily page of the *Nautical Almanac* pictured in Figure 5-1A; it is 137° 43.2′. Hence, the value of the right ascension angle, 360° − SHA, is approximately

$$360° - 137.5° = 222.5°;$$

the SHA is rounded to the nearest half degree for this computation, in keeping with the precision of the graduations of the star base. Since the approximate position of the observer is in south latitudes at the time at which Venus is to be observed, the south sides of the star base plate and red template are used, and the arrow on the template is aligned with 222.5° on the base plate. The declination of Venus tabulated on the daily page of the almanac for the whole hour of GMT nearest the observation time is S 13° 16.1′; this value is also rounded to the nearest half degree, S 13.5°, consistent with the precision of the declination graduations of the red template. Because this declination is of the same name as the side of the star base used in this instance, the planet is plotted opposite the innermost 13.5° point on the declination scale of the template; thus, Venus at this time is almost alongside the third magnitude star Zubenelgenubi. At this point, the Starfinder appears as pictured in Figure 8-3 on the following page.

Finally, the template is removed and the plot of Venus is labeled.

In similar fashion, the remaining three navigational planets could each be plotted on the base plate. After all four have been plotted and labeled, it is a good practice to print the date for which they were plotted somewhere on the base plate, so as to be able to judge when to replot them.

The apparent altitudes and true azimuths of the planets can now be obtained and recorded along with the stars by using the appropriate blue template. With the 35° S template aligned to LHA♈ 145°, the apparent altitude of Venus is read as 18°, and the true azimuth is 097°T.

Use of the Starfinder for Unlisted Stars, the Sun, or the Moon

If the navigator wishes to use the starfinder to obtain the altitude and true azimuth of a star not among the 57 selected navigational stars, the body must first be plotted on the star base plate. To do this, the SHA and declination of the star is first located in an almanac,

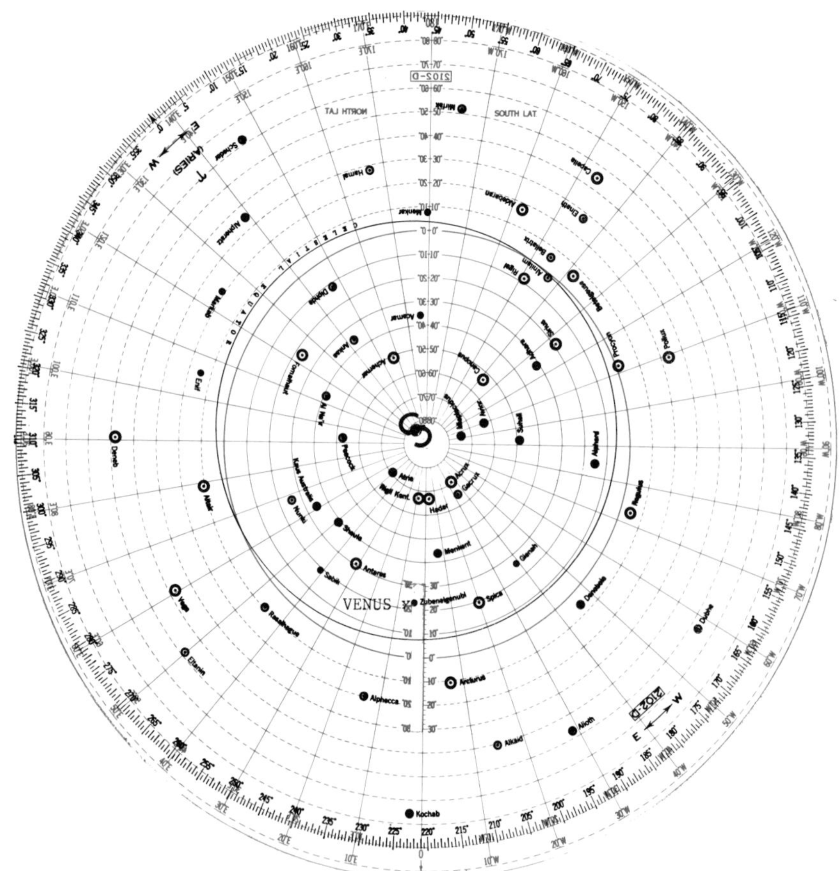

Figure 8-3. Rude Starfinder set up for plotting the planet Venus.

then the body is plotted in exactly the same manner as described above for a planet. Once it has been located on the star base, an unlisted star may be left in its plotted position indefinitely, as it will not change position relative to the other stars.

Although the location of the sun and moon in the heavens is of course no problem for the navigator, it may on occasion be desirable to use the Rude Starfinder to determine their positions at some future time for tactical planning purposes. Should this become necessary, the body of interest is first plotted on the star base. The GHA♈ and the GHA and declination of the body for the time of interest are extracted from an almanac, and the right ascension angle is found by subtracting the GHA of the body from the GHA♈. After the body has been plotted using its declination and the red template, the apparent altitude and true azimuth of the body for the time in question are determined by use of the proper blue template.

Errata to Marine Navigation 2, Second Edition

The following are corrections to the first printing of *Marine Navigation 2, Second Edition*, that should be incorporated in the book prior to its use:

Change caption page 243 Figure 13-5C to read,
"Loran-C LOP plotted for a reading of 9960–Z–60306.0."

Cut below on dotted line, and paste over Figure 8-3 page 162:

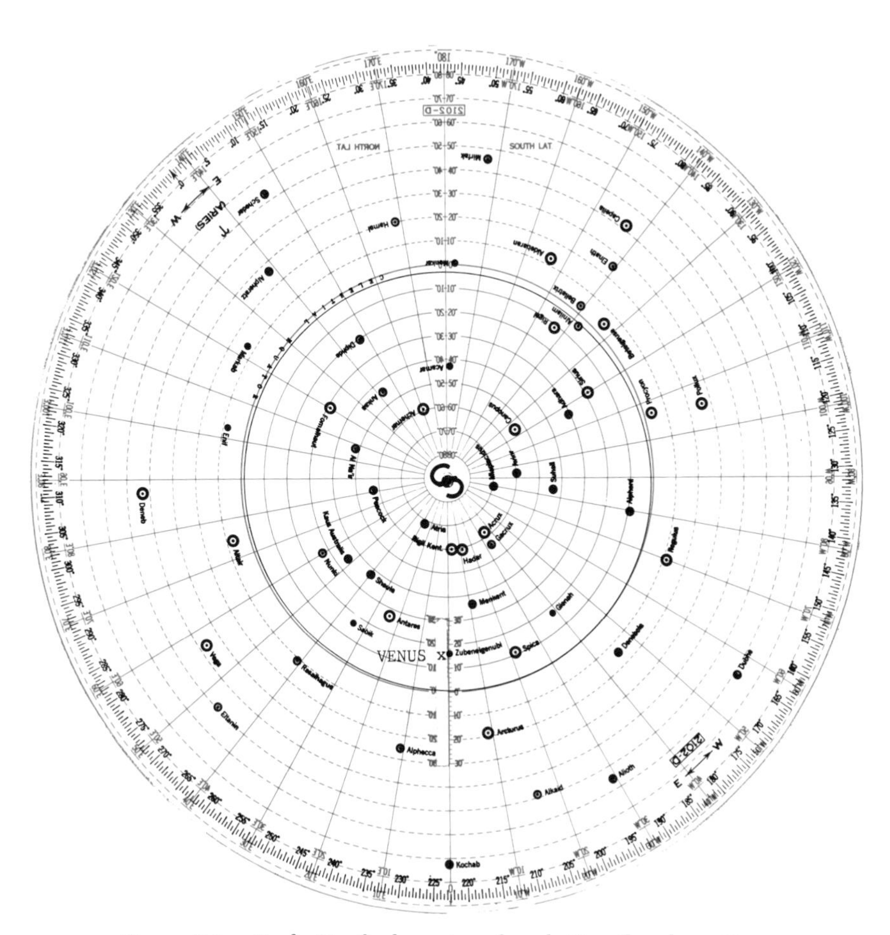

Figure 8-3. Rude Starfinder set up for plotting the planet Venus.

Use of the Starfinder to Identify an Unknown Body

Occasionally the navigator may obtain an observation of an unknown body not included in his preselected list of stars and planets suitable for observation. In such cases, the Rude Starfinder may be used to identify the body, if both its altitude and true azimuth for the time of the sight are recorded. To set up the device for this purpose, both the appropriate blue template and the red template are placed on the proper side of the star base plate, with the red template on top, and all three properly oriented for latitude. The index arrows of both templates are aligned with the proper value of LHA♈ on the base plate. To identify the body, the observed value of the true azimuth and the observed altitude are located on the blue template. If a star appears on the base plate at or near this location, it may be assumed that this was the body observed. If no star appears nearby, however, the red template is used to determine the approximate declination and meridian angle corresponding with the location. Knowing the meridian angle and the ship's longitude, the approximate GHA of the body may be computed, and from that, the approximate SHA. First, the navigator searches the appropriate daily page of the *Nautical* or *Air Almanac* with the GHA and declination to determine if the body is a planet. If it happens that the body is not a planet, the SHA and declination are used with the list of additional stars in the back of the *Nautical Almanac* to identify the body and obtain its exact GHA and declination.

Summary

This chapter has examined the use of the Rude Starfinder, 2102-D, in obtaining predicted values of the apparent altitude and true azimuth for compilation of a list of selected celestial bodies to be observed at a given location and time. The device is best suited for use with the 57 selected stars tabulated in the *Nautical* and *Air Almanacs,* but if necessary, it can be used to determine the positions of the navigational planets, unlisted stars, and even the sun or moon. It can also be used in conjunction with the *Nautical* or *Air Almanac* to identify an unknown body, if the true azimuth as well as the altitude is recorded at the time of the observation.

9

The Determination of Latitude at Sea

Earlier in this book, the process of determining a celestial fix by the altitude-intercept method of sight reduction was examined in some detail. In this chapter, a special case solution of the celestial triangle to produce a type of LOP called the *latitude line* will be discussed. This LOP may be combined with other LOPs that were obtained simultaneously to form a celestial fix, or with a nonsimultaneous LOP to produce a running fix. If no other LOP is available, the latitude LOP may be used in conjunction with the DR plot to form a highly accurate estimated position.

The Celestial Triangle at Meridian Transit of a Body

The celestial latitude line of position is a special case solution of the celestial triangle, in which all three vertices of the triangle—the celestial pole, the zenith of the observer, and the position of the body—lie on a single great circle, coincident with a celestial meridian. This condition occurs whenever any celestial body transits the upper or lower branch of the observer's meridian, as shown in Figure 9-1.

Observation of any celestial body at or near the moment of meridian transit produces a highly accurate LOP because its altitude changes most slowly at this time. If a series of observations is commenced shortly before the time of meridian transit and continued until after meridian transit has occurred, a series of nearly identical altitudes results. Reduction of the highest altitude observed in the case of a body at upper transit, or the lowest altitude of a body at lower transit, should therefore yield a highly accurate latitude line in almost every case. The only exception occurs when the navigator's platform is moving rapidly (i.e., in excess of twenty knots or so) in a generally northerly or southerly direction at the time of transit. In such situations, because the vessel is moving either directly toward or away from the GP beneath the transiting body, the vessel motion can cause some distortion in the apparent rate of change of altitude near the time of transit. In these situations, care must be taken to use the altitude recorded when the body bears either due north or south of the observer.

When a very high degree of accuracy in the latitude line is required,

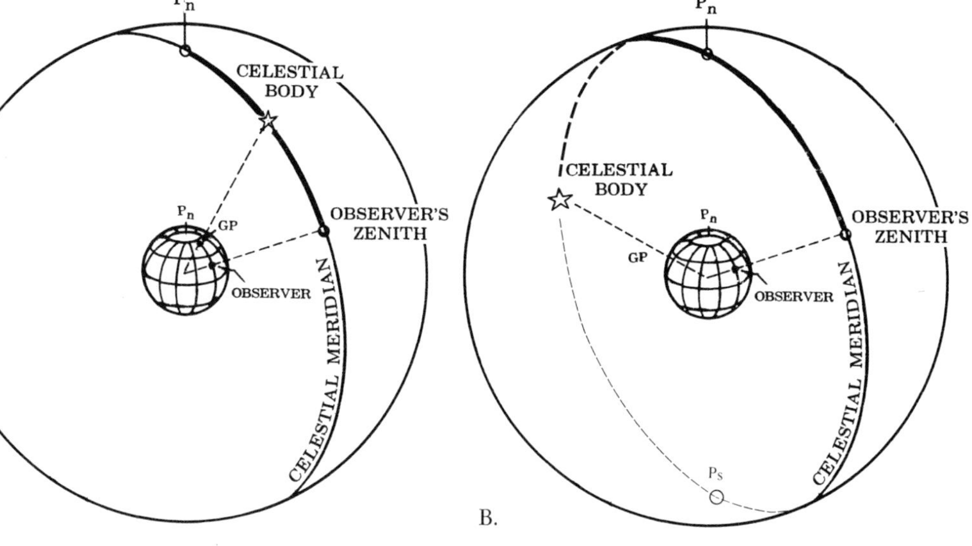

Figure 9-1. Celestial triangle of a body at (A) upper meridian transit, (B) lower meridian transit.

many navigators construct a graph such as that shown in Figure 9-2 to assist in picking off the precise altitude at the moment of transit. Such a series of sights is called a *time sight;* the technique can also be used to enhance the precision of observations made at other times, when extreme accuracy is desired. More information on this advanced technique can be obtained from either *Duttons Navigation and Piloting* or *Bowditch.*

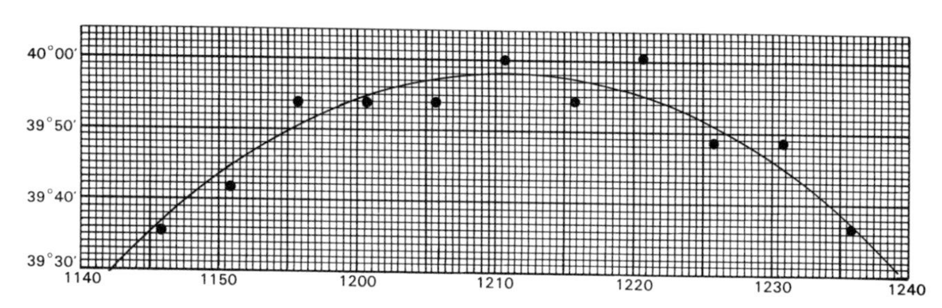

Figure 9-2. A plot of observed altitudes versus time near meridian transit (upper branch).

In addition to the high accuracy of the resulting LOP, a further advantage of any celestial latitude line obtained by observation of a body at meridian transit is that only the celestial triangle itself is solved, with no requirement for the derivation of a related navigational triangle. No sight reduction tables are required to obtain the solution, eliminating the necessity for the selection of an assumed position. Only the declination of the observed body extracted from an almanac for the time of the observation is needed.

In practice, the latitude line is normally obtained by direct observation of one of only two celestial bodies—the pole star Polaris or the sun. Polaris is ideally suited for obtaining a latitude line in north latitudes because it is virtually always at meridian transit, being located approximately at the north celestial pole; the sun is the only celestial body readily observable at transit during daylight hours in mid latitudes.

Obtaining a Latitude Line by Polaris

In the celestial triangle for the star Polaris the position of the star and the north celestial pole P_n are nearly coincident. The sides of the triangle linking the zenith of the observer with the position of the star and with the celestial north pole, 90° − Ho and 90° − Latitude, respectively, are therefore approximately coincident and equal in length, as shown in Figure 9-3A.

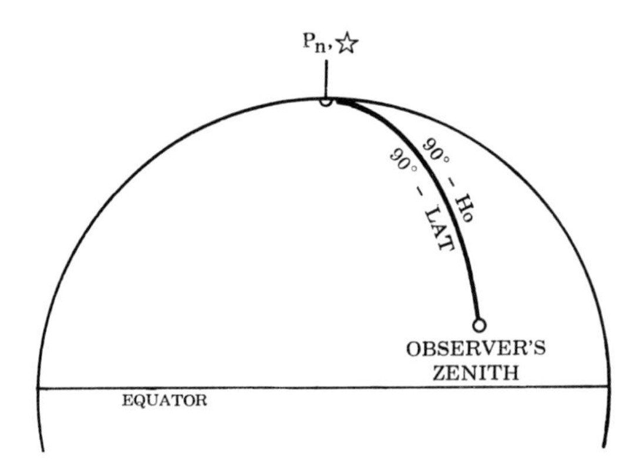

Figure 9-3A. Celestial triangle for Polaris.

If Polaris were in fact exactly coincident with the north celestial pole, the relationship between the two coincident sides of the celestial triangle could be expressed by the following equation:

$$(90^\circ - \text{Ho}) = (90^\circ - \text{Latitude})$$

This relationship is clearly indicated in Figure 9-3B, in which a cross section of the celestial sphere bisected along the observer's celestial meridian is shown.

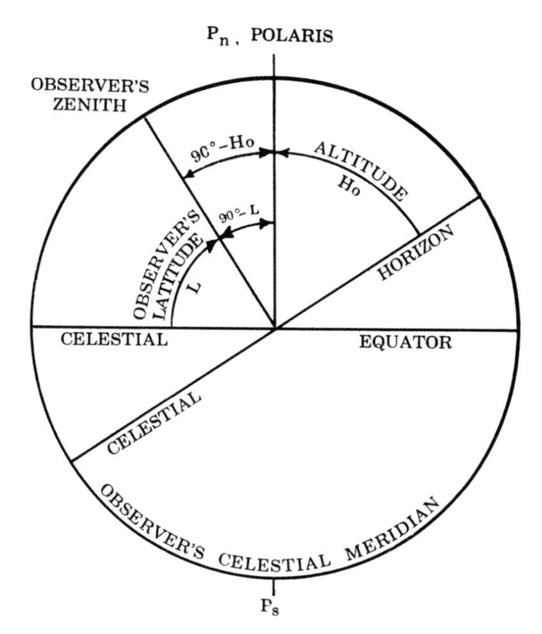

Figure 9-3B. Celestial triangle for Polaris (side view).

Removing the parentheses and combining terms algebraically, the equation above can be simplified to the following expression:

$$\text{Ho} = \text{Latitude}$$

Thus, the observed altitude Ho of Polaris is approximately equal to the latitude of the observer, a fact of great usefulness to the navigator operating at sea in northern latitudes.

In reality Polaris is *not* exactly coincident with the north celestial pole, but rather it is about three-quarters of a degree off to one side; the exact separation varies somewhat over time, as a result of the effects of the earth's precession and nutation, and the aberration effect. Hence, the star has a small diurnal circle of variable radius about the pole, which must be taken into account in order to obtain a latitude line precise to more than about 1° by Polaris. In the days of sail this 1° error could be disregarded, but the precision required by modern-day navigation makes it mandatory to take the distance between Polaris and the north celestial pole into account. Tables are provided in both the *Nautical* and *Air Almanacs* for the purpose of correcting the apparent altitude ha of Polaris for this distance at any given moment. In practice, this correction is usually combined with the refraction and temperature/barometric pressure corrections to form a total correction to the apparent altitude ha, even though the Polaris correction is not, strictly speaking, an altitude correction.

For marine navigation, the Polaris correction tables in the *Nautica* *Almanac* are generally preferred over the corresponding *Air Almana* tables because of their greater precision. An extract from the Polari correction tables of a *Nautical Almanac* appears in Figure 9-4 Another part of the tables, not shown in Figure 9-4, is used for determining the true azimuth of Polaris; it will be discussed in the followin chapter.

The Polaris correction is taken from the "Polaris Tables" in thre parts, which are designated a_0, a_1, and a_2. The a_0 part of the Polari correction can be thought of as a compensation for the componen of the distance between the position of Polaris in its diurnal circl and the north celestial pole, measured along the observer's celestia meridian; the effect, and hence the a_0 correction, varies with th longitude of the observer. The a_1 part is a compensation for the til of the diurnal circle of Polaris with respect to the vertical circle o the observer; the a_1 correction increases as the latitude of the observe increases. The a_2 part compensates for an aberration in the apparen position of Polaris because of bending of the incoming light rays fron the star, occurring as a result of the velocity of the earth in its orbit the orbital velocity, and therefore the a_2 correction, varies with th time of year. The composition of the "Polaris Tables" in the *Nautica* *Almanac* is such that these three parts are always positive; but 1° (60′ is always subtracted from their sum. The total Polaris correction thu obtained can be negative at times.

As can be seen from Figure 9-4, there is one column in the table for every 10° of LHA♈, and one horizontal section for each of th three parts. The three parts of the Polaris correction are all base on the exact LHA♈ determined from the DR position of the observe at the time of the observation. After the exact value of LHA♈ ha been obtained, the 10° LHA♈ column containing this value is entered The a_0 part of the Polaris correction corresponding to the exact LHA♈ is obtained by mental interpolation from the top section of the column The a_1 part is extracted from the second section of the column, usin as an entering argument the tabulated latitude closest to the DR latitude of the observer. The a_2 part is obtained from the third sectio of the same column, using the month in which the observation wa made as the entering argument. No interpolation is necessary in th a_1 and a_2 sections. An example of the determination of a Polari correction is given below.

Even though the Polaris sight reduction is uncomplicated, a sigh reduction form can be of convenience during the process. Such a form is used as the medium on which the following example Polaris sight i worked.

POLARIS (POLE STAR) TABLES

FOR DETERMINING LATITUDE FROM SEXTANT ALTITUDE AND FOR AZIMUTH

L.H.A. ARIES	0°–9°	10°–19°	20°–29°	30°–39°	40°–49°	50°–59°	60°–69°	70°–79°	80°–89°	90°–99°	100°–109°	110°–119°
	a_0	a_0	a_0	a_0	a_0	a_0	a_0	a_0	a_0	a_0	a_0	a_0
°	° ′	° ′	° ′	° ′	° ′	° ′	° ′	° ′	° ′	° ′	° ′	° ′
0	0 14·2	0 10·2	0 07·6	0 06·7	0 07·4	0 09·6	0 13·4	0 18·6	0 25·0	0 32·4	0 40·7	0 49·5
1	13·7	09·8	07·5	06·7	07·5	09·9	13·9	19·2	25·7	33·2	41·5	50·4
2	13·3	09·5	07·3	06·7	07·7	10·3	14·3	19·8	26·4	34·0	42·4	51·3
3	12·8	09·2	07·2	06·7	07·9	10·6	14·8	20·4	27·1	34·8	43·3	52·2
4	12·4	09·0	07·1	06·8	08·1	11·0	15·3	21·0	27·9	35·7	44·1	53·1
5	0 12·0	0 08·7	0 07·0	0 06·8	0 08·3	0 11·3	0 15·8	0 21·7	0 28·6	0 36·5	0 45·0	0 54·0
6	11·6	08·5	06·9	06·9	08·5	11·7	16·4	22·3	29·4	37·3	45·9	54·9
7	11·2	08·2	06·8	07·0	08·8	12·1	16·9	23·0	30·1	38·1	46·8	55·8
8	10·9	08·0	06·8	07·1	09·1	12·5	17·5	23·6	30·9	39·0	47·7	56·7
9	10·5	07·8	06·7	07·2	09·3	13·0	18·0	24·3	31·7	39·8	48·6	57·6
10	0 10·2	0 07·6	0 06·7	0 07·4	0 09·6	0 13·4	0 18·6	0 25·0	0 32·4	0 40·7	0 49·5	0 58·5
Lat.	a_1	a_1	a_1	a_1	a_1	a_1	a_1	a_1	a_1	a_1	a_1	a_1
°	′	′	′	′	′	′	′	′	′	′	′	′
0	0·5	0·6	0·6	0·6	0·6	0·5	0·5	0·4	0·3	0·2	0·2	0·1
10	·5	·6	·6	·6	·6	·5	·5	·4	·3	·3	·2	·2
20	·5	·6	·6	·6	·6	·5	·5	·4	·4	·3	·3	·3
30	·6	·6	·6	·6	·6	·6	·5	·5	·4	·4	·4	·4
40	0·6	0·6	0·6	0·6	0·6	0·6	0·6	0·5	0·5	0·5	0·5	0·5
45	·6	·6	·6	·6	·6	·6	·6	·6	·6	·5	·5	·5
50	·6	·6	·6	·6	·6	·6	·6	·6	·6	·6	·6	·6
55	·6	·6	·6	·6	·6	·6	·6	·6	·7	·7	·7	·7
60	·6	·6	·6	·6	·6	·6	·7	·7	·7	·8	·8	·8
62	0·7	0·6	0·6	0·6	0·6	0·6	0·7	0·7	0·8	0·8	0·9	0·9
64	·7	·6	·6	·6	·6	·7	·7	·8	·8	·9	0·9	0·9
66	·7	·6	·6	·6	·6	·7	·7	·8	·9	0·9	1·0	1·0
68	0·7	0·6	0·6	0·6	0·6	0·7	0·8	0·8	0·9	1·0	1·1	1·1
Month	a_2	a_2	a_2	a_2	a_2	a_2	a_2	a_2	a_2	a_2	a_2	a_2
	′	′	′	′	′	′	′	′	′	′	′	′
Jan.	0·7	0·7	0·7	0·7	0·7	0·7	0·7	0·7	0·7	0·7	0·7	0·7
Feb.	·6	·6	·7	·7	·7	·8	·8	·8	·8	·8	·8	·8
Mar.	·5	·5	·6	·6	·7	·7	·8	·8	·9	·9	·9	0·9
Apr.	0·3	0·4	0·4	0·5	0·6	0·6	0·7	0·8	0·8	0·9	0·9	1·0
May	·2	·3	·3	·4	·4	·5	·6	·6	·7	·8	·8	0·9
June	·2	·2	·2	·3	·3	·4	·4	·5	·5	·6	·7	·8
July	0·2	0·2	0·2	0·2	0·2	0·3	0·3	0·4	0·4	0·5	0·5	0·6
Aug.	·4	·3	·3	·3	·3	·3	·3	·3	·3	·3	·4	·4
Sept.	·5	·5	·4	·4	·3	·3	·3	·3	·3	·3	·3	·3
Oct.	0·7	0·7	0·6	0·6	0·5	0·4	0·4	0·3	0·3	0·3	0·3	0·2
Nov.	0·9	0·9	0·8	0·7	·7	·6	·5	·5	·4	·3	·3	·3
Dec.	1·0	1·0	1·0	0·9	0·8	0·8	0·7	0·6	0·6	0·5	0·4	0·3

Figure 9-4. Excerpt from "Polaris Tables," Nautical Almanac.

Suppose that Polaris was observed at zone time 20-12-09, at which time the DR position of the observer was L 30° 30.5′ N, λ 67° 37.2′ W. The sextant index correction (IC) is +.5′, and the height of eye is 44′.

After entering this information on the sight form, the LHA♈ is determined. To do this, the GHA♈ for the time of the observation is first obtained by combining the tabulated GHA♈ on the daily pages with the proper GHA increment; in this case, the GHA♈ is 102° 18.5′. At this time, the time diagram at the top of the form can be completed to assist in finding the proper value of LHA♈. Since the longitude is west and is less than the GHA♈ in this instance, the exact LHA♈ is found by subtracting the DR longitude from the GHA♈. The form now appears as in Figure 9-5A.

After determining the exact LHA♈, the IC and dip corrections are summed and applied to the sextant altitude hs to obtain the apparent altitude ha of 31° 27.5′. The refraction correction for this altitude is −1.6′; after entering this value on the form, the Polaris Tables shown in Figure 9-4 are entered to find the three parts of the Polaris correction. The entering argument for the tables is the LHA of Aries, 34° 41.3′. The appropriate column is therefore the third one, headed 30°-39°. To find the a_0 part, the upper third of the table containing corrections for each integral degree of LHA from 30° to 40° is used, with interpolation if necessary to arrive at the a_0 correction corresponding to LHA♈ 34° 41.3′. In this case, the tabulated a_0 values for both LHA♈ 34° and 35° are the same, 0° 06.8′, so no interpolation is needed. After recording this 0° 06.8′ a_0 correction on the form, we proceed to the middle third of the table, in the same LHA♈ column, to obtain the a_1 part. The entering argument in the left margin is the tabulated latitude closest to the DR latitude, or 30° in this case. The corresponding a_1 value is 0.6′, which is recorded onto the form. The a_2 part is obtained from the lower third of the table, staying in the same column, opposite the month of observation, December in this case; it is 0.9′. As the final step in finding the total correction to be applied to ha, a_0, a_1, and a_2 are added algebraically with the negative refraction correction, any required temperature-barometric (TB) correction, and the required −60′ constant. In this example, the addition yields a total correction of −53.3′. Applying this correction to ha yields the desired latitude line, 30° 34.2′ N. The sight form as it appears at this point is pictured in Figure 9-5B. The remaining spaces on the form pertain to the determination of the gyro error based on the true azimuth of Polaris; they will be completed in the next chapter.

A plot of the resulting latitude line by Polaris is shown in Figure 9-5C on page 172.

LATITUDE BY POLARIS		
DR Lat.	30-30.5 N	
DR Long.	67-37.2 W	
Date	15 DEC	
ZT	20-12-09	
ZD (W+, E-)	+5	
GMT	01-12-09	
Date (GMT)	16 DEC	
Tab. GHA ♈	99-15.8	
Incr'm't ♈	3-02.7	
Total GHA ♈	102-18.5	
±360 if needed		
DR λ (-W, +E)	67-37.2	
LHA ♈ (Exact)	34-41.3	
IC		
Dip (Ht. ')		
Sum		
hs		
ha		
Refr. Corr.		-
TB (ha < 10°)	+	-
a_0	+	
a_1	+	
a_2	+	
Add'n'l		-60.0
Sub Total	+	-
Total Corr. to ha (±)		
ha		
Latitude	N	
True Azimuth	°T	
Gyro Brg.	°pgc	
Gyro Error	°(E or W)	

NOTES:

Figure 9-5A. Polaris sight form, LHA ♈ computed.

LATITUDE BY POLARIS		
DR Lat.	30-30.5 N	
DR Long.	67-37.2 W	
Date	15 DEC	
ZT	20-12-09	
ZD (W+, E-)	+5	
GMT	01-12-09	
Date (GMT)	16 DEC	
Tab. GHA ♈	99-15.8	
Incr'm't ♈	3-02.7	
Total GHA ♈	102-18.5	
±360 if needed		
DR λ (-W, +E)	67-37.2 W	
LHA ♈ (Exact)	34-41.3	
IC	+.5	
Dip (Ht. 44 ')	-6.4	
Sum	-5.9	
hs	31-33.4	
ha	31-27.5	
Refr. Corr.		- 1.6
TB (ha < 10°)	+	-
a_0	+ 06.8	
a_1	+ .6	
a_2	+ .9	
Add'n'l		-60.0
Sub Total	+ 08.3	- 61.6
Total Corr. to ha (±)	-53.3	
ha	31-27.5	
Latitude	30-34.2 N	
True Azimuth	°T	
Gyro Brg.	°pgc	
Gyro Error	°(E or W)	

NOTES:

Figure 9-5B. Completed determination of latitude by Polaris.

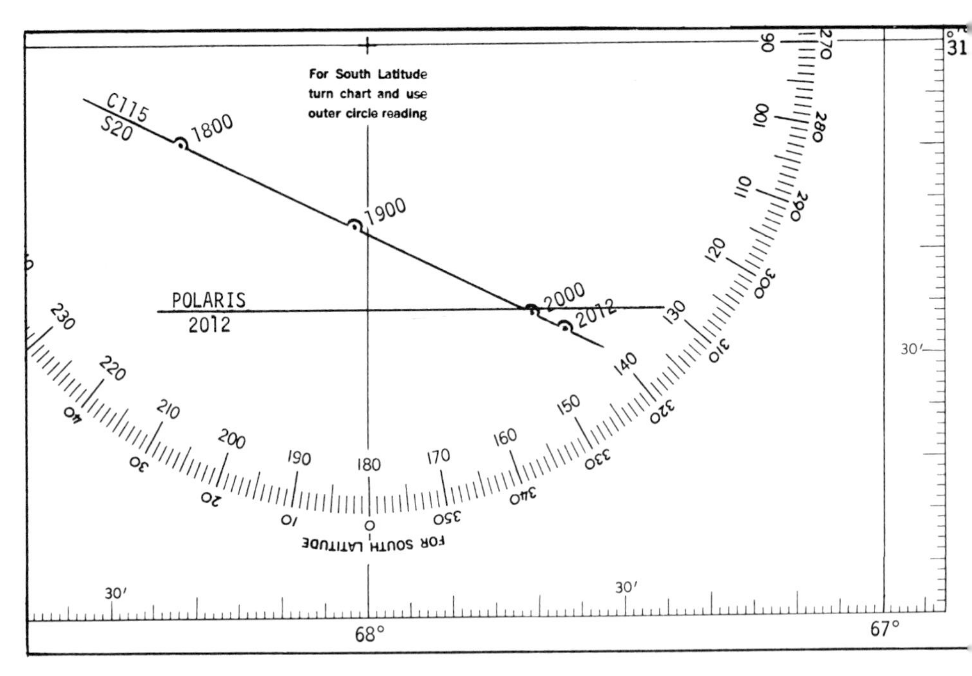

Figure 9-5C. Plot of a latitude line by Polaris.

After the latitude line determined by an observation of Polaris h been plotted, it may be combined with other LOPs to form a celesti fix or running fix by methods discussed in Chapter 7. If no other LC is available, a perpendicular may be dropped to the latitude line fro the DR position at the time of the Polaris observation (or in oth words, the DR longitude may be crossed with the Polaris latitud to obtain an estimated position of high reliability.

Because the star Polaris is below the observer's celestial horiz when he is located south of the terrestrial equator, it follows that t determination of latitude by Polaris is not possible when the observ is in south latitudes. When Polaris is not available for observatio either during daylight hours or because of the position of the observe there is an excellent alternative body available for determination a latitude line—the sun.

Obtaining a Latitude Line by the Sun

Because the sun always completes upper transit above the observe celestial horizon in the mid-latitudes of the world, observation of t sun at this meridian transit is a very convenient method of determini a latitude line. The sun latitude line thus obtained is considered to one of the most accurate LOPs available in the course of a typic day's work in navigation at sea, except possibly when the sun's altitude is extremely high. In practice, most navigators observe the sun at meridian transit as a matter of routine at sea every day the weather conditions permit.

The determination of a latitude line by observation of the sun is more complicated than a similar determination by Polaris, since the geographical position of the sun can be located anywhere within a 47-degree-wide band of latitude centered on the equator, depending on the date and time of day. As may be recalled from Chapter 3, the moment at which the apparent sun transits the upper branch of the observer's meridian each day is referred to in celestial navigation as *local apparent noon,* abbreviated *LAN.* The observed altitude of the sun and its declination at the moment of LAN are the two basic data required to determine the latitude line at LAN.

Since the declination of the sun changes from about north $23\frac{1}{2}°$ to south $23\frac{1}{2}°$ in the course of each year, there are a number of different relationships possible among the elevated celestial pole, the position of the sun, and the zenith of the observer at LAN. The observer's zenith and the position of the sun can both lie in the same celestial hemisphere, or in different hemispheres; if they are in the same hemisphere, the observer's zenith can be either north or south of the position of the sun. Three possible relationships are shown in Figures 9-6A through C on the following page.

In each figure, the upper branch of the observer's principal vertical circle (the vertical circle passing through the north and south points of the observer's celestial horizon, coincident with his celestial meridian) is pictured at the moment of LAN. The horizontal line represents the observer's celestial horizon, with the northernmost point to the left. The observer's zenith (directly over his DR position) is represented by a vertical line labeled Z. The sun's position is indicated by a line labeled with the sun symbol ⊙, the elevated pole by a line labeled P_n or P_s, and the celestial equator by a line labeled Q. The vertex from which all lines originate represents the center of the earth.

In the figures, the latitude of the observer is identical to the angle formed at the earth's center between the celestial equator Q and the observer's zenith Z. The angle formed between the sun's position and the observer's zenith may be recognized as the 90° − Ho side of the celestial triangle; for purposes of the latitude determination at LAN, this angle is referred to by its alternate name, the *zenith distance.* It is abbreviated by a lower case *z*, and labeled with a suffix to indicate the direction of the observer's zenith from the body at the moment of meridian transit. The suffix N is used to denote north, and the suffix S, south.

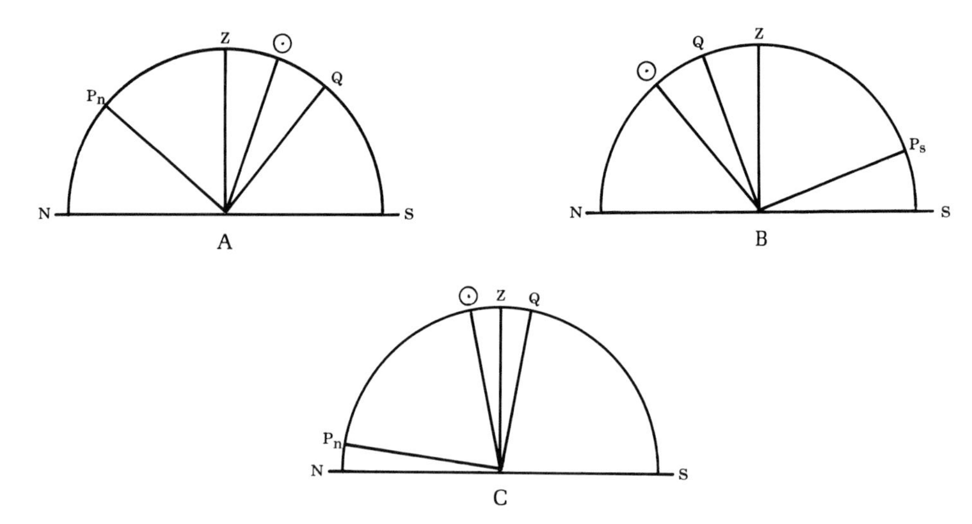

Figure 9-6. A. Latitude N, Dec. N; B. Latitude S, Dec. N; C. Latitude N, Dec. greater than Lat.

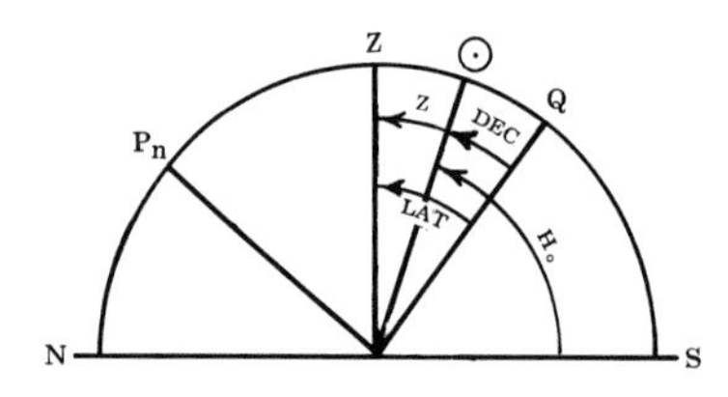

Figure 9-7. Relationship of observer's latitude to zenith distance and declination.

Consider the diagram in Figure 9-6A. If the zenith distance z(N) is indicated, it would be apparent that it is equal to 90° − Ho. Furthermore, if the declination of the sun is indicated, it would be seen that the latitude of the observer in this case is equal to the sum of the zenith distance z plus the declination of the sun. These angles are all identified in Figure 9-7.

If these same angles were indicated on Figures 9-6B and C and on other diagrams representing other possible relationships among the sun, the observer's zenith and the celestial equator at LAN, it would be seen that the latitude of the observer at LAN is always yielded by the sum or difference of the zenith distance z and the declination of the sun. In general, the latitude of the observer can be obtained by an observation of the sun at LAN by applying the following two rules:

> If the zenith distance z and declination are of the *same* name, *add* them for the latitude of the observer.
>
> If the zenith distance z and declination are of *contrary* name, *subtract* the smaller from the larger.

In the latter case, the latitude will have the same name as the remainder.

As an example, if the declination of the sun in Figure 9-7 were N 20° and the observed altitude Ho at LAN were 70°, the zenith distance z would be 90° − 70° or 20° N, and the observer's latitude would be 20° N + N 20°, or 40° N. In Figure 9-6B, if the declination of the sun were N 20° and the Ho were 50°, the zenith distance would be 90° − 50°, or 40° S, since the observer's zenith is located south of the sun. Hence, the latitude would be 40° S − N 20°, or 20° S. In Figure 9-6C, if the declination of the sun were N 20° and the observed altitude Ho were 80°, the latitude of the observer would be 10° N.

Determination of the Time of LAN

Because the declination of the sun at the moment of meridian transit and the zenith distance are the two quantities combined to yield the latitude of the observer, it follows that both the declination and observed altitude of the sun must be determined as precisely as possible for an accurate latitude line. As an aid in obtaining the correct value for the declination of the sun and also to determine when he should plan to make his observation, the navigator will normally precompute the zone time at which the meridian transit of the sun should occur, based on either his stationary position, if his ship is not underway with way on, or on his projected DR track, if his ship is moving. In other words, the navigator will usually *estimate the zone time of local apparent noon* for the position at which he expects to be located at meridian transit of the sun.

The first step in the estimation of the time of LAN at the ship's stationary or anticipated DR position is to obtain the local mean time of meridian passage of the apparent sun by referring to the lower right-hand side of the daily pages of the *Nautical Almanac.* Although the times of meridian passage listed are computed for the Greenwich meridian (and hence are Greenwich mean times), they can also be used with insignificant error as the local mean times of meridian passage at every other meridian, because of the nearly constant rate of motion of the sun about the earth. The data for 15, 16, and 17 December of a typical year are reproduced in Figure 9-8.

Day	SUN Eqn. of Time 00ʰ	12ʰ	Mer. Pass.
	m s	m s	h m
15	05 12	04 58	11 55
16	04 43	04 29	11 56
17	04 14	04 00	11 56

Figure 9-8.
Times of meridian passage of the sun, Nautical Almanac, *15, 16, 17 December.*

After plotting a DR position on a suitably prepared plotting sheet for the local mean time of meridian passage, the navigator must then convert the LMT to the zone time of the phenomenon at the plotted DR position. To perform this conversion, the navigator determines the arc difference between the DR longitude and the central meridian of the time zone by which the ship's clocks are set. Usually but not always the ship will be within this zone. Next, this arc difference is converted to a time difference using the "Conversion of Arc to Time" tables in the almanac. Since the sun will cross the observer's meridian before the central meridian of his time zone if he is located east of the central meridian, the time difference is *subtracted* if the initial DR position is to the *east* of the time zone central meridian. Conversely, if the initial DR position is to the *west* of the central meridian of the time zone being used, the time difference is *added,* since the sun will arrive over the observer's meridian after it has crossed the time zone central meridian. The resulting zone time of meridian passage is referred to as the *first estimate* of the zone time of LAN.

If his ship is stationary, this first estimate of the zone time of LAN should be the time at which the sun will transit the navigator's meridian. If his ship is moving, a second estimate of the time of LAN must be made, to account for the motion of the ship during the time required for the sun to travel between the time zone central meridian and the meridian passing through the first estimate DR position.

To obtain the second and final estimate of the time of LAN, the process described in the preceding paragraph is repeated, using a DR position plotted for the first estimate zone time as a starting point. The longitude difference for this position is found and converted to a time difference, and this time difference is again applied to the tabulated LMT of meridian passage to obtain the second and final estimate.

The following is a summary of the steps necessary to determine the zone time of LAN:

1. Obtain the LMT of meridian passage from the *Nautical Almanac.*

2. Plot a DR position for this time.

3. Determine the difference in longitude between this position and the central meridian of the time zone being used; convert this arc difference to a time difference.

4. Apply this time difference to the LMT of meridian passage, adding if west of the time zone central meridian, subtracting if east. The resulting time is the first estimate of the zone time of LAN.

5. If the ship is stationary, this first estimate should be the actual zone time of LAN. If the ship is moving:

6. Plot another DR position for the first estimate zone time of LAN.

7. Compute and apply the time difference for this position to the tabulated LMT of meridian passage. The resulting time is the second and final estimate of the zone time of LAN.

In practice, the navigator will usually begin observing the sextant altitude of the sun about five minutes prior to the second estimated zone time of LAN, to account for any possible errors in the estimating procedure. The altitude recorded is normally the highest altitude reached by the sun as it transits his meridian. It should be repeated here that if the navigator's vessel is proceeding on a heading having a large north-south velocity component, the motion of the vessel toward or away from the GP of the transiting sun will cause its altitude to appear to be changing at LAN. If the velocity component is in a direction toward the sun, the vessel is approaching the subsolar point, thereby causing the apparent altitude to seem to be increasing at meridian transit. Conversely, if the velocity component is away from the subsolar point, the sun will appear to be sinking at transit. Thus, if the greatest altitude is taken as the observed altitude at LAN under these circumstances, an error is introduced. When such a large north-south velocity component is present, the navigator should either record the altitude observed at the precomputed time of LAN, or record the altitude observed when the true azimuth of the sun is due north or south.

After the actual meridian transit of the sun has been observed and the sextant altitude and exact zone time of LAN have been recorded, the declination of the sun for this time is determined from either the *Nautical Almanac* or *Air Almanac*, the LAN diagram is drawn, and the observer's latitude is determined. As was the case with the observation of Polaris, no sight reduction tables are required to obtain a latitude line by observation of the sun. Nevertheless, a standard sight form for estimating the time of LAN and the reduction of the LAN sun line can be of considerable assistance to the navigator. The following example demonstrates the use of such a form, as prepared at the U.S. Naval Academy.

Determining the Latitude at LAN—an Example

As an example of determining a latitude line by observation of the sun at local apparent noon, suppose that at 1100 ZT on 16 December a ship was located at a DR position L 35° 45.0′ N, λ 54° 45.0′ W, steaming on course 080°T at a speed of 20 knots. The navigator desires to obtain a second estimate of the zone time of LAN.

Working through the procedure set forth in the preceding section, the LMT of meridian passage is first obtained from the *Nautical*

Almanac excerpt shown in Figure 9-8; for 16 December, it is 1156. A DR position is then plotted for this time as shown in Figure 9-10 (p. 181), and the coordinates of this position are recorded on the LAN sight form. The difference between the DR longitude, 54° 22.0′ W, and the central meridian of the time zone, 60° W, is 5° 38.0′:

$$60° - 54° 22.0' = 5° 38.0'$$

Conversion of this arc to time by the tables in the *Nautical Almanac* yields 22 minutes 32 seconds, which may be rounded to the nearest whole minute, 23 minutes. Since the 1156 DR position is to the east of the 60° central meridian, the first estimate of the zone time of LAN is 1156 − 23, or ZT 1133. At this point, the partially completed sight form appears as in Figure 9-9A.

To obtain the second estimate, a DR position for 1133 is plotted, and the DR longitude 54° 31.8′ W is picked off and subtracted from 60° to form the second estimate longitude difference, 5° 28.2′. Converting this arc to time and again subtracting from the LMT of meridian passage yields the second and final estimate of the zone time of LAN, 1134. The form now appears as in Figure 9-9B.

To complete the example, suppose that the actual zone time of the observation recorded when the sun appeared at its greatest altitude was 11-35-05 and the sextant altitude hs of the lower limb was 30° 52.1′. The sextant IC was −.5′, and height of eye is 44′. The true declination for the zone time of observation is found from the *Nautical Almanac* to be S 23° 19.3′. With this information and the DR latitude for the approximate time of the sight, the LAN diagram can now be completed as it appears in Figure 9-9C. After applying the IC, dip, and altitude corrections to hs, an Ho of 30° 59.8′ results. Subtracting this Ho from 90° yields the zenith distance 59° 00.2′ N, labeled "north" because the DR position is north of the GP of the sun. Inasmuch as the zenith distance and declination are of contrary name, the true declination is subtracted from the larger zenith distance to yield the latitude 35° 40.9′ N. The completed sight form now appears as in Figure 9-9C, and the latitude line is shown in Figure 9-10.

As a point of interest, if by chance the navigator is operating in extreme northern latitudes in summer (or in extreme southern latitudes in summer), the sun may be observable across the pole at low altitudes at *lower transit*, i.e., local midnight. In such regions of constant daylight, the equation of time may be used to determine the local mean time of lower transit, from which the local zone time of the transit can be derived by using the same procedure as that for finding the zone time of LAN. The lowest observed altitude is used to determine

LATITUDE AT L.A.N.	Z N — S
FIRST ESTIMATE	
Date	16 DEC
DR Latitude	35-49.0 N
DR λ	54-22.0 W
Central Meridian	60 W
dλ (arc)	5-38.0
dλ (time)	23 m
Mer. Pass. (LMT)	1156
ZT (est) 1st	1133
SECOND ESTIMATE	
DR λ	
Central Meridian	
dλ (arc)	
dλ (time)	
Mer. Pass. (LMT)	
ZT (est) 2nd	
ZT (actual)	
ZD (W+, E−)	
GMT	
Date (GMT)	
Tab. Dec. / d(+ or −)	
d Corr. (+ or −)	
True Dec.	
IC	
Dip (Ht ')	
Sum	
hs (at LAN)	
ha	
Alt. Corr.	
89-60.0	89-60.0
Ho	
Zenith Dist.	
True Dec.	
Latitude	

NOTES:

Figure 9-9A. Sun LAN form, first estimate completed.

FIRST ESTIMATE	
Date	16 DEC
DR Latitude	35-49.0 N
DR λ	54-22.0 W
Central Meridian	60 W
dλ (arc)	5-38.0
dλ (time)	23 m
Mer. Pass. (LMT)	1156
ZT (est) 1st	1133
SECOND ESTIMATE	
DR λ	54-31.8 W
Central Meridian	60 W
dλ (arc)	5-28.2
dλ (time)	22 m
Mer. Pass. (LMT)	1156
ZT (est) 2nd	1134

Figure 9-9B. Sun LAN form, second estimate completed.

LATITUDE AT L.A.N.	Z, Q, Pn, ☉, N, S
FIRST ESTIMATE	
Date	16 DEC
DR Latitude	35-49.0 N
DR λ	54-22.0 W
Central Meridian	60 W
dλ (arc)	5-38.0
dλ (time)	23 m
Mer. Pass. (LMT)	1156
ZT (est) 1st	1133
SECOND ESTIMATE	
DR λ	54-31.8 W
Central Meridian	60 W
dλ (arc)	5-28.2
dλ (time)	22 m
Mer. Pass. (LMT)	1156
ZT (est) 2nd	1134
ZT (actual)	11-35-05
ZD (W+, E–)	+4
GMT	15-35-05
Date (GMT)	16 DEC
Tab. Dec. / d(+ or –)	S 23-19.2 / +.1
d Corr. (+ or –)	+.1
True Dec.	S 23-19.3
IC	–.5
Dip (Ht 44 ')	–6.4
Sum	–6.9
hs (at LAN)	30-52.1
ha	30-45.2
Alt. Corr.	+14.6
89-60.0	89-60.0
Ho	– 30-59.8
Zenith Dist.	59-00.2 N
True Dec.	S 23-19.3
Latitude	35-40.9 N

NOTES:

Figure 9-9C. Completed sun LAN form.

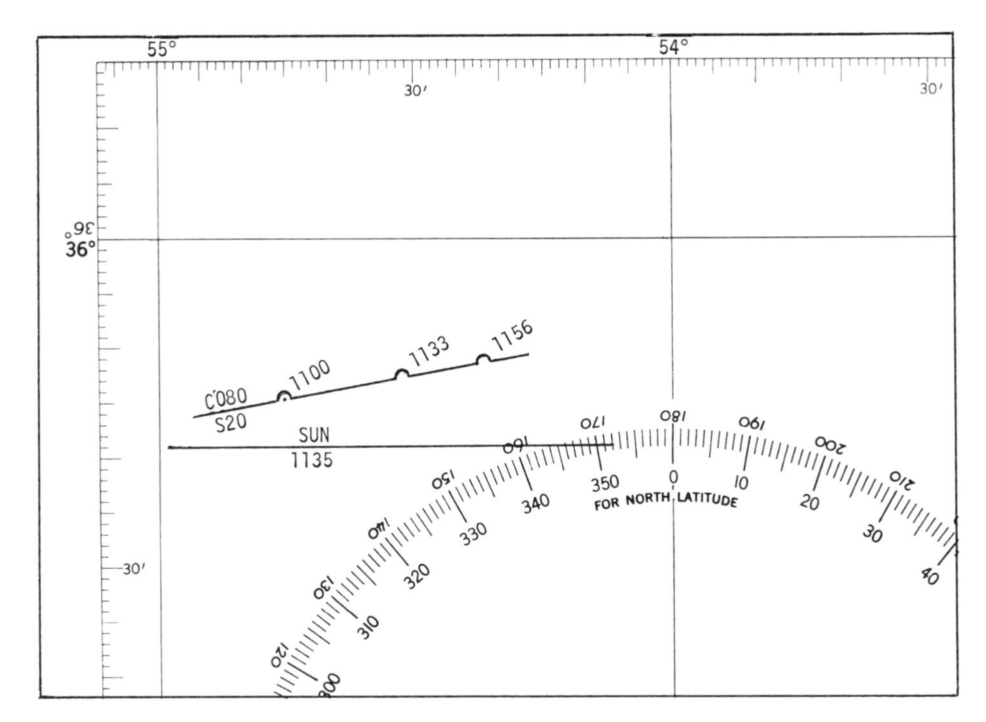

Figure 9-10. Plot of a sun LAN line.

zenith distance, which must then be added to the declination, and the sum subtracted from 180° to obtain the latitude, as indicated in Figure 9-11.

Just as was the case with the Polaris line discussed earlier, the sun latitude line may be combined with other simultaneous or nonsimul-

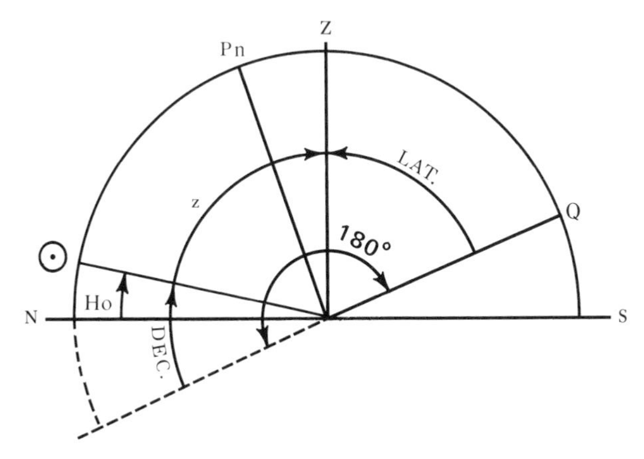

Figure 9-11. Relationship of observer's latitude, zenith distance(z), and declination of the sun at lower transit in northern latitudes.

taneous LOPs to form a fix or running fix, or the line can be crossed with the longitude of the DR position corresponding to the time of the observation to produce an estimated position. In practice, it is standard procedure whenever the weather permits to obtain a running fix at the time of LAN by advancing a morning sun line, and to obtain a second afternoon running fix by advancing the LAN line. These running fixes, together with any electronic fixes, the morning and evening celestial fixes obtained at twilight, and the determination of compass error as set forth in the next chapter, constitute a basic "day's work in navigation" for the navigator at sea. This subject will be discussed in greater detail in Chapter 16.

Summary

This chapter has presented the basic theory and procedures for determining a special type of celestial LOP at sea known as the latitude line, by observing either the star Polaris or the sun. Observation of any celestial body at meridian passage can produce a highly accurate latitude LOP, but Polaris and the sun are particularly well suited for this purpose. Polaris is perpetually near meridian transit, while the sun is the only body always observable at upper transit above the celestial horizon by day in mid-latitudes. Once the latitude line has been derived, it may be combined with other LOPs to form a celestial fix or running fix of high accuracy, or it may be used in conjunction with the DR plot to obtain an estimated position of unusual reliability.

10

Determining the Gyro Error at Sea

Among the duties of the navigator prescribed in *NWP 50, Shipboard Procedures,* is the requirement that the error of the ship's gyrocompass and magnetic compass be checked and recorded daily. The procedures for determining the gyro and compass error in piloting waters are set forth in Chapter 9 of *Marine Navigation 1.* In this chapter, the method of obtaining a gyro error at sea by observation of a celestial body will be discussed. In practice, once the error of the ship's gyrocompass system has been established, the ship's magnetic compasses are then compared with the gyro to obtain the compass errors.

Essentially, gyro error at sea is determined by first observing the azimuth of a selected celestial body by means of the gyro repeater fitted with an azimuth circle, and then comparing this observed azimuth to a true azimuth computed by means of a special case solution of the celestial triangle. Although any celestial body could be used for this purpose, in practice most navigators prefer one of only two bodies—the star Polaris or the sun. As was the case with the latitude line discussed in the last chapter, Polaris is a favorite body for compass error determinations in north latitudes because of its location near the north celestial pole, while the sun is usually the only body available for observation for this purpose during daylight hours.

Observing the Azimuth of Polaris or the Sun

To observe the azimuth of Polaris, the sun, or any other celestial body, an instrument introduced in Chapter 7 of *Marine Navigation 1* is used—the *azimuth circle.* The inexperienced navigator should review the description of the design and function of this device before proceeding further.

When the azimuth of the star Polaris is to be recorded, the procedure used is somewhat similar to that for the observation of the bearing of a terrestrial object with the azimuth circle. The star is first aligned in the black-coated reflector incorporated in the far sight vane, then the gyro azimuth is read from the portion of the compass card directly beneath the sight vane. Care should be taken to ensure that the azimuth circle is level at the moment the sight is recorded, to obtain an accurate reading.

To observe the azimuth of the sun, the mirror and prism assemblies at right angles to the sight vanes on the instrument are used, as explained in Chapter 7, *Marine Navigation 1*. A narrow reflected beam of sunlight is cast across the proper portion of the compass card, and the bearing illuminated on the card by the beam is the gyro azimuth of the sun recorded. An alternative method of recording the azimuth of the sun, if its face is partially obscured by clouds, is to orient the azimuth circle in such a way that the image of the sun appears centered in the far sight vane reflector. The gyro bearing read beneath the vane then represents the sun's azimuth. As before, the azimuth circle should be level at the time of the observation.

Because of the construction of the azimuth circle, azimuths of celestial bodies are most accurately read when the body under observation is less than 20° in altitude; the most accurate observation possible occurs when the body is located on the celestial horizon of the observer. Hence, when the sun is to be used to determine gyro error, the optimum time of observation is near the time of sunrise and sunset. Polaris, because of its location near the north celestial pole, is always at about the same position both in altitude and azimuth above the observer's horizon in north latitudes. Obtaining a meaningful gyro error by Polaris in latitudes above 60° north is virtually impossible because of its high altitude in these higher latitudes.

Determining the Gyro Error by Polaris

The true azimuth of Polaris is tabulated in the *Nautical Almanac* "Polaris Tables" for latitudes between 0° and 65° north, in a horizontal section located beneath the a_0, a_1, and a_2 correction sections discussed in the last chapter. An extract from this Polaris azimuth section for LHA♈ between 0° and 119° appears in Figure 10-1.

Entering arguments for this azimuth section are the same LHA♈ used as a basis for entering the top three sections of the table, and the computed latitude by Polaris determined earlier. Mental interpolation is used where necessary to obtain the true azimuth of Polaris precise to the nearest tenth of a degree.

After the true azimuth of Polaris has been extracted from the Polaris azimuth table, it is then compared with the observed azimuth. The difference between the observed and true azimuth is the gyro error. The error is usually rounded to the nearest half degree, in keeping with the precision of the gyro repeater, and it is labeled either E (east) or W (west), depending on whether the true azimuth is greater than or less than the observed azimuth. The last few spaces of the Polaris

274 POLARIS (POLE STAR) TABLES, 19__

FOR DETERMINING LATITUDE FROM SEXTANT ALTITUDE AND FOR AZIMUTH

L.H.A. ARIES	0°– 9°	10°– 19°	20°– 29°	30°– 39°	40°– 49°	50°– 59°	60°– 69°	70°– 79°	80°– 89°	90°– 99°	100°– 109°	110°– 119°
Lat.	AZIMUTH											
°	°	°	°	°	°	°	°	°	°	°	°	°
0	0·4	0·2	0·1	359·9	359·8	359·6	359·5	359·4	359·3	359·2	359·2	359·1
20	0·4	0·3	0·1	359·9	359·8	359·6	359·5	359·4	359·2	359·2	359·1	359·1
40	0·5	0·3	0·1	359·9	359·7	359·5	359·4	359·2	359·1	359·0	358·9	358·9
50	0·6	0·4	0·1	359·9	359·7	359·4	359·2	359·0	358·9	358·8	358·7	358·7
55	0·7	0·4	0·2	359·9	359·6	359·4	359·1	358·9	358·8	358·6	358·5	358·5
60	0·8	0·5	0·2	359·9	359·6	359·3	359·0	358·8	358·6	358·4	358·3	358·3
65	0·9	0·6	0·2	359·8	359·5	359·1	358·8	358·5	358·3	358·1	358·0	357·9

Latitude = Apparent altitude (corrected for refraction) $- 1° + a_0 + a_1 + a_2$

The table is entered with L.H.A. Aries to determine the column to be used; each column refers to a range of 10°. a_0 is taken, with mental interpolation, from the upper table with the units of L.H.A. Aries in degrees as argument; a_1, a_2 are taken, without interpolation, from the second and third tables with arguments latitude and month respectively. a_0, a_1, a_2 are always positive. The final table gives the azimuth of *Polaris*.

Figure 10-1. Excerpt from "Polaris Azimuth Tables," Nautical Almanac.

sight reduction form introduced in the last chapter are intended for the computation of the gyro error by Polaris.

As an example of the determination of a gyro error by Polaris, suppose that at the moment of the observation of Polaris (see Figure 9-5B), the gyro bearing of the star was observed and recorded on the sight form as 000.5° per gyrocompass (pgc). To find the true azimuth of Polaris, the LHA♈ 30° – 39° column of the Polaris Tables is followed down into the bottom "Azimuth" section, shown in Figure 10-1. Since the tabulated azimuth corresponding with all latitudes except 65° N is 359.9°T, the exact azimuth corresponding to the computed altitude 30° 34.2′ N is 359.9°T. Comparison of this value with 000.5° pgc results in a gyro error of .6° W, since the true azimuth is less than the observed azimuth. In practice, this value would be rounded off to .5° W. The completed gyro error determination appears on the sight form in Figure 10-2.

As was mentioned earlier, after the gyro error has been found the magnetic compass can then be compared with the gyrocompass to ascertain the magnetic compass error.

Latitude	30-34.2 N
True Azimuth	359.9 °T
Gyro Brg.	000.5 °pgc
Gyro Error	.6 °(E or (W))

NOTES:

Figure 10-2. Completed gyro error computation, Polaris sight form.

Determining the Gyro Error by the Sun

The conditions under which the star Polaris must be observed are not always conducive to accuracy in the determination of the gyro error. Polaris is a second magnitude star, and is therefore not easily observed with the azimuth circle during twilight. Furthermore, dusk or darkness make it difficult to level the azimuth circle, and the altitude of the star is often in excess of 20°, which is considered the maximum altitude for an accurate azimuth measurement. Hence, the determination of the gyro error by observation of the sun is generally preferred by most navigators. The two methods by which this may be accomplished—the sun amplitude sight and the sun azimuth sight—will be discussed below.

The Sun Amplitude Sight

As was stated earlier in this chapter, observation of the sun for the purpose of obtaining a gyro error is best performed when it is centered on the celestial horizon of the observer, either in the act of rising or setting. An observation of any celestial body at this position is termed an *amplitude sight.* There are two main advantages in observing the sun at this position. Not only is the bearing of the sun easy to observe at this time using an azimuth circle, but its true azimuth can be found without having to use a sight reduction table, by the use of an *amplitude table* described below. Because of the combined effects of refraction and dip, in practice the assumption is usually made that the center of the sun is on the celestial horizon when its lower limb is about two-thirds of a solar diameter above the visible sea horizon. Although in some cases this assumption leads to a small error in the observed azimuth, any such error is insignificantly small in low and mid-latitudes. In high latitudes, the sun must be observed when its disk is centered on the visible horizon to ensure accuracy in the observed azimuth. In this case, a correction to the amplitude angle to compensate for the dip of the visible horizon is obtained from a separate table.

The sun, like all celestial bodies, appears to rise in the east and set in the west, as a result of the effects of the earth's rotation. When the declination of the sun is zero at the time of the vernal or autumnal equinoxes, it seems to an observer anywhere on earth that the sun is located at the easternmost point of his celestial horizon at sunrise, bearing 090°T from him. This effect is illustrated in Figure 10-3.

In this figure, AP_1, AP_2, and AP_3 are assumed positions lying along a meridian experiencing sunrise at the time of equinox. Because of the great distance of the earth from the sun, it would seem to an observer at each of the three positions that the sun were bearing 090°T

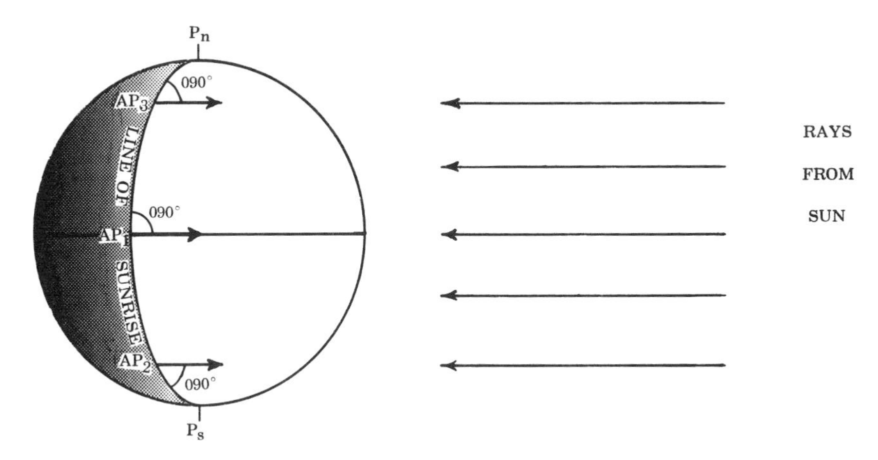

Figure 10-3. Sun at 0° Dec. bears 090° T at sunrise everywhere on earth.

from him at the moment at which this body appeared above his celestial horizon. By the same reasoning, it could be shown that at sunset the true bearing of the sun as it crossed the celestial horizon to the west would appear to be 270°T from all three positions.

If the sun were not at equinox, but rather had a declination of, say, 20° north, it should be obvious that the sun would no longer appear to rise bearing 090°T and set bearing 270°T. It would in fact subtend an angle between its apparent position and the easternmost or westernmost point of the observer's celestial horizon called the *amplitude angle*. This amplitude angle may be defined as the horizontal angular distance measured in a northerly or southerly direction from the prime vertical—that vertical circle passing through the east and west points of the observer's celestial horizon—to the apparent position of the body on the celestial horizon. The angle is given the prefix E (east) if the body is rising, or W (west) if it is setting, and the suffix N (north) if it has northerly declination, or S (south) if it has southerly declination. If the observer were located on the equator, the amplitude angle would be identical in size to the declination; thus, if the sun had 20° north declination at sunrise, the amplitude angle at the equator would be E 20° N. As the distance of the observer from the equator increases, however, the size of the amplitude angle also increases, reaching a maximum of 90° when the sum of the degrees of declination and the observer's latitude equal 90. In the case of the sun having declination N 20° at sunrise, the maximum value E 90° N occurs at latitude 70° north and south. Figures 10-4A and 10-4B may help to visualize this.

Figure 10-4A depicts the earth as it might be viewed from a position in space directly opposite the great circle line along which sunrise is being observed from three different positions on earth, AP_1, AP_2,

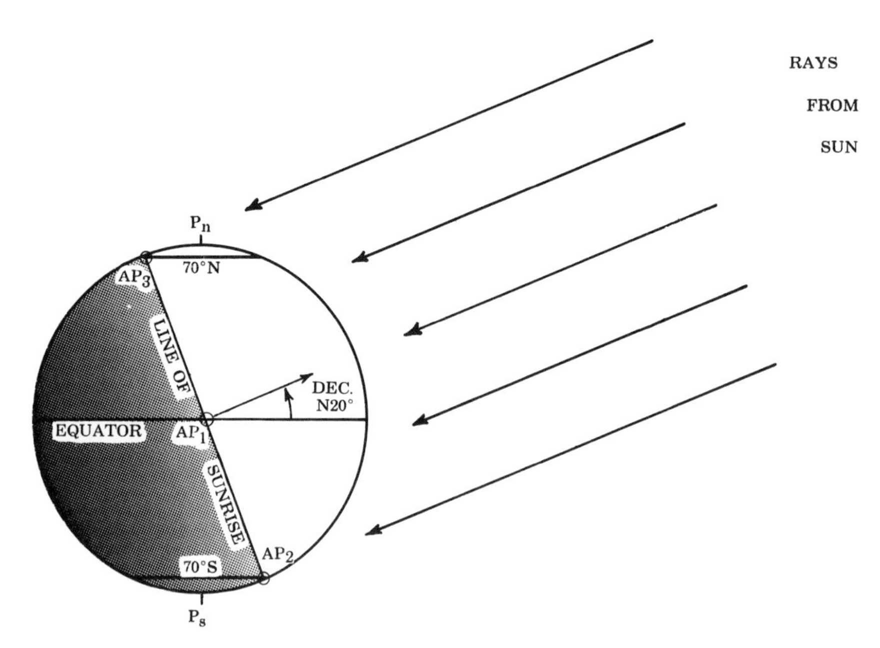

Figure 10-4A. Representation of sunrise great circle, with sun at declination N 20°.

and AP_3; Figure 10-4B represents the plane of the celestial horizon for each of these positions at sunrise. At AP_1, located on the equator, the sun appears to form an amplitude angle equal in size to its declination, E 20° N, at sunrise, which occurs here at about 0600 ZT. At AP_3 located at 70° N, however, sunrise occurs much earlier, at about 0000 ZT. Because the sun is visible across the north pole at that time, its amplitude angle is E 90° N, equivalent to a true bearing of due north. At AP_2, sunrise does not occur until about 1200 ZT, when the sun again forms an amplitude angle of E 90° N, or 000° true azimuth.

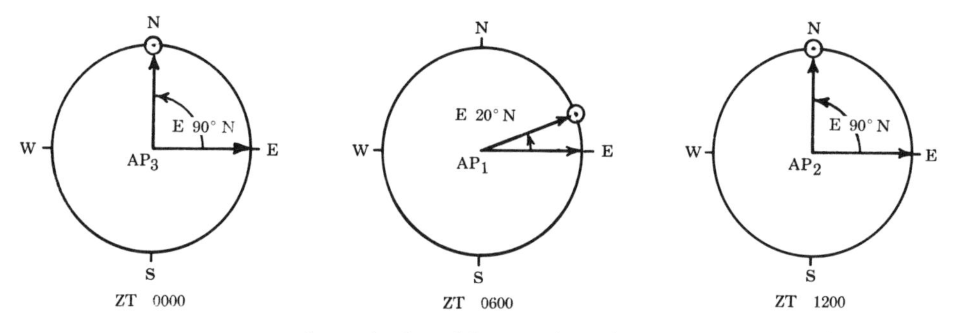

Figure 10-4B. Plane of celestial horizon from three positions at sunrise.

Consequently, all latitudes north of 70° N on this date would experience 24 hours of sunlight, while all positions south of 70° S would experience no daylight whatsoever. The subject of sunrise and sunset will be discussed in greater detail in the next chapter.

From the preceding discussion, it follows that the size of the amplitude angle depends on the declination of the sun and the latitude of the observer. If the sun is observed when its center is located on the celestial horizon, Table 27 in the *American Practical Navigator* (Bowditch) may be used to obtain the value of the amplitude angle. An excerpt from this table for declinations between 18° and 24° appears in Figure 10-5.

1296

TABLE 27
Amplitudes

Latitude	Declination													Latitude
	18°0	18°5	19°0	19°5	20°0	20°5	21°0	21°5	22°0	22°5	23°0	23°5	24°0	
°	°	°	°	°	°	°	°	°	°	°	°	°	°	°
0	18. 0	18. 5	19. 0	19. 5	20. 0	20. 5	21. 0	21. 5	22. 0	22. 5	23. 0	23. 5	24. 0	0
10	18. 3	18. 8	19. 3	19. 8	20. 3	20. 8	21. 3	21. 8	22. 4	22. 9	23. 4	23. 9	24. 4	10
15	18. 7	19. 2	19. 7	20. 2	20. 7	21. 3	21. 8	22. 3	22. 8	23. 3	23. 9	24. 4	24. 9	15
20	19. 2	19. 7	20. 3	20. 8	21. 3	21. 9	22. 4	23. 0	23. 5	24. 0	24. 6	25. 1	25. 6	20
25	19. 9	20. 5	21. 1	21. 6	22. 2	22. 7	23. 3	23. 9	24. 4	25. 0	25. 5	26. 1	26. 7	25
30	20. 9	21. 5	22. 1	22. 7	23. 3	23. 9	24. 4	25. 0	25. 6	26. 2	26. 8	27. 4	28. 0	30
32	21. 4	22. 0	22. 6	23. 2	23. 8	24. 4	25. 0	25. 6	26. 2	26. 8	27. 4	28. 0	28. 7	32
34	21. 9	22. 5	23. 1	23. 7	24. 4	25. 0	25. 6	26. 2	26. 9	27. 5	28. 1	28. 7	29. 4	34
36	22. 5	23. 1	23. 7	24. 4	25. 0	25. 7	26. 3	26. 9	27. 6	28. 2	28. 9	29. 5	30. 2	36
38	23. 1	23. 7	24. 4	25. 1	25. 7	26. 4	27. 1	27. 7	28. 4	29. 1	29. 7	30. 4	31. 1	38
40	23. 8	24. 5	25. 2	25. 8	26. 5	27. 2	27. 9	28. 6	29. 3	30. 0	30. 7	31. 4	32. 1	40
41	24. 2	24. 9	25. 6	26. 3	26. 9	27. 6	28. 3	29. 1	29. 8	30. 5	31. 2	31. 9	32. 6	41
42	24. 6	25. 3	26. 0	26. 7	27. 4	28. 1	28. 8	29. 5	30. 3	31. 0	31. 7	32. 5	33. 2	42
43	25. 0	25. 7	26. 4	27. 2	27. 9	28. 6	29. 3	30. 1	30. 8	31. 6	32. 3	33. 0	33. 8	43
44	25. 4	26. 2	26. 9	27. 6	28. 4	29. 1	29. 9	30. 6	31. 4	32. 1	32. 9	33. 7	34. 4	44
45	25. 9	26. 7	27. 4	28. 2	28. 9	29. 7	30. 5	31. 2	32. 0	32. 8	33. 5	34. 3	35. 1	45
46	26. 4	27. 2	27. 9	28. 7	29. 5	30. 3	31. 1	31. 8	32. 6	33. 4	34. 2	35. 0	35. 8	46
47	26. 9	27. 7	28. 5	29. 3	30. 1	30. 9	31. 7	32. 5	33. 3	34. 1	35. 0	35. 8	36. 6	47
48	27. 5	28. 3	29. 1	29. 9	30. 7	[illegible]	32. 4	33. 2	34. 0	34. 9	35. 7	36. 6	37. 4	48
49	[illegible]	[illegible]	[illegible]	[illegible]	[illegible]	[illegible]	[illegible]	[illegible]	[illegible]	[illegible]	[illegible]	[illegible]	[illegible]	49

Figure 10-5. Excerpt from Table 27, Bowditch.

The table is entered using as arguments the declination of the sun at the time of the amplitude sight, as determined from an almanac, and the DR latitude of the observer for the time of the observation. The magnitude of the amplitude angle corresponding to the exact declination and DR latitude is found by interpolation, and the proper prefix and suffix are added. Finally, the amplitude angle is converted to a true azimuth of the rising or setting sun, and this computed azimuth is compared with the observed azimuth to obtain the gyro error.

The conversion of an amplitude angle to a true azimuth is very similar to the conversion of a prefixed and suffixed azimuth angle to true azimuth. The base direction for the amplitude conversion is 090°T if the prefix is east (E), or 270°T if the prefix is west (W). If the suffix is north (N), the value of the amplitude angle is subtracted from 090°T for rising bodies, or added to 270°T for setting bodies. Conversely, if the suffix is south (S), the value of the amplitude angle is added to 090°T or subtracted from 270°T.

As an example of the use of the amplitude sight of the sun to obtain a gyro error, suppose that at 0600 ZT on 16 December the gyro bearing of the rising sun observed when the lower limb was two-thirds of a diameter above the visible horizon was 116° pgc. The 0600 ZT DR position was L 22° 30.0′ N, λ 58° 30.0′ W.

The first step in obtaining the gyro error is to obtain the declination of the sun for the time of the observation. The DR longitude places the ship in the +4 time zone, so the GMT of the observation is 1000 16 December. The declination of the sun for this time from the *Nautical Almanac* (see Figure 5-1B) is S 23° 18.6′. The amplitude angle can now be determined by interpolation in Table 27 of Bowditch, using this declination and the DR latitude as entering arguments. To simplify the computations, both arguments can be expressed to the nearest tenth of a degree with no loss in the accuracy of the result.

For purposes of the interpolation, the four values of the amplitude angle for the tabulated arguments bracketing the exact calculated arguments are extracted from the excerpt shown in Figure 10-5 and written as shown below:

	Declination		
	23.0°	S 23.3°	23.5°
20°	24.6		25.1
Latitude 22.5°			
25°	25.5		26.1

The interpolation is conducted in two steps, once for latitude, and again for declination; the order of the steps in immaterial. Thus, interpolating first for the exact latitude, we have:

	Declination		
	23.0°	S 23.3°	23.5°
20°	24.6		25.1
Latitude 22.5°	*25.1*		*25.6*
25°	25.5		26.1

Next, for declination:

	Declination		
	23.0°	S 23.3°	23.5°
Latitude 22.5°	25.1	*25.4*	25.6

Since it is sunrise, the prefix is E, and because the declination is south, the suffix is S. Thus, the resultant amplitude angle is E 25.4° S. Converting this amplitude angle to a true azimuth yields a value of 90° + 25.4° or 115.4°T. Finally, this true azimuth is compared to the observed azimuth 116.0° pgc to obtain the resultant gyro error, .6° W. In practice, the error is usually rounded to the nearest half degree, or .5° W in this case.

1297

TABLE 28

Correction of Amplitude as Observed on the Visible Horizon

Latitude	Declination													Latitude
	0°	2°	4°	6°	8°	10°	12°	14°	16°	18°	20°	22°	24°	
°	°	°	°	°	°	°	°	°	°	°	°	°	°	°
0	0. 0	0. 0	0. 0	0. 0	0. 0	0. 0	0. 0	0. 0	0. 0	0. 0	0. 0	0. 0	0. 0	0
10	0. 1	0. 1	0. 1	0. 1	0. 1	0. 1	0. 1	0. 1	0. 1	0. 1	0. 1	0. 1	0. 1	10
15	0. 2	0. 2	0. 2	0. 2	0. 2	0. 2	0. 2	0. 2	0. 2	0. 2	0. 2	0. 2	0. 2	15
20	0. 3	0. 3	0. 3	0. 3	0. 3	0. 3	0. 3	0. 3	0. 3	0. 3	0. 3	0. 3	0. 3	20
25	0. 3	0. 3	0. 3	0. 3	0. 3	0. 4	0. 3	0. 3	0. 3	0. 3	0. 3	0. 3	0. 3	25
30	0. 4	0. 4	0. 4	0. 4	0. 5	0. 4	0. 4	0. 4	0. 4	0. 4	0. 4	0. 5	0. 5	30
32	0. 4	0. 4	0. 4	0. 4	0. 5	0. 4	0. 4	0. 4	0. 4	0. 4	0. 5	0. 5	0. 5	32
34	0. 5	0. 5	0. 5	0. 5	0. 5	0. 5	0. 5	0. 5	0. 5	0. 5	0. 5	0. 5	0. 5	34
36	0. 5	0. 5	0. 5	0. 5	0. 5	0. 5	0. 5	0. 5	0. 6	0. 5	0. 6	0. 6	0. 6	36
38	0. 6	0. 6	0. 6	0. 6	0. 6	0. 6	0. 6	0. 6	0. 6	0. 6	0. 6	0. 6	0. 6	38
40	0. 6	0. 6	0. 6	0. 6	0. 6	0. 6	0. 6	0. 6	0. 6	0. 6	0. 7	0. 7	0. 7	40
42	0. 6	0. 6	0. 6	0. 6	0. 7	0. 7	0. 7	0. 7	0. 7	0. 7	0. 7	0. 7	0. 7	42
44	0. 7	0. 7	0. 7	0. 6	0. 6	0. 7	0. 7	0. 7	0. 8	0. 8	0. 8	0. 8	0. 9	44
46	0. 7	0. 7	0. 7	0. 7	0. 7	0. 8	0. 8	0. 8	0. 8	0. 8	0. 8	0. 9	0. 9	46
48	0. 8	0. 8	0. 8	0. 8	0. 8	0. 8	0. 8	0. 8	0. 9	0. 9	1. 0	1. 0	1. 0	48
50	0. 8	0. 8	0. 8	0. 8	0. 9	0. 9	0. 9	0. 9	0. 9	1. 0	1. 0	1. 1	1. 0	50
51	0. 8	0. 8	0. 8	0. 8	0. 9	0. 9	0. 9	0. 9	0. 9	1. 0	1. 1	1. 1	1. 1	51
52	0. 9	0. 9	0. 9	0. 9	0. 9	0. 9	1. 0	1. 0	1. 0	1. 1	1. 1	1. 1	1. 3	52
53	0. 9	0. 9	0. 9	0. 9	0. 9	0. 9	1. 0	1. 0	1. 0	1. 1	1. 2	1. 2	1. 3	53
54	1. 0	1. 0	1. 0	1. 0	1. 0	1. 0	1. 1	1. 1	1. 1	1. 2	1. 2	1. 3	1. 3	54
55	1. 0	1. 0	1. 0	1. 0	1. 1	1. 1	1. 0	1. 2	1. 2	1. 2	1. 3	1. 3	1. 4	55
56	1. 0	1. 0	1. 0	1. 0	1. 1	1. 1	1. 2	1. 2	1. 2	1. 3	1. 3	1. 4	[illegible]	[illegible]
57	[illegible]	1. 1	1. 1	1. 1	1. 1	[illegible]	[illegible]	1. 2	1. 3	[illegible]	[illegible]	[illegible]	[illegible]	[illegible]

Figure 10-6. "Correction of Amplitude," Table 28, Bowditch.

Had the amplitude observation of the sun been made when its center was located on the visible horizon, a small correction to the resulting observed azimuth would have been necessary. This correction is tabulated in Table 28 of Bowditch, reproduced in part in Figure 10-6. The correction extracted for sun observations is always applied in the direction away from the elevated pole, as per the instructions at the bottom of the table.

The Sun Azimuth Sight

At times between sunrise and sunset, the gyro error may be obtained by computing the true azimuth of the sun by means of any one of a number of special tables designed for azimuth computations, or by use of the *Sight Reduction Tables No. 229.* In practice, *Tables No. 229* are generally preferred for obtaining the true azimuth, inasmuch as these tables are widely available, and they are useful for reducing a standard sight as well. It would also be possible to obtain a true azimuth accurate to the nearest whole degree using Volumes II and III of *Tables No. 249,* but this precision is insufficient to yield a gyro error accurate to the nearest half degree. *Tables No. 249,* therefore, are not suitable for the gyro error determination.

To use *Tables No. 229* for computation of the exact azimuth of the sun, it is necessary to interpolate in the tables for the exact values of DR latitude, LHA, and declination of the sun for the time of the azimuth observations. The DR latitude is, of course, obtained from the DR plot, while the LHA and declination of the sun are derived from either the *Air Almanac* or, preferably, the *Nautical Almanac.* A sight form designed for use in performing the required triple interpolation in *Tables No. 229* is helpful; a form prepared for use at the U.S. Naval Academy appears in Figure 10-7.

Entering arguments for the tables for the exact azimuth determination are the whole degrees of DR latitude, LHA, and declination lower than the exact calculated values. The tabulated azimuth angle Z corresponding with these integral arguments is first extracted and entered on the form as "Tab. Z". Next, the amount of change in the value of the tabulated azimuth angle, or "Z difference," is found between this tabulated azimuth angle and the value corresponding to the next higher integral degree of each of the three entering arguments. The Z difference for a one-degree increase of latitude, for example, is found by comparing the tabulated azimuth with the value contained in the next adjacent latitude column to the right, while keeping the LHA and declination the same. Likewise, the other two Z differences are found by successive comparisons of the tabulated azimuth angle for the next higher degrees of LHA and declination. All three Z differences are recorded on the exact azimuth form.

Using the exact minutes of each of the three entering arguments, each Z difference is then interpolated for the exact value of its corresponding argument; the result of each such interpolation is the correction necessary to the tabulated azimuth angle for the minutes of the

EXACT AZIMUTH USING TABLES 229

Body ______

DR L ______

DR λ ______

Date (L) ______

ZT ______

ZD (+ or -) ______

GMT ______

Date (G) ______

Tab GHA ______

Inc'mt ______

GHA ______

DR λ ______

LHA ______

d(+/-)

Tab Dec ______

d corr ______

Dec ______

	EXACT Deg	Min	Z DIFF. (+ or -)	CORR. (+ or -)
LAT				
LHA				
DEC				
			Total (±)	

Tab Z ______

Exact Z ______

Exact Zn ______

Gyro/Compass Brg ______

Gyro/Compass Error ______

NORTH LAT

LHA greater than 180° .... Zn = Z
LHA less than 180° .... Zn = 360° - Z

SOUTH LAT

LHA greater than 180° .... Zn = 180° - Z
LHA less than 180° .... Zn = 180° + Z

Figure 10-7. USNA "Exact Azimuth by Tables No. 229" form.

argument. All three corrections are added algebraically to form the total correction to the tabulated azimuth angle. This total correction is then applied to obtain the exact azimuth angle corresponding to the exact values of the three entering arguments. As the final step, this exact azimuth angle is converted to a true azimuth, using the conversion formula printed on the azimuth sight form.

Comparison of the true azimuth thus computed with the observed azimuth of the sun yields the gyro error.

The above procedure for determining the gyro error by observation of the azimuth of the sun may be summarized by the following steps:

1. Obtain and record the DR latitude, exact LHA, and declination of the sun for the time of the azimuth observation.

2. Using the integral degrees of these three quantities as entering arguments in *Tables No. 229,* extract the corresponding tabulated azimuth angle Z.

3. Obtain and record the three Z differences between this tabulated Z and the tabulated values for the next higher degree of each of the entering arguments.

4. Interpolate each Z difference for the correction corresponding to the exact minutes of its entering argument.

5. Add all three corrections algebraically to form the total correction to tabulated Z.

6. Apply the correction to obtain the exact azimuth angle.

7. Convert the exact azimuth angle to a true azimuth.

8. Compare the true azimuth with the observed azimuth to obtain the gyro error.

As an example of the determination of a gyro error by the sun azimuth sight, suppose that at 11-12-06 on 16 December the azimuth of the sun was observed bearing 161° pgc. The ship's DR position for this time was L 30° 15.0′ N, λ 64° 33.0′ W.

Working through the stepwise procedure enumerated above, the *Nautical Almanac* is first consulted to obtain the LHA and declination of the sun for the GMT of the observation, 15-12-06 on 16 December. After the resulting LHA and declination of the sun have been entered on the form along with the DR latitude, it should appear as in Figure

EXACT AZIMUTH USING TABLES 229

Body	SUN	
DR L	30-15.0 N	
DR λ(+E-W)	64-33.0 W	
Date (L)	16 DEC	
ZT	11-12-06	
ZD (+ or −)	+4	
GMT	15-12-06	
Date (G)	16 DEC	
Tab GHA	46-06.1	
Inc'mt	3-01.5	
GHA	49-07.6	
DR λ	64-33.0 W	
LHA	344-34.6	
		d(+/−)
Tab Dec	S 23-19.2	+.1
d corr	0	
Dec	S 23-19.2	

	EXACT Deg	EXACT Min	Z DIFF. (+ or −)	CORR. (+ or −)
LAT	30	15.0 N		
LHA	344	34.6		
DEC	S 23	19.2		
			Total (±)	

Tab Z

Exact Z

Exact Zn

Gyro/Compass Brg

Gyro/Compass Error

NORTH LAT

LHA greater than 180° .... Zn = Z
LHA less than 180° .... Zn = 360° − Z

SOUTH LAT

LHA greater than 180° .... Zn = 180° − Z
LHA less than 180° .... Zn = 180° + Z

Figure 10-8A. Exact azimuth form prepared for use with Tables No. 229.

10-8A, ready for use with *Tables No. 229.* Note that the GHA in this case was less than the westerly DR longitude, necessitating the addition of 360° to the GHA to produce a positive LHA.

The tabulated azimuth angle Z from *Tables No. 229* corresponding with the integral arguments latitude 30°N, LHA 344°, and declination S 23° is 162.0. The Z difference between this value and the tabulated value for the same LHA and declination but a latitude of 31° N is +.2. The Z difference for 30° N latitude, S 23° declination, but LHA 345° is +1.1, and for 30° N latitude, LHA 344°, and declination S 24°, the Z difference is +.3. Interpolation of the +.2 latitude Z difference for 15.0′ of latitude yields a latitude Z correction of +.1. In similar fashion, interpolation for the remaining LHA and declination corrections results in respective values of +.6 and +.1. Adding the three corrections yields a total Z correction of +.8. Applying this +.8 to the tabulated azimuth figure forms the exact azimuth angle 162.8°, which is converted to a true azimuth 162.8°T. As the final step, comparison with the observed azimuth, 161° pgc, yields a gyro error of 1.8° E, which for practical purposes is rounded off to 2.0° E. The completed azimuth form appears in Figure 10-8B.

EXACT AZIMUTH USING TABLES 229

Body	SUN	
DR L	30-15.0 N	
DR λ(+E-W)	64-33.0 W	
Date (L)	16 DEC	
ZT	11-12-06	
ZD (+ or -)	+4	
GMT	15-12-06	
Date (G)	16 DEC	
Tab GHA	46-06.1	
Inc'mt	3-01.5	
GHA	49-07.6	
DR λ	64-33.0 W	
LHA	344-34.6	d(+/-)
Tab Dec	S 23-19.2	+.1
d corr	0	
Dec	S 23-19.2	

	EXACT Deg	EXACT Min	Z DIFF. (+ or -)	CORR. (+ or -)
LAT	30	15.0 N	+.2	+.1
LHA	344	34.6	+1.1	+.6
DEC	S 23	19.2	+.3	+.1
			Total (±)	+.8
			Tab Z	162.0
			Exact Z	162.8
			Exact Zn	162.8° T
			Gyro/Compass Brg	161° pgc
			Gyro/Compass Error	1.8° E

NORTH LAT

LHA greater than 180° .... Zn = Z
LHA less than 180° .... Zn = 360° - Z

SOUTH LAT

LHA greater than 180° .... Zn = 180° - Z
LHA less than 180° .... Zn = 180° + Z

Figure 10-8B. Completed exact azimuth form.

Before leaving the subject of the azimuth sight, it should be stressed that the true azimuth of *any* celestial body can be found using *Tables No. 229,* for comparison with the observed azimuth of the body to obtain a compass error. The sun, however, is the body most commonly used for this purpose, since observation of its azimuth is very convenient in the latitude regions in which most ships operate.

Summary

This chapter has examined the procedures most commonly used to determine the gyro error of a ship operating at sea beyond piloting waters. In lower northern latitudes, observation of the azimuth of Polaris at twilight is a very convenient method of obtaining the gyro error, while an amplitude or azimuth observation of the sun yields excellent results by day in either northern or southern latitudes. Computation of the true azimuth of Polaris requires only the Polaris tables in the *Nautical Almanac.* The amplitude and thence the true azimuth of the sun at sunrise or sunset can be quickly obtained using Tables 27 and 28 in the *American Practical Navigator* (Bowditch), and the true azimuth of the sun at other times may be derived using the *Sight Reduction Tables for Marine Navigation, No. 229.*

11

Twilight and Other Rising and Setting Phenomena

There are several phenomena associated with the rising and setting of the sun and other celestial bodies of significance and consequent interest to the navigator. The most important of these are twilight, sunrise and sunset, and moonrise and moonset. *Twilight* is the period before sunrise when darkness is giving way to daylight, and after sunset, when the opposite progression takes place. Morning twilight ends at *sunrise*, defined as the first appearance of the sun's upper limb above the visible horizon, and evening twilight begins at *sunset*, or the disappearance of the sun's upper limb below the horizon. *Moonrise* and *moonset* are defined similarly to sunrise and sunset, by the contact of the upper limb of the moon with the visible horizon.

Twilight is of special interest to the navigator, as this is the only time when the visible horizon is still light enough to be clearly defined, while the navigational stars and planets are bright enough to be observed with a marine sextant. Sunrise and sunset are only slightly less important. Not only do the times of these respective events signify the ending and beginning of twilight, but as was discussed in the last chapter, they also indicate the approximate time at which a sun amplitude sight may be observed. Moonrise and moonset do not have much navigational significance, but nevertheless these phenomena can be of interest to the Navy navigator for tactical planning applications.

Because of the regular rate of increase of the Greenwich hour angle (GHA) of the sun and the very small rate of change in its declination with time, phenomena related to the rising and setting of the sun are tabulated in the *Nautical Almanac* for the central day of each three-day period covered. Mean times of sunrise and sunset and the darker limits of civil and nautical twilight are given as they occur along the Greenwich meridian at various latitudes between 72° N and 60° S. On the other hand, because the motion of the moon is very irregular both in GHA and declination, tabulations of the mean times of moonrise and moonset along the Greenwich meridian between latitudes 72° N and 60° S are required for each day. Because of the regular rate of increase of the GHA of the sun (approximately 15 degrees per hour) the tabulated mean times of sun-associated phenomena at

Greenwich (i.e., the GMTs) can be used as the local mean times (LMTs) of the phenomena at all other meridians. But because of the nonuniform motion of the moon, the tabulated GMTs of moonrise and moonset are valid without correction only along the Greenwich meridian.

Because of the inclination of the earth's axis, at certain times of the year the sun is either continually above or continually below the celestial horizon at the higher tabulated latitudes in the *Nautical Almanac*. This results in either constant daylight, twilight, or night being experienced at these latitudes throughout the three-day period. The symbol //// is shown in place of a time to indicate twilight throughout the night, the symbol ⚌ indicates constant daylight, and the symbol ▬ indicates constant night. If the moon is continually above or below the celestial horizon at a tabulated latitude, this fact is also denoted by the symbols ▭ and ▬, respectively. The phase and age of the moon are given at the bottom of each right-hand daily page, along with its time of meridian passage at Greenwich. A set of tabulated data for twilight, sunrise and sunset, and moonrise and moonset from the *Nautical Almanac* for 15, 16, and 17 December of a typical year appears in Figure 11-1.

In the *Air Almanac*, tables for predicting the darker limits of civil twilight and sunrise and sunset are located in the back of each volume, while tables for moonrise and moonset similar to those in the *Nautical Almanac* appear in the daily pages. The *Nautical Almanac* tabulations are generally preferred by most surface navigators, however, and their use will be demonstrated in this chapter.

Predicting the Darker Limit Time of Twilight

As mentioned above, twilight is the period of incomplete darkness occurring just before sunrise or just after sunset. There are two kinds of twilight of concern in celestial navigation, differentiated by the position of the sun below the horizon at the darker limit:

> *Civil twilight* is the period extending from sunrise or sunset at the lighter limit to the time at which the center of the sun is 6° below the celestial horizon at the darker limit.
>
> *Nautical twilight* is the period extending from sunrise or sunset at the lighter limit to the time at which the center of the sun is 12° below the celestial horizon at the darker limit.

Civil twilight is characterized by a fairly sharp horizon and a light sky wherein only the brighter first and second magnitude stars and navigational planets are visible; during the darker stages of nautical twilight, the horizon may become very vague and the sky quite dark,

Lat.	Twilight Naut.	Twilight Civil	Sunrise	Moonrise 15	Moonrise 16	Moonrise 17	Moonrise 18
°	h m	h m	h m	h m	h m	h m	h m
N 72	08 20	10 48	■	□	□	18 22	20 42
N 70	08 01	09 48	■	□	16 01	18 55	20 56
68	07 45	09 14	■	□	17 08	19 18	21 08
66	07 32	08 49	10 28	15 14	17 44	19 36	21 17
64	07 21	08 29	09 47	16 14	18 09	19 51	21 25
62	07 12	08 13	09 19	16 47	18 29	20 03	21 31
60	07 03	08 00	08 58	17 11	18 45	20 13	21 37
N 58	06 56	07 48	08 40	17 31	18 58	20 22	21 42
56	06 49	07 38	08 26	17 47	19 10	20 30	21 47
54	06 43	07 29	08 13	18 01	19 20	20 36	21 51
52	06 38	07 21	08 02	18 13	19 29	20 43	21 54
50	06 32	07 14	07 52	18 23	19 37	20 48	21 58
45	06 21	06 58	07 32	18 45	19 54	21 00	22 05
N 40	06 10	06 44	07 15	19 03	20 07	21 10	22 10
35	06 01	06 33	07 01	19 18	20 19	21 18	22 16
30	05 52	06 22	06 49	19 31	20 29	21 26	22 20
20	05 36	06 04	06 28	19 52	20 47	21 38	22 28
N 10	05 20	05 46	06 09	20 11	21 02	21 49	22 34
0	05 03	05 29	05 52	20 29	21 16	22 00	22 41
S 10	04 44	05 11	05 34	20 46	21 30	22 10	22 47
20	04 22	04 51	05 15	21 05	21 45	22 21	22 53
30	03 53	04 26	04 53	21 26	22 02	22 33	23 01
35	03 34	04 11	04 41	21 39	22 12	22 40	23 05
40	03 11	03 53	04 26	21 53	22 23	22 48	23 10
45	02 41	03 30	04 08	22 10	22 36	22 58	23 16
S 50	01 56	03 01	03 45	22 31	22 52	23 09	23 22
52	01 28	02 46	03 34	22 41	23 00	23 14	23 26
54	00 44	02 28	03 22	22 52	23 08	23 20	23 29
56	////	02 06	03 08	23 04	23 17	23 26	23 33
58	////	01 37	02 52	23 19	23 28	23 33	23 37
S 60	////	00 49	02 31	23 36	23 40	23 41	23 41

Lat.	Sunset	Twilight Civil	Twilight Naut.	Moonset 15	Moonset 16	Moonset 17	Moonset 18
°	h m	h m	h m	h m	h m	h m	h m
N 72	■	13 02	15 30	□	□	13 37	12 45
N 70	■	14 02	15 49	□	14 25	13 03	12 28
68	■	14 36	16 05	□	13 16	12 38	12 15
66	13 22	15 01	16 18	13 29	12 40	12 19	12 04
64	14 04	15 21	16 29	12 29	12 13	12 03	11 55
62	14 31	15 37	16 39	11 55	11 53	11 50	11 47
60	14 53	15 50	16 47	11 30	11 36	11 39	11 40
N 58	15 10	16 02	16 54	11 10	11 22	11 29	11 34
56	15 25	16 12	17 01	10 53	11 10	11 20	11 28
54	15 38	16 21	17 07	10 39	10 59	11 13	11 23
52	15 49	16 29	17 13	10 27	10 49	11 06	11 19
50	15 58	16 37	17 18	10 16	10 41	11 00	11 15
45	16 19	16 53	17 30	09 53	10 23	10 46	11 06
N 40	16 36	17 06	17 41	09 34	10 08	10 35	10 58
35	16 50	17 18	17 50	09 19	09 55	10 25	10 52
30	17 02	17 29	17 59	09 05	09 44	10 17	10 46
20	17 23	17 48	18 15	08 42	09 25	10 02	10 36
N 10	17 42	18 05	18 31	08 22	09 08	09 49	10 28
0	17 59	18 22	18 48	08 03	08 52	09 37	10 20
S 10	18 17	18 40	19 07	07 44	08 36	09 25	10 11
20	18 36	19 00	19 29	07 23	08 19	09 12	10 02
30	18 58	19 25	19 58	06 59	07 59	08 57	09 52
35	19 11	19 40	20 17	06 45	07 47	08 48	09 46
40	19 26	19 58	20 40	06 29	07 34	08 38	09 39
45	19 44	20 21	21 10	06 09	07 18	08 26	09 31
S 50	20 06	20 50	21 56	05 45	06 58	08 11	09 22
52	20 17	21 06	22 23	05 33	06 49	08 04	09 17
54	20 29	21 23	23 08	05 19	06 39	07 57	09 12
56	20 43	21 45	////	05 03	06 26	07 48	09 07
58	21 00	22 15	////	04 44	06 13	07 39	09 01
S 60	21 20	23 04	////	04 21	05 56	07 27	08 54

Day	SUN Eqn. of Time 00ʰ	SUN Eqn. of Time 12ʰ	SUN Mer. Pass.	MOON Mer. Pass. Upper	MOON Mer. Pass. Lower	MOON Age	MOON Phase
	m s	m s	h m	h m	h m	d	
15	05 12	04 58	11 55	01 50	14 16	17	●
16	04 43	04 29	11 56	02 40	15 04	18	
17	04 14	04 00	11 56	03 27	15 49	19	

Figure 11-1. Rising and setting phenomena predictions, Nautical Almanac, *15, 16, 17 December.*

with the dimmer second and third magnitude stars and planets becoming visible.

The prediction of the zone time of the beginning of morning civil or nautical twilight and the ending of evening civil or nautical twilight for either a stationary or moving ship is somewhat similar to the estimation of the zone time of LAN discussed in Chapter 9. Because the LMT of the darker limits of twilight varies with the latitude of the observer, it is necessary to adjust the tabulated mean time of twilight phenomena for the difference in latitude between the latitude tabulated in the *Nautical Almanac* and the latitude of the observer. The adjusted mean time of the twilight phenomenon of interest is then converted to local zone time by applying the arc-time difference in longitude between the observer's longitude and the standard meridian of his time zone, as was done in the LAN problem.

In practice, the navigator is usually most concerned with the prediction of the zone time of the darker limit of civil twilight, inasmuch as this time occurs about midway in the period suitable for celestial observations. If the ship is stationary in a position such as an anchorage, the time of the twilight phenomena can be determined with great accuracy using the *Nautical Almanac*. If the ship is moving, the time can be predicted within an accuracy of one or two minutes based on the DR plot.

Regardless of whether the ship is stationary or moving, the procedure for estimating the zone time of a darker limit of civil or nautical twilight is basically the same. As the first step, the mean time of the event of interest at the nearest tabulated latitude in the daily pages of the almanac smaller than the DR latitude (for a moving ship) or the ship's position (for a stationary ship) is extracted. Next, a correction to this time for the difference between the ship's latitude and the lower tabulated latitude is obtained from Table I of the "Tables for Interpolating Sunrise, Moonrise, Etc." in the back of the *Nautical Almanac*. This table is reproduced in Figure 11-2.

The entering arguments for Table I are the difference in latitude between the two tabulated values bracketing the ship's latitude (called the *tabular interval* in Table I), the closest tabulated difference to the actual difference between the ship's latitude and the lower of the two tabulated latitudes, and the closest tabulated time difference between the mean times of the event of interest at the upper and lower tabulated latitudes. The sign of the resulting LMT correction is determined by inspection of the daily predictions.

Application of this latitude LMT correction to the listed mean time of the phenomenon at the smaller tabulated latitude yields the LMT of the phenomenon at all three meridians of interest: Greenwich, the observer's meridian, and the standard meridian of his time zone.

TABLES FOR INTERPOLATING SUNRISE, MOONRISE, ETC.

TABLE I—FOR LATITUDE

Tabular Interval			Difference between the times for consecutive latitudes															
10°	5°	2°	5m	10m	15m	20m	25m	30m	35m	40m	45m	50m	55m	60m	1h 05m	1h 10m	1h 15m	1h 20m
° ′	° ′	° ′	m	m	m	m	m	m	m	m	m	m	m	m	h m	h m	h m	h m
0 30	0 15	0 06	0	0	1	1	1	1	1	2	2	2	2	2	0 02	0 02	0 02	0 02
1 00	0 30	0 12	0	1	1	2	2	3	3	3	4	4	4	5	05	05	05	05
1 30	0 45	0 18	1	1	2	3	3	4	4	5	5	6	7	7	07	07	07	07
2 00	1 00	0 24	1	2	3	4	5	5	6	7	7	8	9	10	10	10	10	10
2 30	1 15	0 30	1	2	4	5	6	7	8	9	9	10	11	12	12	13	13	13
3 00	1 30	0 36	1	3	4	6	7	8	9	10	11	12	13	14	0 15	0 15	0 16	0 16
3 30	1 45	0 42	2	3	5	7	8	10	11	12	13	14	16	17	18	18	19	19
4 00	2 00	0 48	2	4	6	8	9	11	13	14	15	16	18	19	20	21	22	22
4 30	2 15	0 54	2	4	7	9	11	13	15	16	18	19	21	22	23	24	25	26
5 00	2 30	1 00	2	5	7	10	12	14	16	18	20	22	23	25	26	27	28	29
5 30	2 45	1 06	3	5	8	11	13	16	18	20	22	24	26	28	0 29	0 30	0 31	0 32
6 00	3 00	1 12	3	6	9	12	14	17	20	22	24	26	29	31	32	33	34	36
6 30	3 15	1 18	3	6	10	13	16	19	22	24	26	29	31	34	36	37	38	40
7 00	3 30	1 24	3	7	10	14	17	20	23	26	29	31	34	37	39	41	42	44
7 30	3 45	1 30	4	7	11	15	18	22	25	28	31	34	37	40	43	44	46	48
8 00	4 00	1 36	4	8	12	16	20	23	27	30	34	37	41	44	0 47	0 48	0 51	0 53
8 30	4 15	1 42	4	8	13	17	21	25	29	33	36	40	44	48	0 51	0 53	0 56	0 58
9 00	4 30	1 48	4	9	13	18	22	27	31	35	39	43	47	52	0 55	0 58	1 01	1 04
9 30	4 45	1 54	5	9	14	19	24	28	33	38	42	47	51	56	1 00	1 04	1 08	1 12
10 00	5 00	2 00	5	10	15	20	25	30	35	40	45	50	55	60	1 05	1 10	1 15	1 20

Table I is for interpolating the L.M.T. of sunrise, twilight, moonrise, etc., for latitude. It is to be entered, in the appropriate column on the left, with the difference between true latitude and the nearest tabular latitude which is *less* than the true latitude; and with the argument at the top which is the nearest value of the difference between the times for the tabular latitude and the next higher one; the correction so obtained is applied to the time for the tabular latitude; the sign of the correction can be seen by inspection. It is to be noted that the interpolation is not linear, so that when using this table it is essential to take out the tabular phenomenon for the latitude *less* than the true latitude.

Figure 11-2. Table I, "Sunrise, Moonrise" interpolation tables, Nautical Almanac.

Next, this initial estimate of the LMT of the phenomenon is converted to a zone time by application of an arc-time correction corresponding to the difference between the observer's longitude and the standard meridian of his time zone, just as was done in estimating the time of LAN. The resulting zone time is the time of the darker limit of twilight if the observer's ship is stationary; if it is moving, this time becomes the "first estimate" of the zone time of the event.

In the case of a moving ship, a DR position for the first estimate time is plotted. A second and final estimate of the zone time of the twilight phenomenon is then made by repeating the procedure described above, using the first estimate DR position as the observer's position.

Inasmuch as the procedure for predicting the zone time of the darker limit of twilight is virtually identical to the procedure used for predicting the time of sunrise or sunset, an example of which will be given in the next section of this chapter, no example of a twilight solution will be given herein.

Predicting the Time of Sunrise and Sunset

As was mentioned at the beginning of this chapter, the prediction of the zone time of sunrise and sunset is of importance not only because these phenomena mark the end of morning and the beginning of evening twilight, but also because one of the most accurate types of azimuth sights possible—the amplitude sight—can be made near these times.

The prediction of sunrise and sunset is accomplished in virtually the same manner as the prediction of the time of a darker limit of twilight. As is the case with predictions of the latter type, the determination of times of sunrise and sunset can be made either for a stationary position or a moving ship. As an aid to the inexperienced navigator, the procedure for predicting the zone time of a darker limit of twilight or the zone time of sunrise or sunset using the *Nautical Almanac* is summarized as follows:

1. Obtain the mean time of the phenomenon of interest at Greenwich on the day in question for the tabulated latitude smaller than the DR latitude (moving ship) or the ship's position (stationary ship).

2. In the case of a moving ship, plot and record a DR position for this time.

3. Using this position, determine a first estimate of zone time of the phenomenon as follows:

   A. Find the latitude LMT correction from Table I corresponding to the difference between the ship's latitude and the lower tabulated latitude.

   B. Apply this correction to the tabulated mean time of the phenomenon.

   C. Convert the resulting adjusted LMT of the phenomenon to zone time by applying the arc-time difference between the ship's longitude and the standard meridian of the time zone by which the ship's clocks are set. In so doing, add the difference if west of the standard meridian, subtract if east.

For a stationary ship, this first estimate is the predicted zone time of the phenomenon. In the case of a moving ship, the following additional steps to find a second estimated time are required:

4. Plot a DR position for the zone time of the first estimate.

5. Using this position, determine the second estimate of zone time of the phenomenon:

   A. Find a second latitude correction from Table I.

B. Apply this correction to the tabulated time of the phenomenon at the lower tabulated latitude.

C. Convert the resulting second estimate of LMT to zone time by determining and applying a second longitude arc-time difference for the first estimate DR longitude.

If the first estimate of the zone time of the phenomenon for a moving ship places the ship above the next higher latitude for which tabulations are printed in the *Nautical Almanac,* care must be taken when finding the second estimate to use the listed time corresponding to this next higher latitude as the base time to which the second estimate Table I and arc-time corrections are applied. Conversely, if the first estimate places the ship below the latitude used as an entering argument for the initial estimate of the time of the phenomenon, the second estimate Table I and longitude corrections should be applied to the tabulated mean time corresponding to the next lower tabulated latitude.

As an example of the application of this procedure to find the predicted zone time of sunrise for a moving ship, suppose that at 0600 ZT on 16 December a ship is located at a DR position L 31° 42.0′ N, λ 65° 26.3′ W, steaming on course 140°T at 20 knots. The navigator desires to predict the time of sunrise, assuming the ship remains on the same course and speed.

Using the procedure outlined above, the first step is to extract the tabulated LMT of sunrise at the approximate DR position of the ship. Since the DR plot indicates that the ship is between 30° and 35° N, the mean time tabulated in the *Nautical Almanac* for 16 December for latitude 30° N is extracted; it is 0649. A DR position is plotted and recorded for this time at L 31° 29.3′ N, λ 65° 14.0′ W. At this point, the DR plot appears as shown in Figure 11-3A next page.

For this DR position, the tabular interval is 5°, the difference in latitude is 31° 29.3′ − 30° = 1° 29.3′, and the difference in time between the tabulations for 30° and 35° N is 12 minutes. Entering Table I with the closest tabulated arguments, a latitude LMT correction of +3 minutes is extracted. Applying this correction to the tabulated mean time for 30° N, 0649, yields an adjusted LMT of 0652.

Next, this adjusted LMT of sunrise is converted to zone time by application of the arc-time difference corresponding to the difference in longitude between the 0649 DR position and 60° W, the standard meridian of the time zone in this case. The arc difference is 5° 14.0′, so the corresponding time correction taken from the "Conversion of ARC to Time" tables in the almanac is 21 minutes; the sign of the correction is positive, since the ship is located to the west of the standard meridian. Application of this correction to the LMT adjusted

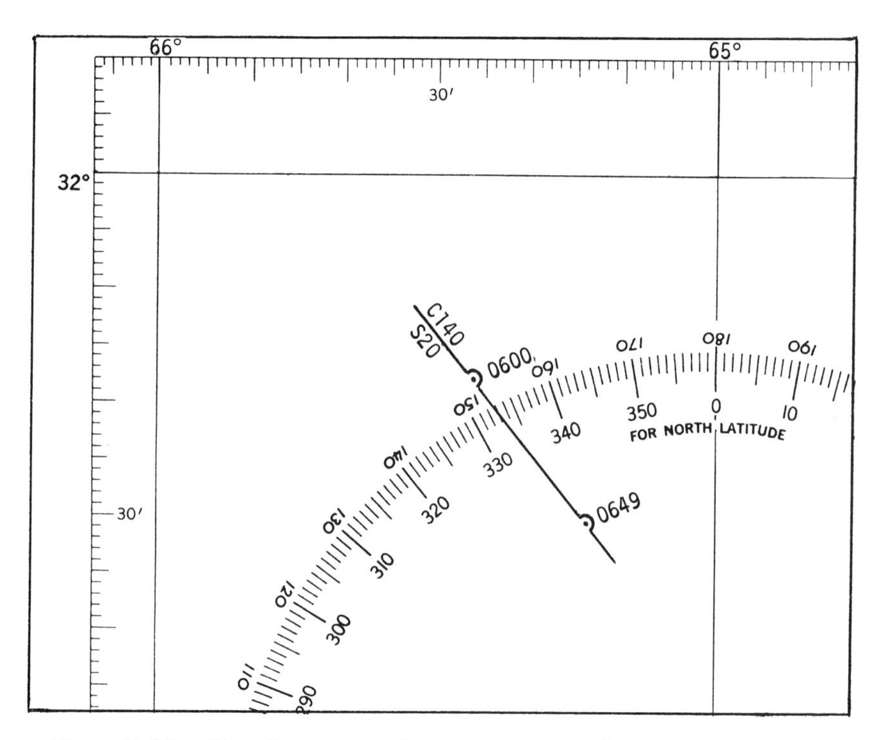

Figure 11-3A. Plot of a sunrise solution on a moving ship.

for latitude, 0652, yields a first estimate of zone time of sunrise of 0713.

To find the second estimate of the zone time of sunrise, a DR position is first plotted for 0713; it is L 31° 23.0′ N, λ 65° 07.8′ W. Now the correction procedure is repeated using this position as a basis. The LMT latitude correction from Table I is +3, resulting in a second estimate of LMT of sunrise identical to the first, 0652. The arc-time longitude correction for the 0713 DR position is again 21 minutes, so the second and final estimate of zone time of sunrise is 0713. The completed DR plot appears in Figure 11-3B.

As an aid to the inexperienced navigator, the sequence of the estimation procedure for this example is shown in tabular form below:

Tab GMT (LMT)	0649
0649 DR L.	31° 29.3′ N
Table I corr.	3 m
Adjusted LMT	0652
0649 DR λ	65° 14.0′ W
dλ	21 m
1st Est. ZT	0713

Page 30, 2nd para, line 3. Change to read, ". . . Hc of 44° 45.5′ was computed . . ."

Page 159, 2nd para, line 3. Change to read, ". . . on page 141. . ."

Page 239. Figure 13–3. Change label in center of figure to read "1000 μsec"

Insert as fifth enter under **Basic PRR/Basic PRI**," 11.15/89,700

Last para line 5. Change to read, ". . . master-secondary X pair . . ."

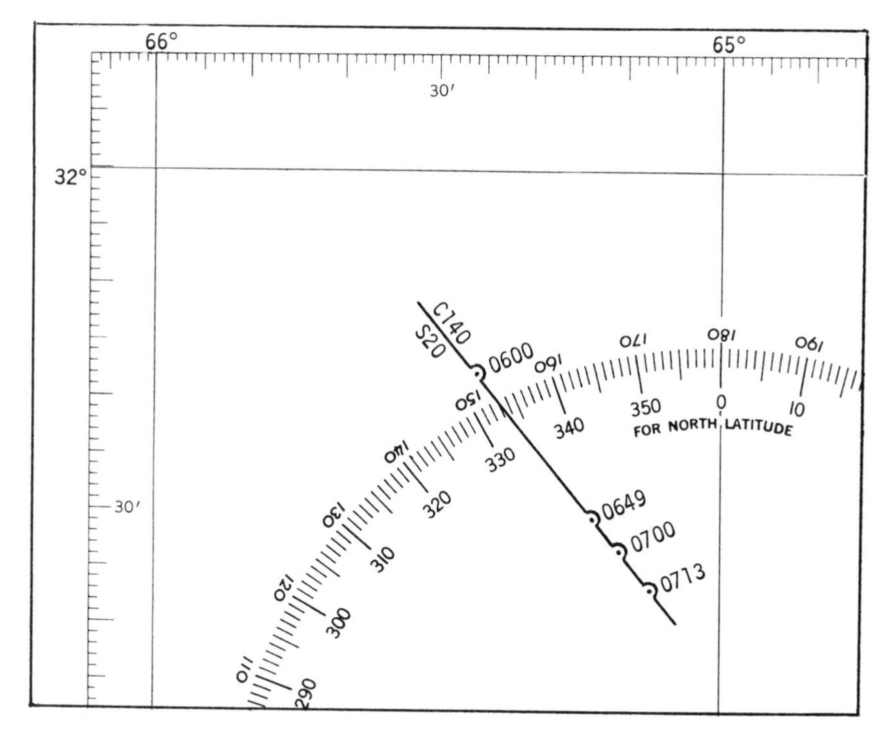

Figure 11-3B. Completed plot of a sunrise solution on a moving ship.

0713 DR L.	31° 23.0′ N
Table I corr.	3 m
Tab GMT (LMT)	0649
Adjusted LMT	0652
0713 DR λ	65° 07.8′ W
dλ	21 m
2nd Est. ZT	0713

Predicting the Time of Moonrise and Moonset

Although the times of moonrise and moonset do not have much navigational significance, the navigator occasionally will have need to predict the times of these phenomena for other purposes, such as operational or tactical planning. Additionally, when the moon is at or near its full phase and its altitude is low, the horizon beneath it may be illuminated to a degree sufficient to permit sextant altitude observations of the moon itself or other celestial bodies in the same vicinity. Prediction of the times of moonrise and moonset may indicate when the possibility for such an auxiliary sight opportunity may exist during a given night.

Prediction of the time of moonrise or moonset is more complicated than a similar prediction for phenomena associated with the sun because the rate of change of GHA and to a lesser extent the declination of the moon are not constant. Whereas the GHA of the sun increases at the rate of approximately 15° per hour, the GHA of the moon increases at an irregular rate, usually faster but at times slower than that of the sun. Thus, in the *Nautical Almanac* daily pages, predictions of the mean times of moonrise and moonset along the Greenwich meridian must be given for each date covered by the three-day period. For convenience, the data for the first day on the following page are also given (Figure 11-1). Because the rate of change of the GHA of the moon is so irregular, the tabulated mean times of its rising and setting at the Greenwich meridian cannot be considered the same as the LMTs of these phenomena at other meridians as in the case of the sun-related phenomena. Hence, after the tabulated GMT of moonrise or moonset has been adjusted for the exact latitude of the observer by use of Table I, shown in Figure 11-2, the resulting LMT at the Greenwich meridian must then be corrected for the longitude of the observer by use of a special longitude LMT correction table designed for use with the moon. This table, designated Table II, is located in the back of the *Nautical Almanac* beneath Table I, and is shown in Figure 11-4. Its use will be explained below. After the tabulated time of the lunar event of interest at Greenwich has been corrected for the exact latitude and longitude of the observer using Tables I and II, the resulting adjusted LMT must then be converted to zone time in the usual fashion.

As was the case with the other rising and setting phenomena discussed earlier, the time of moonrise or moonset may be predicted for either a stationary or moving ship. The first step in the prediction procedure is to choose the proper columns of tabulations in the *Nautical Almanac;* two adjacent days' columns must always be used because of the irregular motion of the moon. If the observer is located in *east* longitudes on the day for which the moonrise or moonset prediction is to be made, the columns containing tabulations for the day in question and for the *preceding* day are used. Conversely, if the observer is in *west* longitudes, the columns containing tabulations for the day in question and for the *following* day are used.

For a stationary ship, after the proper data columns have been selected, the difference in mean time of the phenomenon resulting from the difference between the observer's latitude and the closest tabulated latitude smaller than his latitude is first taken into account by the use of Table I of Figure 11-2. Corrections are found from Table I both for the mean time at the tabulated latitude on the day in question and on the preceding (λE) or following (λW) day. After these correc-

TABLE II—FOR LONGITUDE

Long. East or West	10m	20m	30m	40m	50m	60m	1h + 10m	20m	30m	1h + 40m	50m	60m	2h 10m	2h 20m	2h 30m	2h 40m	2h 50m	3h 00m
	Difference between the times for given date and preceding date (for east longitude) or for given date and following date (for west longitude)																	
°	m	m	m	m	m	m	m	m	m	m	m	m	h m	h m	h m	h m	h m	h m
0	0	0	0	0	0	0	0	0	0	0	0	0	0 00	0 00	0 00	0 00	0 00	0 00
10	0	1	1	1	1	2	2	2	2	3	3	3	04	04	04	04	05	05
20	1	1	2	2	3	3	4	4	5	6	6	7	07	08	08	09	09	10
30	1	2	2	3	4	5	6	7	7	8	9	10	11	12	12	13	14	15
40	1	2	3	4	6	7	8	9	10	11	12	13	14	16	17	18	19	20
50	1	3	4	6	7	8	10	11	12	14	15	17	0 18	0 19	0 21	0 22	0 24	0 25
60	2	3	5	7	8	10	12	13	15	17	18	20	22	23	25	27	28	30
70	2	4	6	8	10	12	14	16	17	19	21	23	25	27	29	31	33	35
80	2	4	7	9	11	13	16	18	20	22	24	27	29	31	33	36	38	40
90	2	5	7	10	12	15	17	20	22	25	27	30	32	35	37	40	42	45
100	3	6	8	11	14	17	19	22	25	28	31	33	0 36	0 39	0 42	0 44	0 47	0 50
110	3	6	9	12	15	18	21	24	27	31	34	37	40	43	46	49	0 52	0 55
120	3	7	10	13	17	20	23	27	30	33	37	40	43	47	50	53	0 57	1 00
130	4	7	11	14	18	22	25	29	32	36	40	43	47	51	54	0 58	1 01	1 05
140	4	8	12	16	19	23	27	31	35	39	43	47	51	54	0 58	1 02	1 06	1 10
150	4	8	13	17	21	25	29	33	38	42	46	50	0 54	0 58	1 03	1 07	1 11	1 15
160	4	9	13	18	22	27	31	36	40	44	49	53	0 58	1 02	1 07	1 11	1 16	1 20
170	5	9	14	19	24	28	33	38	42	47	52	57	1 01	1 06	1 11	1 16	1 20	1 25
180	5	10	15	20	25	30	35	40	45	50	55	60	1 05	1 10	1 15	1 20	1 25	1 30

Table II is for interpolating the L.M.T. of moonrise, moonset and the Moon's meridian passage for longitude. It is entered with longitude and with the difference between the times for the given date and for the preceding date (in east longitudes) or following date (in west longitudes). The correction is normally *added* for west longitudes and *subtracted* for east longitudes, but if, as occasionally happens, the times become earlier each day instead of later, the signs of the corrections must be reversed.

xxxii

Figure 11-4. Table II, moonrise/moonset longitude interpolations, Nautical Almanac.

tions have been applied to both days' times, the correction to LMT for the observer's longitude is then found from Table II in Figure 11-4. Entering arguments for this table are the observer's longitude and the difference between the two LMTs of the phenomenon corrected for latitude as described above. Interpolation to the nearest minute may be required in some cases in which the time difference is large. The correction obtained from Table II is normally added for west longitudes and subtracted for east longitudes, but if the tabulated times on the successive days of interest become earlier instead of later, the sign of the correction must be reversed.

After the LMT correction for longitude from Table II has been obtained, it is applied to the LMT adjusted for latitude on the day in question to obtain the LMT of the phenomenon at the stationary position. This LMT must then be converted to zone time by application of the arc-time difference between the observer's meridian and the standard meridian of his time zone.

In the case of a moving ship, the tabulated time of moonrise or moonset at the tabulated latitude smaller than the approximate DR latitude of the ship on the day in question is used as a starting point. The exact DR position corresponding to this LMT is plotted and used as the basis for finding the two latitude corrections as explained before. The longitude correction from Table II is then found and applied to the DR position time corrected for latitude to form a "first estimate" of the LMT of moonrise or moonset. This first estimate in LMT is then converted to zone time, using as a basis for this conversion the DR position plotted for the tabulated mean time of the phenomenon on the given day. A second estimate is then found by repeating the interpolation procedure, using as a basis the first estimate zone time DR position. This second estimate of zone time of the phenomenon is then considered the approximate time of the event at the projected position of the moving ship.

The procedure for predicting the time of moonrise or moonset may be summarized as follows:

1. Obtain the GMT of the phenomenon of interest on the day in question for the tabulated latitude smaller than the DR latitude (moving ship) or the ship's position (stationary ship).

2. In the case of a moving ship, plot a DR position for this time, using the GMT as local mean time.

3. Using this position, determine a first estimate of zone time of moonrise or moonset as follows:

A. Find a latitude correction to LMT from Table I for both the day in question and the preceding day, if in east longitude, and the following day, if in west longitude.

B. Apply these corrections to the tabulated times corresponding to the nearest tabulated latitudes smaller than the ship's latitude.

C. Find the difference between these two adjusted LMTs.

D. Find the longitude correction to LMT from Table II using as entering arguments this time difference and the ship's longitude.

E. Apply the longitude correction to the LMT adjusted for latitude on the day in question to find the LMT of moonrise or moonset at the ship's stationary or DR position.

F. Convert this LMT to zone time, using the arc-time difference between the initial DR longitude or the ship's stationary position and the standard meridian of the observer's time zone.

The result at this point is the first estimate of the zone time of the moon phenomenon. For a stationary ship, this first estimate is the predicted time of moonrise or moonset. For a moving ship, the following additional steps are required to determine a second and final estimate:

4. Plot a DR position for the zone time of the first estimate.

5. Using this position, repeat all parts of step 3 above. The result is the predicted zone time of moonrise or moonset for the moving ship.

As an example of the prediction of the time of moonset for a moving ship, suppose that in the early morning of 16 December a ship's 0700 DR position was L 35° 03.5′ S, λ 162° 01.0′ E, steering on course 230°T at a speed of 20 knots. The navigator desires to predict the time of moonset if the ship remains on the same course and speed. Following through the procedure set forth above, the first step is to determine the LMT of moonset on 16 December at the approximate DR position. Since the time of moonset tabulated for latitude 35° south (the lower tabulated latitude) on 16 December in the *Nautical Almanac* is 0747, a DR position is plotted for this time for use in obtaining the first estimate of the zone time of moonset; the plot is shown in Figure 11-5A.

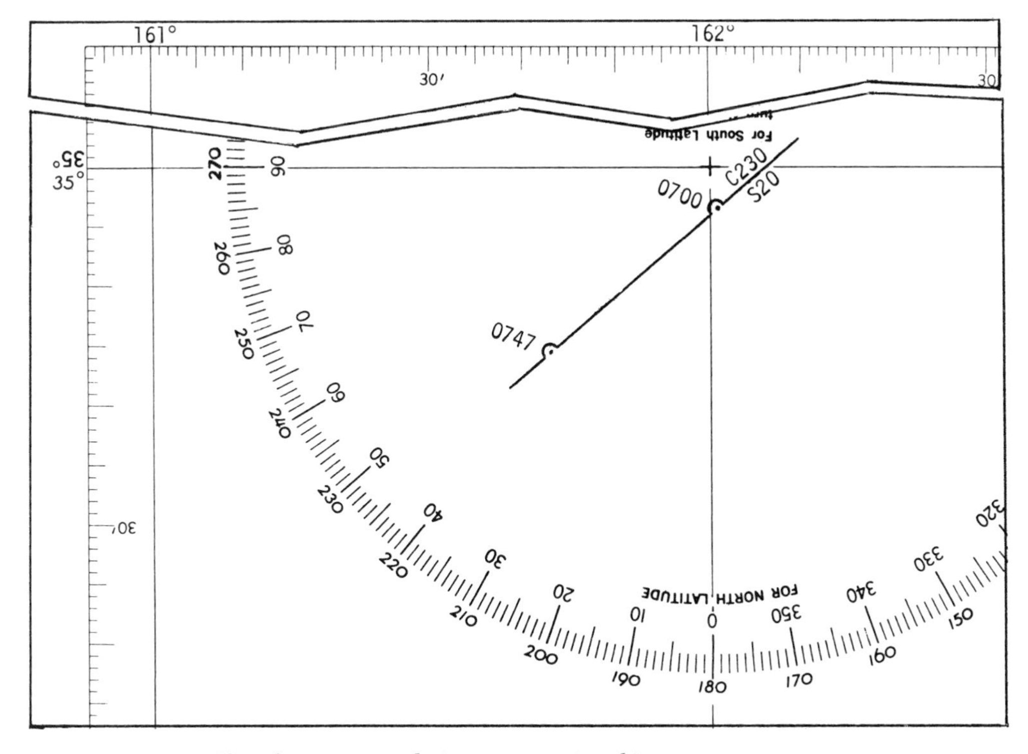

Figure 11-5A. Plot of a moonset solution on a moving ship.

The coordinates of the 0747 DR position are L 35° 15.8′ S, λ 161° 43.0′ E. Since the ship is in east longitude, the columns of the almanac containing moonset time for 15 and 16 December will be used to obtain latitude and longitude LMT corrections. The times of moonset at 35° S on 15 and 16 December are recorded (see Figure 11-1) as 0645 and 0747, respectively. The latitude difference between the 0747 DR latitude and 35° S is 15.8′, the tabular interval is 5°, and the differences between the listed times at 35° S and 40° S on 15 and 16 December are −16 minutes and −13 minutes, respectively. The resulting latitude LMT corrections from Table I of Figure 11-2 are each −1 minute. Applying these corrections to the tabulated times for 15 and 16 December yields values of LMT corrected for latitude of 0644 and 0746, respectively.

Next, Table II is entered for the longitude LMT correction, using as arguments the difference between the two partially corrected LMTs, 0746 − 0644, or 62 minutes, and the 0747 DR longitude. After mental interpolation in the table, a correction of −28 minutes results. Application of this correction to the adjusted LMT 0746 results in a first estimate of the LMT of moonrise of 0746 − 28, or 0718. To convert this LMT to zone time, an arc-time difference is found for the differ-

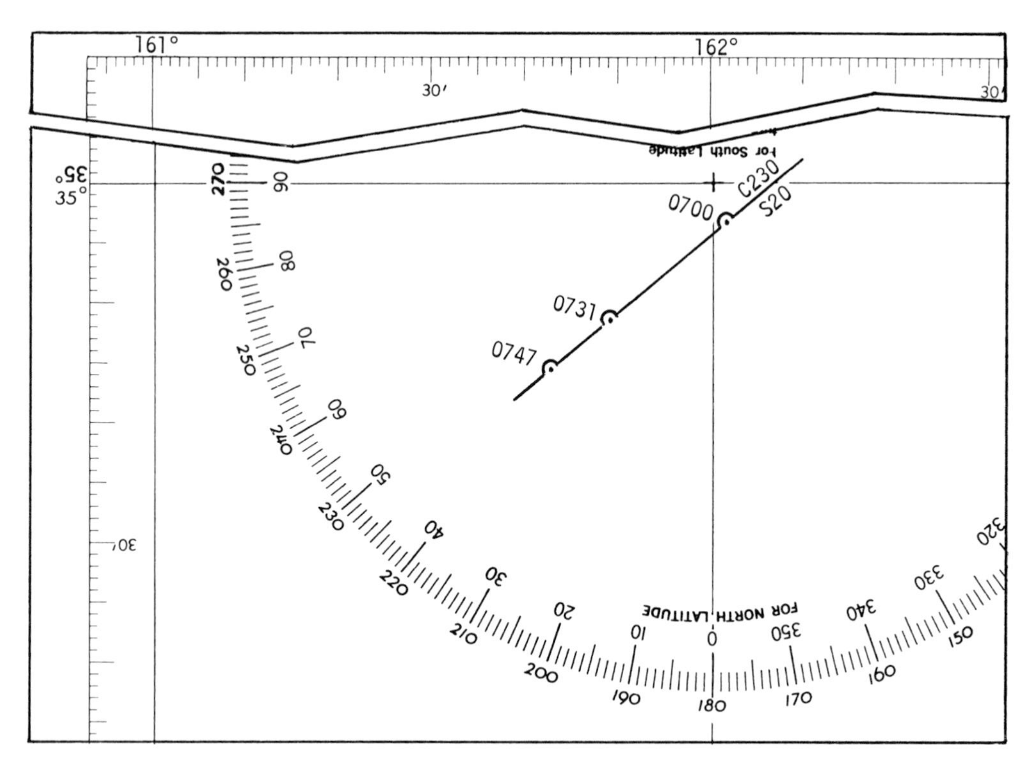

Figure 11-5B. Plot of a moonset solution on a moving ship.

ence in longitude between the 0747 DR position and the time zone standard meridian, 165° E in this case. The arc difference is 3° 17.0′, and the corresponding time correction is 13 minutes; it is added, since the ship is west of the time zone standard meridian. Thus, the first estimate of the zone time of moonset is 0731.

To find the second estimate, a DR position is plotted for 0731, L 35° 12.0′ S, λ 161° 48.0′ E, and the procedure is repeated. After applying the latitude LMT corrections from Table I to the tabulated times for 35° S for 15 and 16 December, partially corrected LMTs of 0644 and 0746 again result. The longitude LMT correction is again −28 minutes, resulting in an LMT corrected for latitude and longitude of 0718. To correct this LMT to zone time, the arc-time correction for the difference between the 0731 DR longitude and 165° E is found; it is again +13 minutes. Hence, the second and final estimate of the zone time of moonset in the morning of 16 December for this ship is 1718 + 13 or ZT 0731. The completed plot is shown in Figure 11-5B, and the sequential list of the computations for this example is given below.

Tab GMT 16 Dec	0747
0747 DR L.	35° 15.8′ S
Tab I corr	−1
Adjusted LMT 16 Dec	0746
Tab GMT 15 Dec	0645
Tab I corr	−1
Adjusted LMT 15 Dec	0646
Time diff	62 m
0747 DR λ	161° 43.0′ E
λ corr Tab II	−28 m
1st Est LMT	0718
dλ, 0747 DR	13 m
1st Est ZT	0731
0731 DR L	35° 12.0′ S
Tab GMT 16 Dec	0747
Tab I corr	−1
Adjusted LMT 16 Dec	0746
Tab GMT 15 Dec	0645
Tab I corr	−1
Adjusted LMT 15 Dec	0644
Time diff	62 m
0731 DR λ	161° 48.0′ E
λ corr Tab II	−28 m

2nd Est LMT	0718
dλ, 0731 DR	13 m
2nd Est ZT	0731

Summary

This chapter concludes the portion of *Marine Navigation 2* dealing with the practice of celestial navigation by discussing the procedures used to predict the times of certain rising and setting phenomena. The determination of the times of the darker limits of morning and evening celestial twilight are of greatest concern to the navigator, followed by the prediction of the times of sunrise and sunset and, finally, moonrise and moonset. The procedures for prediction of the times of sun-related rising and setting phenomena are relatively uncomplicated, while the procedures for estimating the times of the phenomena of least significance in celestial navigation—moonrise and moonset—are fairly involved because of the irregular rate of change of the GHA and declination of the moon.

12

Introduction to Electronic Navigation

The previous chapters of this book have each been concerned with some aspect of the practice of celestial navigation at sea. In the days of sail, the principles of celestial navigation just described were sufficient by themselves to safely direct the voyages of sailing vessels. Because of the relatively slow speed of even the fastest sailing ship, unfavorable weather conditions over part of a transit would not unduly hazard the vessel if a good dead reckoning plot were maintained. Even if several days elapsed without a good celestial fix being obtained, the ship could not stray an excessive distance from her planned track because of her slow speed. A divergence of a few tens of miles between the actual track and the DR track in times of overcast or stormy weather on the open sea was hardly significant for a sailing voyage of several thousand miles.

With the advent of the airplane and the deep-draft steamship, however, and more recently the missile-carrying submarine, the capability of a regular determination of position regardless of the environmental conditions prevailing over the open sea became more and more critical. Not only are the effects of any divergence from the planned track amplified by the increased speed of airborne travel, but the deep draft and huge displacement of modern merchant and combatant ships make close adherence to the preplanned track a necessity for both economy of operations and safe navigation.

The first practical use of electronic navigation occurred in the early years of this century, when primitive shore-based radio direction finding equipment was used to establish a bearing to a ship, and later to an airplane, transmitting a CW signal. Once determined, the bearing was then transmitted to the ship or plane via radiotelegraph. At some locations, several of these shore stations were linked by telephone lines, enabling a rough determination of position to be made by crossed bearings, which position could then be transmitted to the ship or plane. This technique is still in use today in more refined form in many coastal areas of the world.

In the 1930s, improved radio direction finding techniques and equipment led to the establishment of navigational aids for both contemporary air and surface navigation called *radiobeacons*. These aids consisted essentially of small radio transmitters located along coastal areas; they were designed to provide radio bearings that could be used in lieu of visual bearings at sea in foul weather. Position-finding by these beacons became known as *radio navigation*. Continued refinements in the state of the art of electronics technology and a better understanding of electromagnetic wave propagation led to the subsequent development of radar, sonar, and longer-range radio navigation systems, with the result that modern electronic navigational aids bear little resemblance to the early devices employed in radio navigation. The term radio navigation is still applied to any electronic navigation system employing radio waves as the means of position-finding; the term *electronic navigation* has come into wide use to describe not only radio navigation systems, but also all other systems dependent upon an electronic device for the establishment of a position.

Modern electronic navigation systems can be classified both by their range and their scope. The range classifications most generally used are short range, medium or intermediate range, and long range, with the exact limits between each being rather indefinite. The scope of a system is used here to mean either a self-contained or externally supported system, and either an active (i.e., transmitting) or passive (i.e., not transmitting) mode of operation.

The short-range systems in most common use at the present time include radiobeacons, radar, and Decca; medium-range systems include Decca* and certain types of extended-range radar; and long-range systems include Loran-C, Consol, and Omega. All of these systems depend on active radiofrequency (RF) transmissions, and with the exception of radar, all are externally supported with respect to a ship operating at sea. In addition to these systems, there is another category of systems that do not conveniently fit into any range classification. These are the so-called *advanced navigation systems*, of which four are currently of prime importance: the Transit satellite navigation system, the shipboard inertial navigation system, the doppler sonar system, and the NavStar Global Positioning System. The second of these is the only self-contained, passive electronic navigation system currently available.

Before specifically discussing any of these systems, it is necessary to review the basic theory of the electromagnetic wave upon which most of them depend for their operation. This chapter reviews the fundamental characteristics and behavior of the electromagnetic wave, and discusses the way in which these waves are employed in radio naviga-

*Decca is considered to be both a short- and medium-range system.

tion. The radio navigation systems listed above are each the subject of a separate section of Chapter 13, and Chapter 14 deals with the four advanced navigation systems mentioned. Radar is discussed in Chapter 10 of *Marine Navigation 1.*

The Electromagnetic Wave

For navigational purposes it is not necessary to be more than generally aware of the methods by which an electromagnetic wave is propagated, but it is very important to understand what forces affect such a wave once it has left the generating antenna(s). Very briefly, an electromagnetic wave is produced by a rapidly expanding and collapsing magnetic field, which is in turn produced by alternately energizing and deenergizing an electronic circuit especially designed for the generation of such waves. In the science of electronics, such a generating circuit is referred to as an *oscillator.* An amplifier of some type is generally used to boost the power of the oscillator output, and an antenna is used to form the outgoing wave.

An electromagnetic wave, because of the methods by which it is propagated, always resembles a sine wave in appearance; it can be characterized by its wavelength, frequency, and amplitude, each of which are illustrated in Figure 12-1 on the next page. In the figure, one complete electromagnetic wave or cycle is shown, and the terms used to describe it are defined as follows:

> A *cycle* is one complete sequence of values of the strength of the wave as it passes through a point in space.
>
> The *wavelength,* abbreviated in electronics by the symbol λ, is the length of a cycle expressed in distance units, usually either meters or centimeters.
>
> The *amplitude* is the wave strength at any particular point along the wave.
>
> The *frequency,* abbreviated by f, is the number of cycles repeated during one second of time. If the time frame shown in Figure 12-1 were one second long, for example, it could be said that the frequency of the wave depicted is one cycle per second.

In the vacuum of space, an electromagnetic wave is theorized to travel at a velocity approaching the speed of light, or 300,000,000 meters per second. Frequency and wavelength are related by the formula

$$\lambda = \frac{300{,}000{,}000}{f}$$

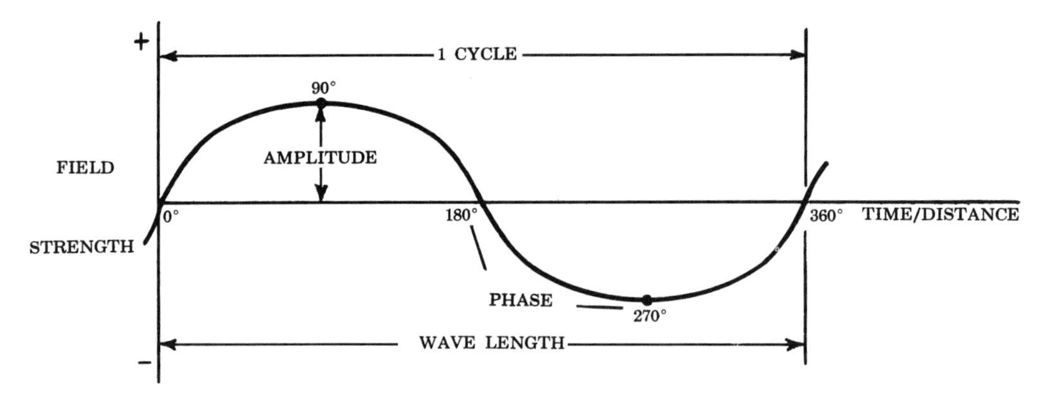

Figure 12-1. Electromagnetic wave terminology.

where λ is the wavelength in meters and f the frequency in cycles per second. Hence, every specific electromagnetic frequency is radiated at a specific wavelength.

In recent years, the term *Hertz,* abbreviated *Hz,* has come to be used in place of *cycles per second,* in honor of the German pioneer in electromagnetic radiation, Heinrich Hertz. One Hertz is defined as one cycle per second. Frequency is expressed in terms of numbers of thousands (kilo-) millions (mega-) or billions (giga-) of Hertz. For example, 10,000 cycles per second is expressed as 10 kiloHertz, usually written 10 kHz, or 2.5 million cycles per second as 2.5 megaHertz, written as 2.5 MHz.

One additional term associated with the electromagnetic wave has great import in electronic navigation. The *phase* of a wave is the amount by which a cycle has progressed from a specified origin. For most purposes, it is stated in degrees, with a complete cycle being 360 degrees in length. Two waves can be partially compared by measuring their phase difference. For example, two waves having crests one-quarter of a cycle apart can be described as being 90 degrees "out of phase."

The behavior of an electromagnetic wave is dependent upon its frequency and corresponding wavelength. For descriptive purposes, electromagnetic frequencies can be arranged in ascending order to form a "frequency spectrum" diagram, shown in Figure 12-2. As can be seen by an inspection of Figure 12-2, electromagnetic waves are classified as *audible waves** at the lower end of the spectrum, *radio waves* from about 5 kHz to 3×10^5 MHz, and *visible light* and various

* That is, audible *frequency* waves; to be heard, such electromagnetic waves must be transformed into sound waves by a receiver.

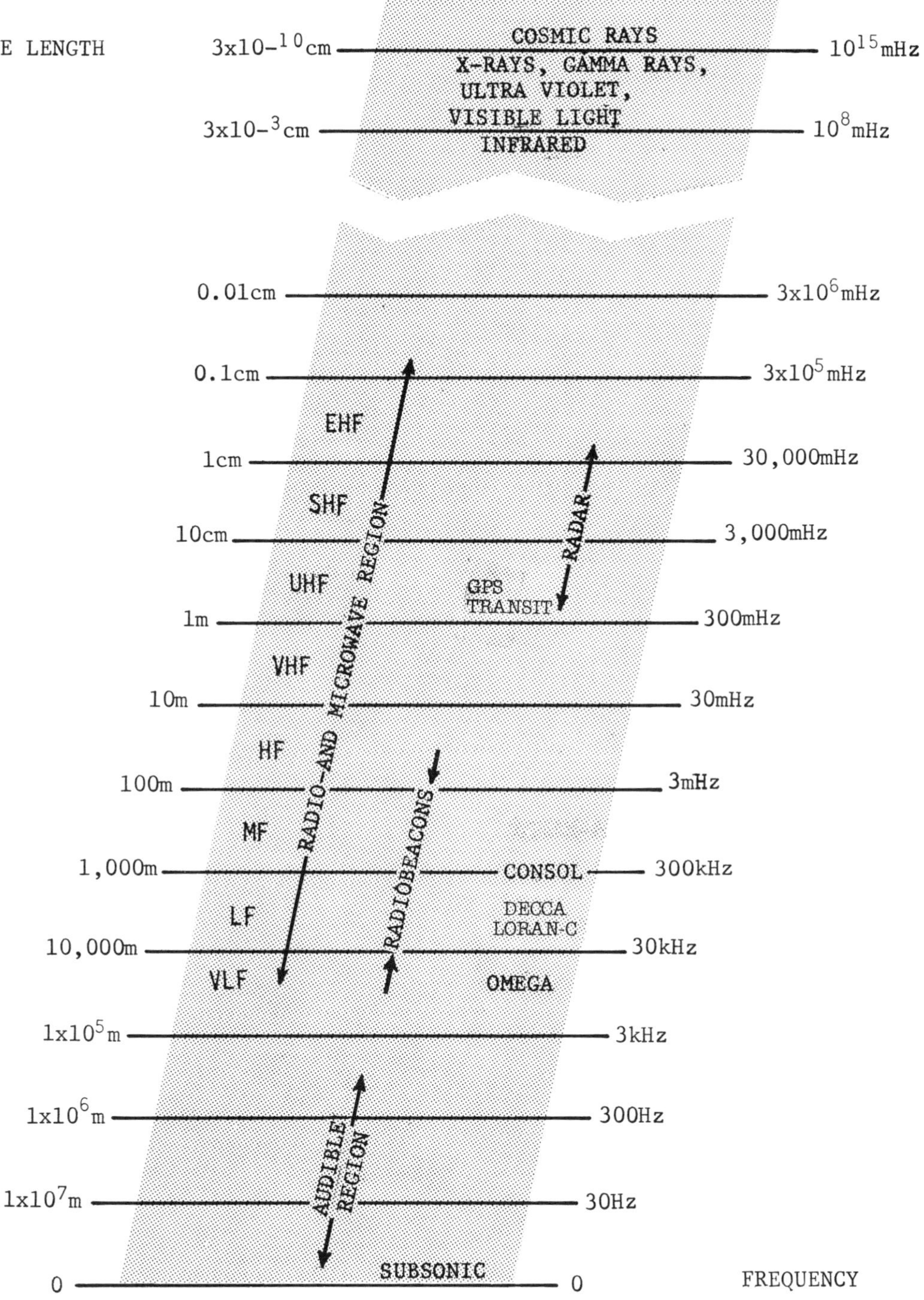

Figure 12-2. The electromagnetic wave frequency spectrum.

other types of *rays* at the upper end of the spectrum. In electronic navigation, frequencies within the radio wave spectrum are used.

For ease of reference, the radio wave spectrum is further broken down into eight so-called *bands* of frequencies, as indicated in Figure 12-2. These eight bands are described as follows:

The *Very Low Frequency* (*VLF*) band includes all radio frequencies less than 30 kHz. The primary navigational use of this band is for the Omega system.

The *Low Frequency* (*LF*) band extends from 30 to 300 kHz. The navigational systems Decca, Loran-C, and Consol, and most radiobeacons, use this band.

The *Medium Frequency* (*MF*) band extends from 300 kHz to 3 MHz, and is used by some Consol stations.

The *High Frequency* (*HF*) band includes frequencies from 3 to 30 MHz, and is chiefly used for long-range communication.

The *Very High Frequency* (*VHF*) band extends from 30 to 300 MHz and is used for short- to medium-range communication.

The *Ultra High Frequency* (*UHF*) band includes frequencies between 300 and 3000 MHz. It is used for short-range communications and radar.

The *Super High Frequency* (*SHF*) band from 3000 to 30,000 MHz and the *Extremely High Frequency* (*EHF*) band from 30,000 to 300,000 MHz are used almost exclusively for precise line-of-sight radar.

A series of electromagnetic waves transmitted at constant frequency and amplitude is called a *continuous wave,* abbreviated CW. This wave cannot be heard except at the very lowest frequencies, where it may produce a high-pitched hum in a receiver. Because an unmodified continuous wave cannot convey much information, in electronic navigation the wave is often modified or *modulated* in some way. When this is done, the basic continuous wave is referred to as a *carrier wave.*

In practice there are three methods by which a carrier wave may be modulated to convey information; these are amplitude, frequency, and pulse modulation. In *amplitude modulation,* abbreviated AM, the amplitude of the carrier wave is modified in accordance with the amplitude of a modulating wave, usually but not always an audible frequency. The top drawing in Figure 12-3 illustrates this type of modulation. In the receiver, the signal is demodulated by removing the modulating wave, which in the case of a voice radio is then

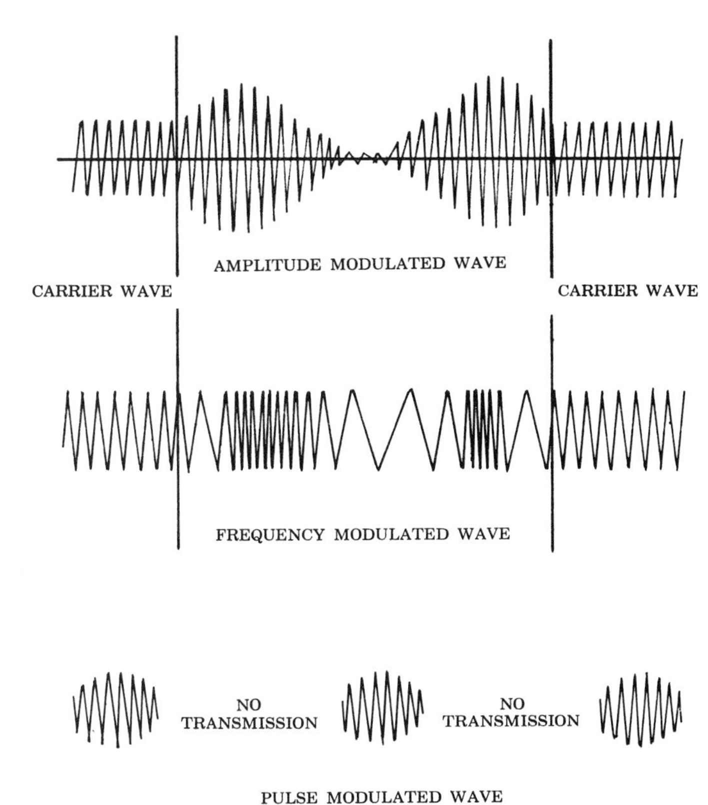

Figure 12-3. Amplitude, frequency, and pulse modulation of a carrier wave.

amplified and related to the listener by means of a speaker. This type of modulation is, of course, quite common, being the usual form of modulation in the commercial radio broadcast band.

In *frequency modulation*, abbreviated FM, the frequency of the carrier wave instead of the amplitude is altered in accordance with the frequency of the modulating wave, as shown in Figure 12-3 (center). This type of modulation is used for FM commercial radio broadcasts and the sound portion of television broadcasts.

Pulse modulation is different from either amplitude or frequency modulation in that there is usually no impressed modulating wave employed. In this form of modification, the continuous wave is actually broken up into very short bursts or "pulses," separated by relatively long periods of silence during which no wave is transmitted. As stated in Chapter 10 of *Marine Navigation 1*, this is the transmission used in most types of marine navigational and surface search radar; it is also used in some common long-range radio navigation aids, most notable of which is Loran. Figure 12-3 (bottom) depicts a pulse-modulated wave.

As was mentioned in the first part of this section, in electronics the device that generates an electromagnetic wave is termed an oscillator. In the case of radio waves, the output of the oscillator is boosted in power by an amplifier, then modulated as described above by a modulator unit. In voice radio, this modulator incorporates a microphone that converts an audible wave into a modulating wave. The modulated radio wave is then passed through a second amplifier and finally transmitted into space by means of the antenna. These components, in addition to a power supply and a device to control the frequency of the wave originated by the oscillator, are collectively referred to as a *transmitter.* A block diagram of an AM radio transmitter appears in Figure 12-4.

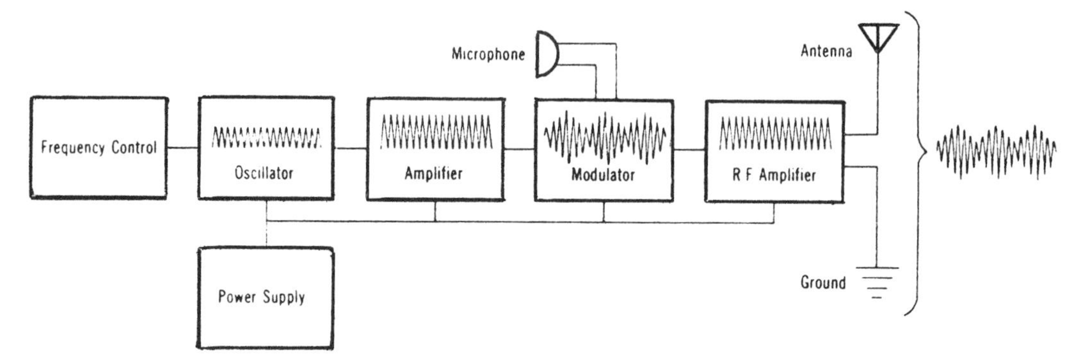

Figure 12-4. Block diagram of components of an AM radio transmitter.

A radio *receiver* is a device that is designed to convert a radio wave into a form suitable to convey information. It should be able to select carrier waves of a desired frequency, demodulate the wave, amplify it as necessary, and present the information conveyed in a usable form. The output of a receiver may be presented audibly by earphones or a loud speaker, or visually on a dial or cathode ray tube (CRT) display. A receiver can be thought of as incorporating three components—an antenna to convert an incoming radio wave into an electric current, a demodulator to separate a modulating wave from the carrier wave, and an output display device to present the output from the demodulator in a usable form.

Radio receivers differ mainly in the six characteristics listed below:

Type. The type of signal which they will receive, ie, AM, FM, or pulse.

Frequency range. The range of frequencies to which they can be tuned.

Selectivity. The ability to separate waves of a desired frequency from others of nearly the same frequency.

Sensitivity. The ability to detect and extract information from weak signals against a background of random radio-frequency noise.

Stability. The ability to resist drift from a frequency to which the receiver is tuned.

Fidelity. The accuracy with which the characteristics of the original modulating wave are reproduced.

The Behavior of Radio Waves in the Earth's Atmosphere

When a radio wave is generated by a transmitter located within the earth's atmosphere, the wave travels outward in all directions much like a light wave from an unshielded light source. Because the radio and light wave are both forms of an electromagnetic wave, differing only in frequency and wavelength, their behavior is very similar in several respects.

Radio waves, like light waves, may be reflected from a surface they strike. In both cases, the quality of the reflection depends on the irregularity of the surface as compared to the wavelength of the impinging wave. Thus, a sea of ten-foot waves would form a good reflecting surface for a radio wave hundreds of meters in length, but a very poor reflecting surface for a wave of a few centimeters. Whereas most of the energy within a wave striking a high-quality reflective surface is retained in the reflected wave, most of the energy striking a nonreflective surface is absorbed by the material in the form of heat.

Electrons within the molecules composing the earth's atmosphere and its crust are excited by the energy of a passing electromagnetic wave; as these electrons collide with each other and with other adjacent molecules, the energy is converted to heat and lost to the original wave. This effect, called *absorption,* is inversely proportional to the frequency of the wave; at lower radio frequencies the effect is maximized, due to the fact that at these frequencies the time available for each part of the wave to affect the electrons is relatively long. Because of the lower frequencies and longer wavelengths of a radio wave as compared to a light wave, the absorption effect is much more pronounced for the radio wave. The lower the frequency of the radio wave, the more energy it must have to proceed through the atmosphere to a given distance from the transmitter.

When an electromagnetic wave encounters an obstruction of opaque material (i.e., impervious to radio waves), the area behind the obstruction is shadowed, since the waves that would otherwise reach the area are blocked. This blockage is not complete, however, because the portion of the wave to either side of the obstruction begets a secondary series of waves that travel into the shadow zone. This effect, called *diffraction,* is illustrated in Figure 12-5. In the case of a light wave, it accounts for a shadow behind an obstruction being less than absolutely black, and in the case of a radio wave, for the reception of a weak radio signal within such a shadow zone.

If two or more electromagnetic waves arrive simultaneously at the same point in space, *interference* may result. The amount of this interference depends on the phase and frequency relationship of the waves involved. If two waves of the same frequency but 180° out of phase arrive simultaneously, each wave cancels the other, resulting in a *null* at that point.

Under certain conditions, a portion of the electromagnetic energy in a radio wave may be reflected back toward the earth's surface from the *ionosphere,* a layer of charged particles about 90 to 400 kilometers high; such a reflected wave is called a *sky wave.* The portion of the

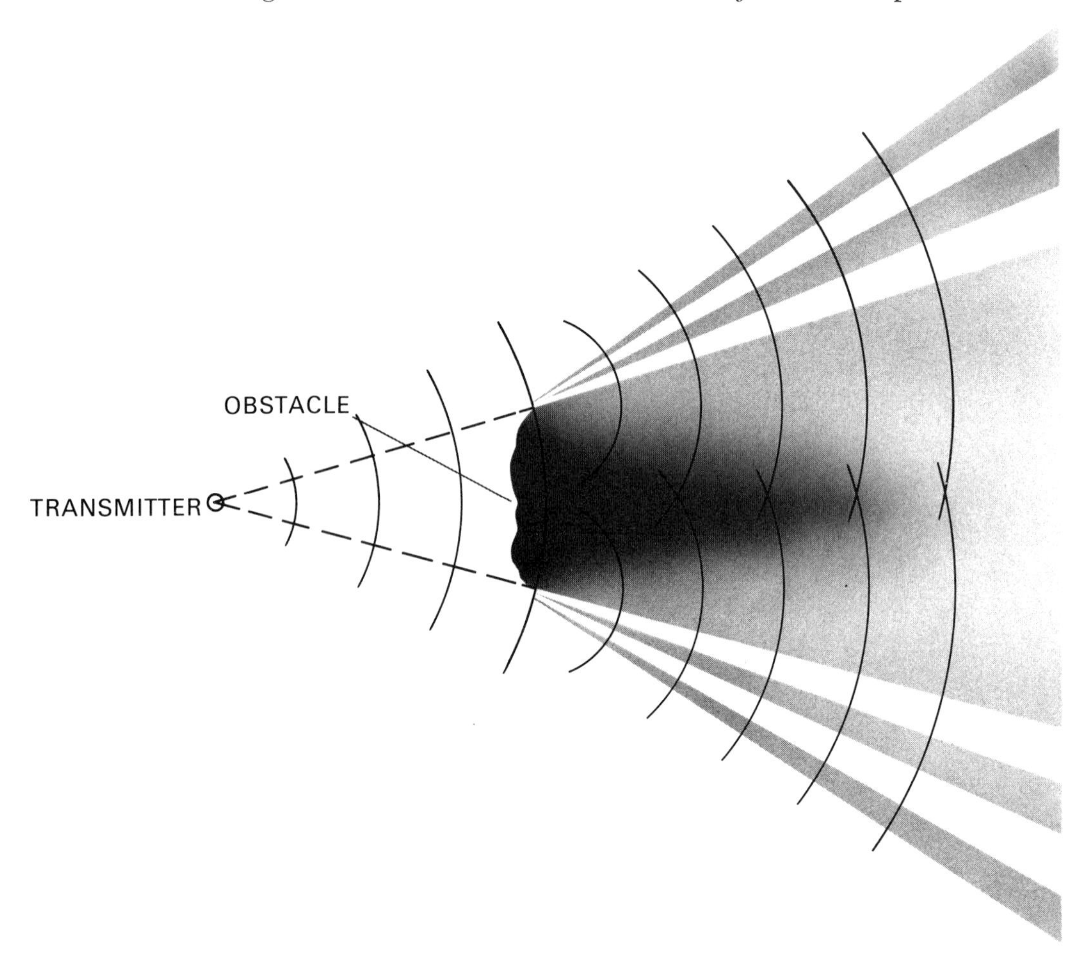

Figure 12-5. Diffraction effect of an electromagnetic wave.

earth's upper atmosphere in daylight is subjected to continual bombardment by ultraviolet rays of the sun. These high-energy light waves cause electrons in the gas molecules of the upper atmosphere to become excited and to free themselves from their molecules, forming ionized layers. These layers reach their maximum intensity when the sun is highest.

There are four such *ionospheric layers* of importance in the study of radio wave propagation, pictured in Figure 12-6; they are designated the D-layer, E-layer, and F_1 and F_2 layers, each of which is described on the following page:

> The *D-layer* is the ionized layer closest to the earth's surface, about 60 to 90 kilometers high. It is considerably less dense than the other layers, and appears to be formed only during the daylight hours.
>
> The *E-layer* is located about 110 kilometers above the earth's surface. Its density is greatest in the area immediately beneath the sun, and it persists throughout the night with decreased intensity.
>
> The *F_1-layer* occurs only in daylight regions of the upper atmosphere, usually between 175 and 200 kilometers above the earth's surface.

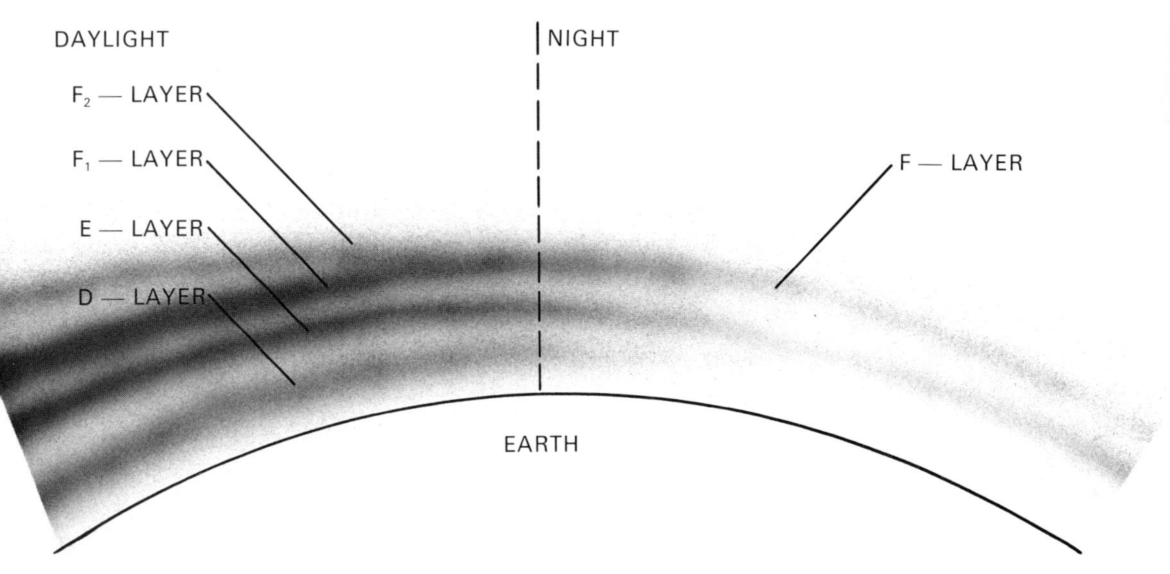

Figure 12-6. The four layers of the ionosphere of importance in radio wave propagation.

The *F_2-layer* is found at altitudes between 250 to 400 kilometers. Its strength is greatest by day, but due to the very low density of the atmosphere at this height, the freed electrons in this layer persist several hours after sunset. There is a tendency for the F_1 and F_2 layers to merge thereafter to form the so-called *F-layer,* which is ordinarily the only layer of importance to radio wave propagation after dark.

All layers of the ionosphere are somewhat variable, with the main patterns seeming to be diurnal, seasonal, and by sunspot cycle. The layers may either be conducive to the sky wave transmission of a radio wave to a desired area of reception, or they may hinder or even entirely prevent such transmission, depending on the frequency of the wave, its angle of incidence, and the height and density of the various layers at the time of the transmission. In general, frequencies in the MF and HF bands are most suitable for ionospheric reflection during both day and night, with frequencies in the upper LF and lower VHF bands producing usable sky waves only at night. Frequencies outside of these limits either will not produce any sky waves or the sky waves they do form are so weak as to be unusable. In times of abnormal sunspot activity the normal ionospheric reflection patterns may be disrupted to such an extent that no sky waves of any frequency are produced.

Because of the higher resistance of the earth's crust as compared to the atmosphere, the lower portion of a radio wave radiated parallel to the earth's surface is slowed somewhat, causing the wave to bend toward earth, just as a light wave from a star is bent upon entering the earth's atmosphere. A wave of this type that tends to follow the earth's curvature is termed a *ground wave.* Again, the amount of curvature resulting from this cause is inversely proportional to the frequency of the wave; the lower the frequency, the more it will tend to bend in conformity to the earth's shape. The ultimate range of such a ground wave is dependent upon the absorption effect.

Combining the effects of sky wave and ground wave transmission for a given radio wave of suitable frequency, a radiation pattern of the type depicted in Figure 12-7 may result. In this figure, ray *1* of the transmitted radio wave is radiated at an angle too great to permit its reflection by the ionosphere as a sky wave. Hence, it penetrates the ionosphere and escapes into space. Ray *2* impinges upon the ionosphere at the steepest angle permitting a reflected sky wave under these conditions. The resultant sky wave, called a "one-hop" wave, reaches the earth's surface at point P_1; no sky waves could be received within the distance of P_1 from the transmitter, since the angle of

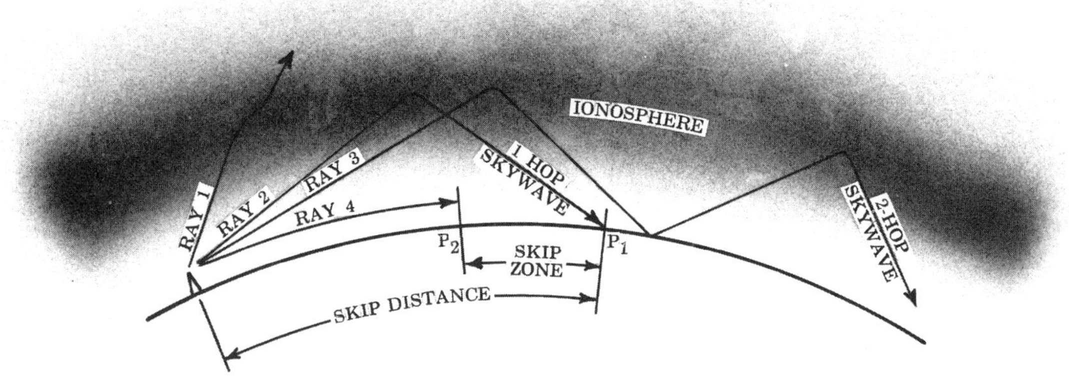

Figure 12-7. A sky wave and ground wave pattern for a given radio wave transmission.

incidence of ray 2 on the ionosphere is the maximum possible angle that will produce a sky wave. Ray *3* strikes the ionosphere at a lesser angle, is reflected in turn by the earth's surface, and is once again reflected by the ionosphere; such a doubly reflected wave is termed a "two-hop" sky wave. Ray *4* is a ground wave. It can penetrate the atmosphere only as far as point P_2 before absorption weakens the wave to such an extent that it is not receivable beyond this point. Hence, this particular radio wave cannot be received between points P_1 and P_2; such a zone between the maximum range of the ground wave and the minimum range of a sky wave is termed a *skip zone.* The term *skip distance* is sometimes applied to the minimum range at which a sky wave may be received.

Occasionally under relatively rare atmospheric conditions a phenomenon known as *surface ducting* may occur, extending the range of a ground wave well beyond its normal limits of reception. A surface duct is formed between the earth's surface and the bottom of a layer of air within which an extreme temperature inversion exists, shown in Figure 12-8 on the following page. Because the width of the duct must be large as compared to the wavelength of the radio wave to be effective, surface ducting is usually associated with the higher radio and radar frequencies. The phenomenon is most common in tropical latitudes, especially in the Pacific regions, where a surface duct once formed may persist for several hours or even days.

Short and Medium Range Navigation Systems

Most short- to medium-range radio navigation systems are designed to provide either a bearing to a shore-based transmitter site, as in the case of radiobeacons, or a range and bearing from the transmitter

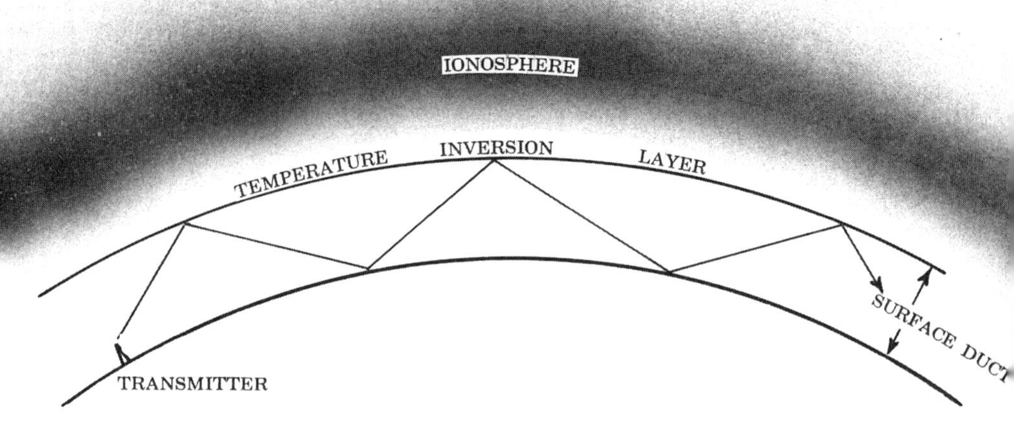

Figure 12-8. Formation of a surface duct.

to a natural or man-made navigation aid, in the case of radar. Decca, which may be considered both a short- and medium-range system, employs a type of hyperbolic lattice pattern to determine position; principles of its operation are discussed in the following section of this chapter. Since radar as a navigation system is discussed in some detail in Chapter 10 of *Marine Navigation 1*, it will not be dealt with here.

Essentially, a radiobeacon is a single transmitter, frequently combined with visual navigational aids such as a buoy or permanent light structure. Radiobeacons transmit a continuous wave at rather low power, usually modulated by an audible Morse Code character or characters for identification. The transmitted signal is received by a shipboard receiver incorporating a radio-direction-finding (RDF) capability, and used very much like an observed terrestrial bearing. The bearings of signals received simultaneously from two or more different radiobeacons may be crossed to form a fix, or a single RDF bearing may be crossed with any other type of LOP to obtain a fix or running fix.

Radiobeacons and their use will be discussed in further detail in the first section of the following chapter.

Long-Range Hyperbolic Navigation Systems

Most contemporary long-range electronic navigation systems are called *hyperbolic systems* because the line of position that they yield is a segment of a hyperbola rather than a radial line, as in the case of a terrestrial or radiobeacon LOP. Suppose that two transmitting stations designated M (Master) and S (Slave) located some distance apart both transmit a short radio wave pulse simultaneously at fixed time intervals. If the speed of propagation of the ground waves were assumed constant, a series of equally spaced concentric circles could be drawn surrounding each of the stations to represent the distance a single pulse would travel during a fixed time interval; the spacing between each circle is constant, since the pulse would travel a uniform distance during each time interval. Figure 12-9 illustrates a series of such circles.

A line drawn connecting the two stations M and S is called the *base line.* If a perpendicular line AB were drawn bisecting this base line, simultaneously transmitted pulses from both stations would arrive along this bisector at the same time. Thus, the bisector would represent a locus of all points at which the time difference in receipt of the master and slave station signals was zero.

Now, suppose that we wished to locate a second locus of points at which the slave signal arrived one time unit ahead of the master pulse. Returning to the base line, if we moved one-half the spacing of a circle from the bisector toward the slave station, we would also be moving one-half of the time interval away from the master. Hence,

Figure 12-9. Construction of a hyperbolic navigation pattern.

at this point on the base line, the slave station pulse would be received one time unit before the master station pulse. The locus of all other points at which an identical time difference could be measured would form the hyperbola labeled CD in the figure.

In similar fashion, other hyperbolas could be constructed, each of which would represent the locus of all points at which a specific time difference between the times of arrival of the master and slave pulses would be measured. A family of such hyperbolas appears in Figure 12-9.

As can be seen by an inspection of Figure 12-9, the divergence between adjacent hyperbolas representing successive time differences increases as the distance of the receiver from the base line increases, and the curvature of the hyperbolas approaches a straight line. In fact, at extreme ranges, the branches of the hyperbolas become asymptotic to straight lines converging at the midpoint of the base line. Thus, any hyperbolic system is inherently more accurate on or near the base line, with the precision of any LOP obtained decreasing with increased distance from the base line. Near the base line, any change in the position of the receiver would correspond to a relatively large change in the time difference of receipt of the master and slave pulses, but at longer ranges the change in time difference because of a similar change in position is relatively small.

The time difference measured anywhere on either base-line extension is a constant equal to the time required for a pulse to traverse the length of the base line. Moreover, the time difference is nearly constant in the vicinity of the extensions, with a large change in position in a direction normal to the extensions being necessary to produce a measurable change. Hence, the hyperbolic system subtended by a single pair of stations is unusable in the parabolic areas centered on the base-line extensions.

Up to this point in the discussion of the hyperbolic system it has been assumed that two pulses were transmitted simultaneously by both the master and slave stations. In actuality in such a pulsed hyperbolic system the pulses are not transmitted simultaneously, but rather the slave station transmits its pulse at a fixed interval called a *coding delay* after the mean time of arrival of the master pulse at the slave or *secondary* station; this is done so that the master pulse will always precede the slave pulse, no matter where the receiver is located in the hyperbolic grid pattern. An additional variable delay can also be included for security purposes during wartime. In such a system the time difference measured along the perpendicular bisector of the base line, line AB in Figure 12-9, is no longer zero but rather the interval corresponding to the sum of the time delay plus the transmission time between the master and slave stations. This type of system is employed in Loran-C.

As an alternative to a short pulse, the circular transmission patterns shown in Figure 12-9 can also be achieved by use of continuous waves, with the leading edges of each cycle representing a given time and corresponding distance interval. The corresponding hyperbolas then represent loci of constant phase differences called *isophase lines,* with zero phase differences occurring half a wavelength apart along the base line. In this type of system the spaces between the hyperbolas are referred to as *lanes;* the position of the receiver within a lane is determined by measuring the *phase difference* between the master and slave signals. A disadvantage of this type of system is that it is not possible to distinguish one lane from another by use of the phase comparison alone. Hence, the lanes must either be counted as they are traversed from some fixed position, or the lane must be identified by other means. Consol, Decca, and Omega all employ either a continuous wave or an interrupted continuous wave (i.e., a relatively long CW transmission as compared to a pulse) to establish their hyperbolic grid patterns.

Regardless of the type of transmission, in hyperbolic navigation systems the user's distance from the stations can seldom exceed about six times the length of the base line for good accuracy in the line of position. In systems in which the base line is very short compared to the distances over which the system is to be radiated, the pattern is referred to as a *collapsed* or *degenerated hyperbolic type,* and only the straight asymptotic part of the hyperbola is used. Consol is a system representative of this type. In general, the shorter the base line, the more the character of the LOP resembles a bearing.

In electronic navigation, as in visual piloting and celestial navigation, a fix is obtained by the intersection of two or more simultaneous or near simultaneous LOPs. In the case of the hyperbolic system, at least two pairs of stations must be available to the user in order to obtain a fix. In cases in which only one hyperbolic LOP is available, it may be crossed with a simultaneous LOP obtained by any other means to determine a fix. Such single hyperbolic LOPs, when available, are often used to check the accuracy of a celestial fix, especially when only two or three bodies are observed.

Accuracy of an Electronic Fix

In electronic navigation it is frequently desirable to use a standard measure by which the accuracy of a fix produced by a given system can be expressed. In radio navigation the measure most often used for this purpose is the *root mean square* (r.m.s.) error, which can be thought of as the standard deviation (σ in statistics) within which a given position might vary from the actual position 68 percent of the time. A detailed explanation of the computation and use of r.m.s. error is beyond the scope of this text; it will suffice here to regard the r.m.s. accuracy as being the maximum distance of the indicated position from the true position 68 percent of the time.

Summary

This chapter has described the characteristics and behavior of the electromagnetic wave upon which the radio navigation systems to be discussed in the following chapter are based. In general, the behavior of any particular radiated RF signal is dependent on its power, frequency, and corresponding wavelength. The type of modulation used to produce meaningful positioning information to the user depends on the purpose for which the system is intended. Radar and Loran-C use pulsed transmissions, while most radiobeacons, Decca, Consol, and Omega employ continuous waves. Radio navigation systems can provide bearings to the transmitter site, as in the case of radiobeacons and Consol, or hyperbolic lattice patterns, as in the case of Loran-C, Decca, and Omega; radar provides both a range and a bearing from the transmitter to a desired natural or artificial navigation aid. Each of the former systems will be discussed in the following chapter, while radar is the subject of Chapter 10 of *Marine Navigation 1.*

13

Radio Navigation Systems

The preceding chapter reviews the principles of behavior of the electromagnetic wave in the earth's atmosphere, and discusses the general characteristics of the most widely used short-, medium-, and long-range radio navigation systems. This chapter specifically describes radiobeacons, Loran-C, Decca, Consol, and Omega, in order to acquaint the inexperienced navigator with the practical methods of position-finding through the use of each of these systems. However, in electronic navigation, just as in celestial navigation, there is no substitute for practical experience. Working with receivers under actual conditions is the only way to attain proficiency in the use of any electronic navigation system.

Radiobeacons

As previously mentioned, radiobeacons were first employed as coastal navigational aids along the shores of the United States in the 1930s. They consist of single transmitters, often co-located with visual navigational aids. Radiobeacon signals are transmitted at rather low power on frequencies in the LF and MF bands between about 250 to 400 kHz; most signals are limited in range to less than 200 miles, with the majority not receivable beyond about 20 miles. The signals are continuous waves modulated by Morse Code characters for identification. Often radiobeacons located in a given area are grouped on a common frequency, and each transmitter transmits only during a segment of a time-sharing plan. Radio bearings to the site of the transmitter are determined by the use of a radio receiver equipped with a radio direction finding (RDF) capability. There are a number of moderately priced manually operated RDF receivers and several more expensive fully automatic models available commercially for private users. Many Navy navigators use the services of the ECM (electronic countermeasures) personnel in CIC during piloting to obtain radiobeacon bearings.

All RDF receivers, whether manual or automatic, operate on the principle that when a coil is rotated in the field of radiation of a transmitter, the received signal strength is greatest when the coil face is oriented 90° away from the direction of the incoming signal, and weakest or *nulled* when the coil is normal to the signal, facing either directly toward or directly away from it. Thus, particularly on older models of receivers, it is very easy for the inexperienced operator to read an RDF bearing as 180° away from the true bearing of the transmitter. Most newer models, however, have features that tend to negate the possibility of such error. Other potential sources of error include improper calibration of the receiver, the effect of coastal refraction if the path of the signal passes over land or along the shoreline between the beacon and the receiving ship, and varying amounts of ionospheric interference at night. As a general rule, RDF bearings are normally considered accurate only to within ±2° for distances under 150 miles to the transmitter in favorable conditions, and ±5–10° when conditions are unfavorable.

Descriptions of the characteristic signals and normal ranges of all permanent radiobeacons in U.S. and foreign waters are contained in the DMAHTC *Publications No. 117A* and *B, Radio Navigational Aids.* Volume A covers coastal areas contiguous to the North and South Atlantic Oceans, the Mediterranean Sea, and the Arctic Ocean, while Volume B covers coastal areas adjoining the North and South Pacific Oceans, the Indian Ocean, and the Bearing Sea. The various coastal regions covered in the two volumes are divided into areas, with a separate section for each of the areas covered. Within each area, the radiobeacons are listed in geographic order. A sample page from the U.S. Atlantic Coast Section of *Publication 117A* appears in Figure 13-1A.

Brief descriptions of each radiobeacon located along coastal areas of the United States and its possessions are also included in the introductory pages of the *Light List* described in *Marine Navigation 1.* Because of the more complete data in *Publication 117,* however, this publication should normally be consulted whenever RDF bearings to radiobeacons are to be used.

Since radio waves transmitted by radiobeacons travel outward along great circles, radio bearings to the beacons cannot be directly read from a receiver and plotted on a Mercator chart. All RDF bearings must first be converted to rhumb lines before they can be plotted, using a conversion table included in *Publications 117A* and *B.* A portion of this table appears in Figure 13-1B on page 234.

Entering arguments for the table are the difference in longitude to the nearest half degree between the positions of the radiobeacon and the receiver, and the mid-latitude to the nearest whole degree between the two positions. The sign of the correction is determined by the rules included at the bottom of the table.

Corrected radiobeacon bearings are plotted on a Mercator chart using a reciprocal drawn from the radiobeacon to the approximate

RADIOBEACONS

RADIO NAVIGATIONAL AIDS

CANADA–UNITED STATES–Great Lakes–Lake Superior (Cont'd)

1074. **Superior Entry South Breakwater Light Station Marker Radiobeacon, Wis.** **46°42'37"N., 92°00'22"W.**
FREQ.: 316 kHz, A2. Low power, for local use only.
CHARACTERISTIC SIGNAL: Seconds
A series of 0.5 second dashes 13.5
Silent 1.5
Period 15.0
HOURS OF TRANSMISSION: Continuous.
NOTE: Operates during navigation season only. Carrier signal. (See sec. 100A.)

UNITED STATES—Atlantic Coast

All U.S. Marine Radiobeacons are equipped with a continuous carrier (see sec. 100A).

1076. **West Quoddy Head Light Station, ME.** **44°48'54"N., 66°57'04"W.**
FREQ.: 308 kHz, A2. RANGE: 20 miles.
CHARACTERISTIC SIGNAL: Seconds:
WQ (• ▬ ▬ ▬ ▬ • ▬) 50
Long dash 10
Period .... 60 (1 min.)
HOURS OF TRANSMISSION: Continuous.
GROUP SEQUENCE: IV
REMARKS: Grouped with Cape D'Or I (1033.5), Partridge Island II 1033.7, Southwest Head III (1034) and Seal Island VI (1033).

1077. **Great Duck Island Light Station, ME.** **44°08'32"N., 68°14'47"W.**
FREQ.: 286 kHz, A2. RANGE: 50 miles.
CHARACTERISTIC SIGNAL: Seconds:
GD (▬ ▬ • ▬ • •) 50
Long dash 10
Period .... 360 (6 min.)
HOURS OF TRANSMISSION: Continuous.
GROUP SEQUENCE: V
REMARKS: Grouped with Highland I (1087), Nantucket II (1090), Montauk Point III (1098), Ambrose IV (1106) and Manana VI 1079.

1078. **Martinicus Rock Light Station, ME.** **43°47'01"N., 68°51'19"W.**
FREQ.: 314 kHz, A2. RANGE: 20 miles.
CHARACTERISTIC SIGNAL:
MR (▬ ▬ • ▬ •).
HOURS OF TRANSMISSION: Continuous.

1079. **Manana Island Fog Signal Station, ME.** **43°45'48"N., 69°19'38"W.**
FREQ.: 286 kHz, A2. RANGE: 100 miles.
CHARACTERISTIC SIGNAL: Seconds:
MI (▬ ▬ • •) 50
Long dash 10
Period 360 (6min.)
HOURS OF TRANSMISSION: Continuous.
GROUP SEQUENCE: VI
REMARKS: Grouped with Highland I (1087), Nantucket II (1090), Montauk III (1098), Ambrose IV (1106) and Great Duck V (1077).

1080. **The Cuckolds Light Station Marker Radiobeacon, ME.** **43°46'46"N., 69°39'02"W.**
FREQ.: 320 kHz, A2. RANGE: 10 miles.
CHARACTERISTIC SIGNAL:
CU (▬ • ▬ • • • ▬).
HOURS OF TRANSMISSION: Continuous.

Figure 13-1A. Sample page from Pub. 117A, Radio Navigational Aids.

RADIO NAVIGATIONAL AIDS

RADIO DIRECTION-FINDER AND RADAR STATIONS

200F. Radio Bearing Conversion Table

Correction to be applied to radio bearing to convert to Mercator bearing

Difference of longitude

Mid. lat.	0.5°	1°	1.5°	2°	2.5°	3°	3.5°	4°	4.5°	5°	5.5°	6°	6.5°	7°	7.5°	Mid. lat.
°	°	°	°	°	°	°	°	°	°	°	°	°	°	°	°	°
4	...	...	...	...	0.1	0.1	0.1	0.1	0.2	0.2	0.2	0.2	0.2	0.2	0.3	4
5	...	0.1	0.1	.1	.1	.1	.2	.2	.2	.2	.2	.3	.3	.3	.3	5
6	...	.1	.1	.1	.1	.2	.2	.2	.2	.3	.3	.3	.3	.4	.4	6
7	...	.1	.1	.1	.2	.2	.2	.3	.3	.3	.3	.4	.4	.4	.5	7
8	...	.1	.1	.1	.2	.2	.2	.3	.3	.4	.4	.4	.5	.5	.5	8
9	...	.1	.1	.1	.2	.2	.2	.3	.3	[illegible]	[illegible]	.5	.5	.6	.6	[illegible]
[illegible]	[illegible]	[illegible]	[illegible]	[illegible]	.2	.2	.3	[illegible]	[illegible]	[illegible]	[illegible]	[illegible]	.6	[illegible]	[illegible]	[illegible]
[illegible]	[illegible]	.2	[illegible]	[illegible]	[illegible]	.3	[illegible]	[illegible]	1.1	1.2	1.4	[illegible]	[illegible]	[illegible]	[illegible]	[illegible]
31	.1	.2	.4	.5	[illegible]	[illegible]	[illegible]	1.0	1.2	1.3	1.4	1.6	[illegible]	1.8	1.9	31
32	.1	.3	.4	.5	.7	[illegible]	.9	1.1	1.2	1.3	1.4	1.6	1.7	1.8	2.0	32
33	.1	.3	.4	.6	.7	.8	1.0	1.1	1.2	1.4	1.5	1.6	1.8	1.9	2.1	33
34	.1	.3	.4	.6	.7	.8	1.0	1.1	1.2	1.4	1.5	1.7	1.8	2.0	2.1	34
35	.1	.3	.4	.6	.7	.9	1.0	1.2	1.3	1.4	1.6	1.7	1.9	2.0	2.2	35
36	.1	.3	.4	.6	.7	.9	1.0	1.2	1.3	1.5	1.6	1.8	1.9	2.1	2.2	36
37	.2	.3	.4	.6	.8	.9	1.1	1.2	1.4	1.5	1.6	1.8	2.0	2.1	2.2	37
38	.2	.3	.5	.6	.8	.9	1.1	1.2	1.4	1.5	1.7	1.8	2.0	2.2	2.3	38
39	.2	.3	.5	.6	.8	1.0	1.1	1.2	1.4	1.6	1.7	1.9	2.1	2.2	2.4	39
40	.2	.3	.5	.6	.8	1.0	1.1	1.3	1.4	1.6	1.8	1.9	2.1	2.2	2.4	40
41	.2	.3	.5	.6	.8	1.0	1.2	1.3	1.5	1.6	1.8	2.0	2.1	2.3	2.5	41
42	.2	.3	.5	.7	.8	1.0	1.2	1.3	1.5	1.7	1.8	2.0	2.2	2.3	2.5	42
43	.2	.3	.5	.7	.8	1.0	1.2	1.4	1.5	1.7	1.9	2.1	2.2	2.4	2.6	43
44	.2	.4	.5	.7	.9	1.1	1.2	1.4	1.6	1.7	1.9	2.1	2.2	2.4	2.6	44
45	.2	.4	.5	.7	.9	1.1	1.2	1.4	1.6	1.8	2.0	2.1	2.3	2.5	2.6	45
46	.2	.4	.5	.7	.9	1.1	1.3	1.4	1.6	1.8	2.0	2.2	2.3	2.5	2.7	46
47	.2	.4	.6	.7	.9	1.1	1.3	1.5	1.7	1.8	2.0	2.2	2.4	2.6	2.8	47
48	.2	.4	.6	.8	.9	1.1	1.3	1.5	1.7	1.8	2.1	2.2	2.4	2.6	2.8	48
49	.2	.4	.6	.8	1.0	1.1	1.3	1.5	1.7	1.9	2.1	2.3	2.5	2.6	2.8	49
50	.2	.4	.6	.8	1.0	1.1	1.3	1.5	1.7	1.9	2.1	2.3	2.5	2.7	2.9	50
51	.2	.4	.6	.8	1.0	1.2	1.4	1.6	1.8	2.0	2.1	2.3	2.5	2.7	2.9	51
52	.2	.4	.6	.8	1.0	1.2	1.4	1.6	1.8	2.0	2.2	2.4	2.6	2.8	3.0	52
53	.2	.4	.6	.8	1.0	1.2	1.4	1.6	1.8	2.0	2.2	2.4	2.6	2.8	3.0	53
54	.2	.4	.6	.8	1.0	1.2	1.4	1.6	1.8	2.0	2.2	2.4	2.6	2.8	3.0	54
55	.2	.4	.6	.8	1.0	1.2	1.4	1.6	1.8	2.1	2.2	2.4	2.7	2.9	3.1	55
56	.2	.4	.6	.8	1.0	1.2	1.4	1.7	1.9	2.1	2.3	2.5	2.7	2.9	3.1	56
57	.2	.4	.6	.8	1.1	1.2	1.5	1.7	1.9	2.1	2.3	2.5	2.7	2.9	3.2	57
58	.2	.4	.6	.8	1.1	1.3	1.5	1.7	1.9	2.1	2.3	2.6	2.8	3.0	3.2	58
59	.2	.4	.6	.8	1.1	1.3	1.5	1.7	1.9	2.2	2.4	2.6	2.8	3.0	3.2	59
60	.2	.4	.6	.9	1.1	1.3	1.5	1.7	2.0	2.2	2.4	2.6	2.8	3.0	3.2	60

Receiver (latitude)	Transmitter (direction from receiver)	Correction Sign	Receiver (latitude)	Transmitter (direction from receiver)	Correction Sign
North	Eastward	+	South	Eastward	−
North	Westward	−	South	Westward	+

2–5

Figure 13-1B. Portion of "Radio Bearing Conversion Table," Publication 117A.

position of the ship. Additional lines are drawn representing the probable limits of error at the time the bearings were obtained, as illustrated in Figure 13-1C. The intersection of two or more such bearing envelopes indicates the probable ship's position as well as the area within which the ship might lie given the probable bearing errors, as shown in the figure.

Because of the relative inaccuracy of RDF bearing LOPs as compared to LOPs obtained visually or by the use of radar, radiobeacons are not widely used by U.S. Navy navigators. Navigators of small boats and merchant ships not equipped with effective navigational radars, however, make extensive use of these aids whenever they are engaged in coastal navigation and piloting.

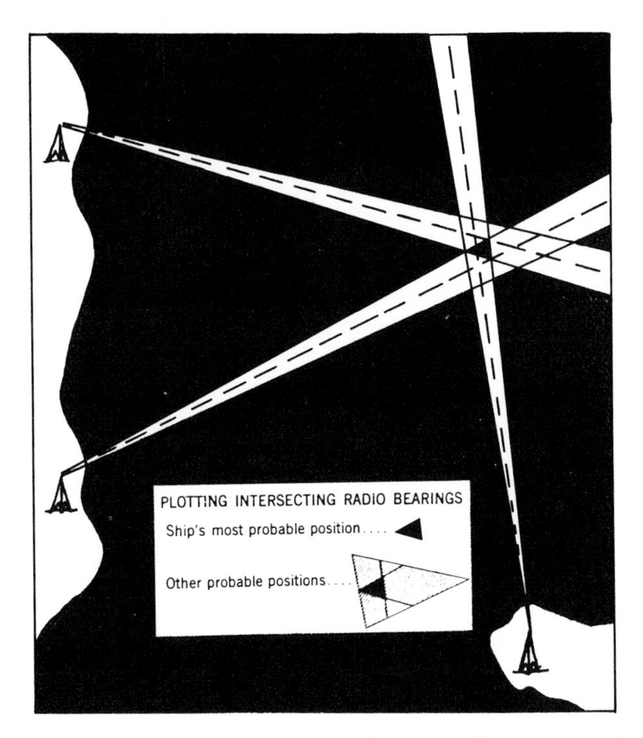

Figure 13-1C. A plot of a ship's position as determined by three radiobeacon bearings.

Loran-C

The Loran (for Long Range Navigation) system was originally developed by the United States in 1940. It was one of the first attempts to implement a long-range hyperbolic navigation system that could provide all-weather fixing information for both ships and aircraft at sea. World War II and the Korean War spurred the establishment of the Loran system. As originally conceived, the system used pairs of master and slave stations transmitting sequential pulsed radio waves in the upper MF band, with frequencies between 1850 and 1950 kHz, to establish a hyperbolic lattice pattern in the area of coverage based on time differences of receipt of the master and the slave station pulse. Each hyperbolic LOP in the pattern represented a locus of points at which a unique time difference could be recorded between receipt of the master station pulse and the pulse of its associated slave. This original system, which came to be called *Loran-A*, featured ground wave coverage out to between 450 and 800 miles from the base line by day, and sky wave ranges to 1,400 miles by night. Between about 1955 and 1970 it was deployed throughout the coastal regions of the northern Atlantic and Pacific Oceans, where it became probably the most widely used electronic navigation aid in the world up to that time.

In the early 1950s, the need for a longer-range electronic navigation system that would extend coverage farther out into the mid-ocean regions began to be felt. Initial attempts to extend the range of Loran-A by simply lowering its carrier frequencies were abandoned because of difficulties encountered with station synchronization and in distinguishing between ground and sky waves as a result of the shorter reflection time of the sky wave at lower frequencies. To overcome these difficulties, a system employing synchronized pulses for both time-difference and phase-comparison measurements was developed. This system became known as Loran-C; the first stations in the system became operational in 1957. The Loran-C system has since been continually improved and expanded, to the point where in the late seventies its ground wave coverage extended over most littoral regions of the northern Atlantic and Pacific Oceans and the Mediterranean, with sky waves receivable over most of the northern hemisphere with the exception of the Indian Ocean. Loran-A was thus rendered redundant, with the result that it began to be phased out of existence in 1977, following a decision by the Department of Transportation to designate Loran-C as the sole government-provided radio-navigation system for civil marine use in U.S. coastal areas. The Coast Guard assumed managerial control of all U.S.-operated Loran-C stations in 1978. In early 1981, the only Loran-A stations still transmitting were those located in the Japanese islands in the Western Pacific, this for the benefit of the many small commercial fishermen there who have become dependent on the system to locate their fishing grounds. The coverage of Loran-C in early 1981 is shown in Figure 13-2.

Both Loran-A and Loran-C originally operated under a master station/slave station concept wherein the receipt of the master pulse at the associated slave station would trigger the sequential transmission of the "slave" pulse. In the late sixties and early seventies, however, atomic time standards were introduced to regulate both transmissions more precisely, and the slave stations in Loran-C were consequently redesignated *secondary* stations; each group of master and associated secondary stations is referred to as a *chain*. All stations in the system transmit a signal on a common carrier frequency in the mid-LF band of 100 kHz, with a band width extending 10 kHz to either side. Because this frequency is relatively low compared to the basic 1900 kHz frequency of the old Loran-A, the range of both ground and sky waves in Loran-C extends considerably farther. The ground-wave range of Loran-C is considered to be about 1,200 miles; one-hop sky waves have a range of about 2,300 miles, and two-hop signals have been received as far as 4,000 miles from the transmitting stations.

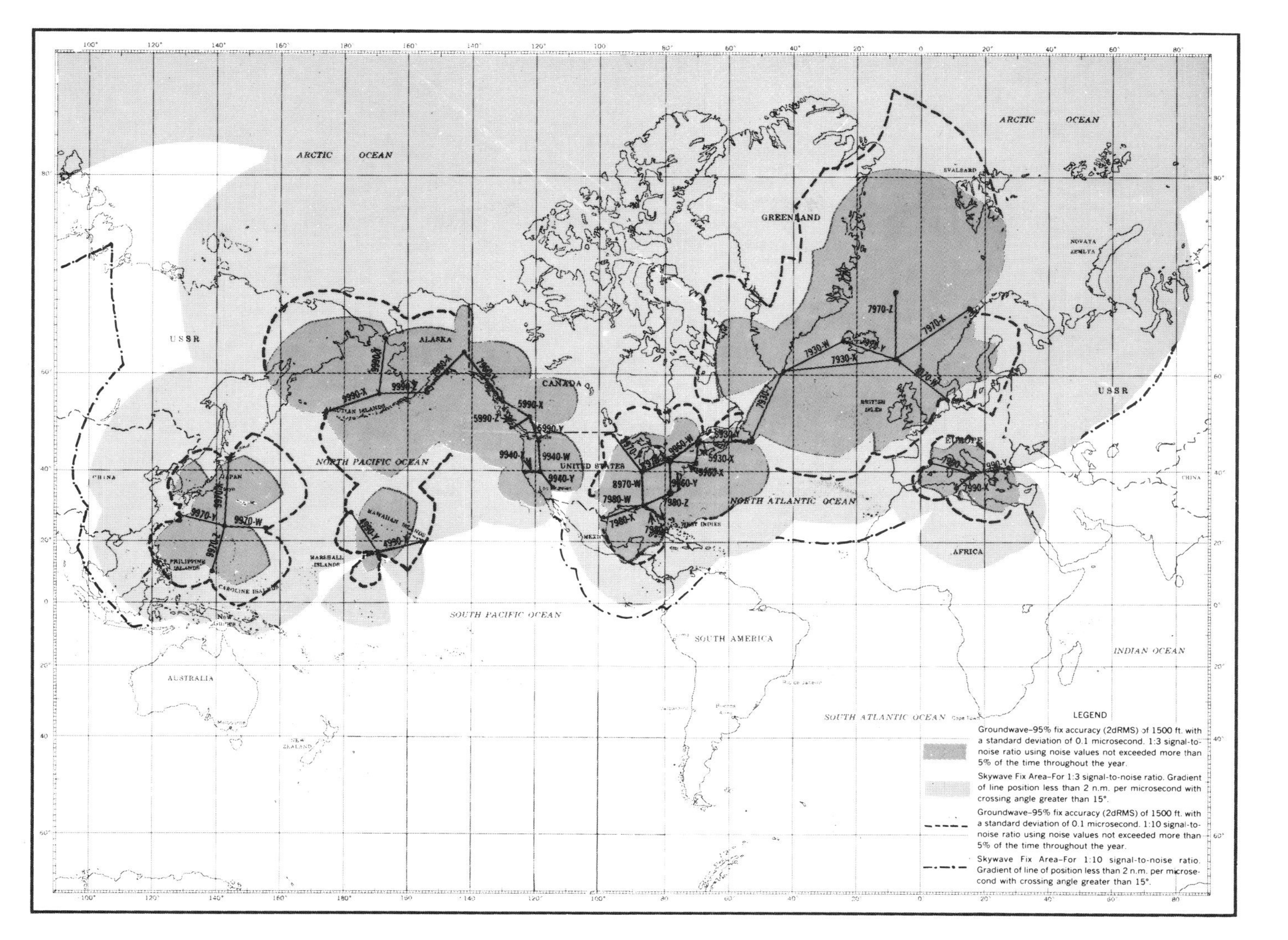

Figure 13-2. Areas of the world covered by Loran-C, 1981.

One-hop sky waves are produced both by day and by night, while two-hop sky waves are formed only at night. The increased range of the Loran-C ground wave, coupled with the use of atomic time standards vice the receipt of the master pulse to regulate the transmission of the secondary station pulses, has permitted base-line distances to be extended from a maximum of about 500 miles in the old Loran-A system, to between 1,000 and 1,500 miles for present-day Loran-C chains. The area of coverage of a particular Loran-C pair, and the sensitivity at longer ranges, are thereby both considerably increased as compared to the old Loran-A. Accuracy of the system varies from about ±700 feet r.m.s. near the base line to ±2,000 feet r.m.s. near the extreme range of the system. The low frequencies and high power (over 1,500 kw in some cases) at which Loran-C signals are transmitted allow ground waves to penetrate the surface layers of sea water, enabling them to be received by submerged submarines.

Unfortunately, the increased advantages of Loran-C as compared to Loran-A are achieved at the cost of increased complexity of both the system itself and the receiver required to use it. This complexity is necessary for two reasons. First, at the frequency at which the Loran-C signal is transmitted, a great deal of power is required to propagate the wave for the long range for which the system is designed. Second, at this frequency sky waves are propagated that arrive a very short time after the arrival of the ground wave at any particular location.

To lessen the large power requirements, in Loran-C a so-called *multipulsed* signal is used. Each pulse transmitted by a master station actually consists of nine separate comparatively weak pulses; the first eight are separated by an interval of 1000 μs, and the ninth by 2000 μs. Each secondary station transmits eight such pulses at 1000 μs intervals, with the extra pulse in the master signal being used for visual master station identification and trouble blink. Within the Loran-C receiver, each of the basic eight weak pulses is electronically superimposed or "integrated" to form "strong" master and secondary pulses of approximately 320 μs duration, which can then be electronically compared to obtain the time-difference measurement.

To eliminate sky-wave interference, the integrated master and secondary pulses are compared at a sampling point exactly 30 μs from their leading edges. The comparison is thus made before any reflected sky waves of the pulse can reach the receiver.

The multipulsed signals of each master station and its associated secondary stations are transmitted in a predetermined sequence, as shown in Figure 13-3. As was the case with Loran-A, the master station signal is allowed to arrive at each of its secondary stations before the secondary station signals are commenced, and a coding delay is included between the secondary transmissions to ensure that they will always be received in the same sequence throughout the area covered by the chain.

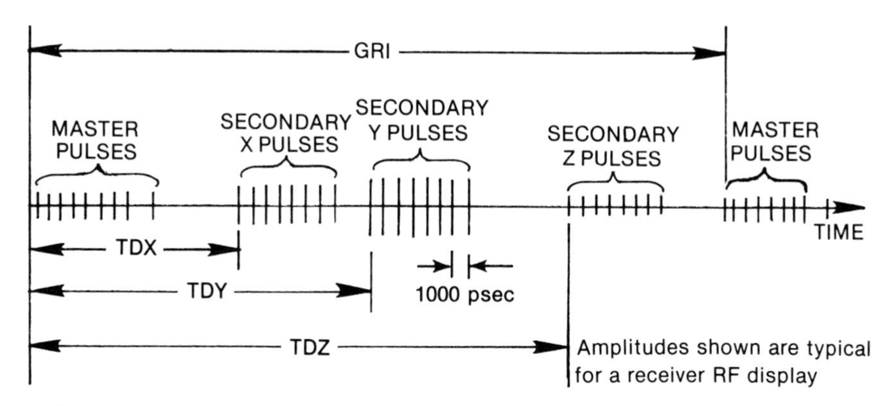

Figure 13-3. Loran-C pulse sequence, for a four-station chain.

Currently there are thirteen Loran-C chains, and four basic pulse repetition rate (PRR)/basic pulse repetition interval (PRI) combinations. Each chain uses a different basic PRR/basic PRI combination, with the basic PRI modified to one of eight specific PRIs, as shown below:

Basic PRR/Basic PRI (μs)	Specific PRI (μs)
10/100,000	0 — Basic rate
$12\frac{1}{2}$/80,000	1 — 100 less than basic rate
$16\frac{2}{3}$/60,000	2 — 200 less than basic rate
20/50,000	3 — 300 less than basic rate
	4 — 400 less than basic rate
	5 — 500 less than basic rate
	6 — 600 less than basic rate
	7 — 700 less than basic rate

The specific PRI used in a given chain is referred to as its *group repetition interval* (*GRI*), often called a *rate*. The first four digits of the GRI are used as the chain designator; the secondary station is identified by suffixing its letter to the GRI. Thus, the code 7970-X designates the master-slave X pair of the Norwegian Sea chain, which uses a basic PRR of $12\frac{1}{2}$ pulses per second transmitted at a specific PRI of 79,700 μs. An observed time difference in microseconds is added as a suffix to the basic code, as for example 7970-X-11347.

Loran-C Receivers

The Loran-C receiver used at present in many U.S. Navy ships is the SPN 32, pictured in Figure 13-4. This receiver is designed in such a way that time-difference readings for any two station pairs receivable at a given location are determined and indicated simultaneously, thus eliminating the necessity to advance Loran-C LOPs. Once it is set up for a given Loran-C chain, the receiver has the capability of continuous automatic monitoring of two time differences until such time as the vessel passes out of range. The receiver must then be manually reset for another chain. Within the receiver, a rough determination of the time difference between receipt of the master signal and each secondary station signal is first made electronically. The exact time difference for each pair is then determined by a comparison of the phase of the integrated master pulse with the phase of each of the various integrated secondary station pulses. These measurements are both made automatically by components within the receiver.

There are a number of Loran-C receivers available commercially, ranging in price from about $1,500 for manually operated models that will track one and in some cases two master-secondary station pairs automatically, to $6,000 for more sophisticated models that feature automatic acquisition and tracking of two pairs simultaneously. Some of the latter types can even be purchased with optional electronic packages that continuously convert the readings received to a latitude and longitude readout.

Figure 13-4. SPN 32 Loran-C receiver.

Plotting the Loran-C Fix

After the time-difference readings for two or more Loran-C station pairs receivable at a particular location have been obtained and recorded, it remains only to plot the corresponding portions of the time-difference hyperbolas on a chart or plotting sheet to obtain the Loran-C fix. Of course, if the user's receiver has a lat/long conversion feature, only the fix position need be plotted. Even in these cases, however, it is prudent to check the actual time-difference readouts against the Loran-C chart occasionally, to be certain that the position being read out is consistent with the charted time-difference hyperbolas in the area in which the position plots. If by chance only one Loran-C line is obtainable in the area in which the vessel is operating, it may be combined with LOPs obtained by other means to form a fix or running fix, in the usual fashion.

The Defense Mapping Agency and the National Ocean Survey both issue series of Loran-C charts upon which the hyperbolic patterns covering the area represented on the chart are overprinted. There is a separate set of color-coded time-difference hyperbolas for each master-secondary station pair receivable within that area. The hyperbolas are printed at intervals of 25, 50, 100, or 200 μs, depending on the scale of the chart and the degree of convergence of the hyperbolic patterns depicted. A portion of a typical Loran-C chart appears in Figure 13-5A on the following page.

It is fairly simple to plot a Loran-C LOP on a chart overprinted with Loran hyperbolas. The navigator merely selects the two time-difference hyperbolas printed on the chart near his DR position that bracket the observed reading for the secondary station in question, and plots a short segment of the hyperbola corresponding to this reading in the appropriate position between them. The segment is a straight line plotted in rough conformance with the trend of the hyperbolas to either side.

To assist in the correct positioning of the segment between the two printed hyperbolas, an aid called a *linear interpolator* is printed in the margin of every Loran chart. An example appears in Figure 13-5B.

The interpolator is used in conjunction with either a pair of dividers or a drawing compass; most navigators prefer the dividers. First, the dividers are spread to the interval appearing between the two printed hyperbolas between which the observed segment is to be drawn in the vicinity of the DR position. Without changing the spread thus established, the dividers are then moved to the linear interpolator, where they are held at a right angle to the bottom base line with one point resting thereon. The dividers are then shifted to the right or left until the top point rests on the top curve of the interpolator.

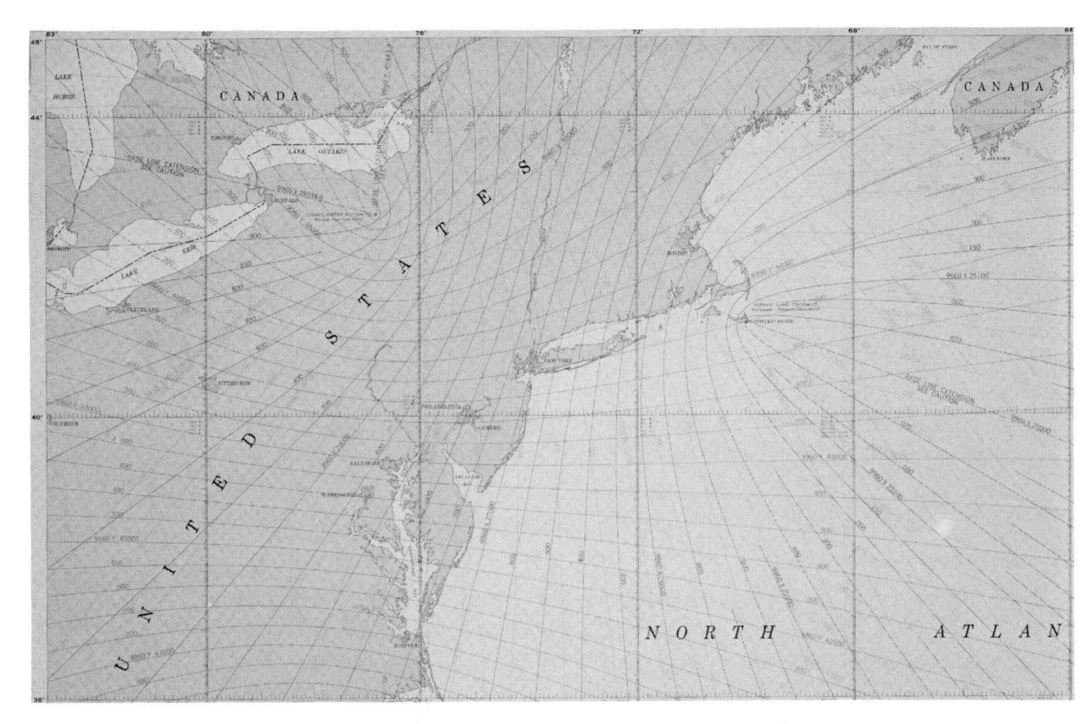

Figure 13-5A. Portion of a Loran-C chart of an area covered by the Northeast U.S. chain (GRI 9960).

The portion of the interpolator between the divider points now represents the charted interval. The top point of the dividers is then compressed downward across the interpolator scale until the point rests

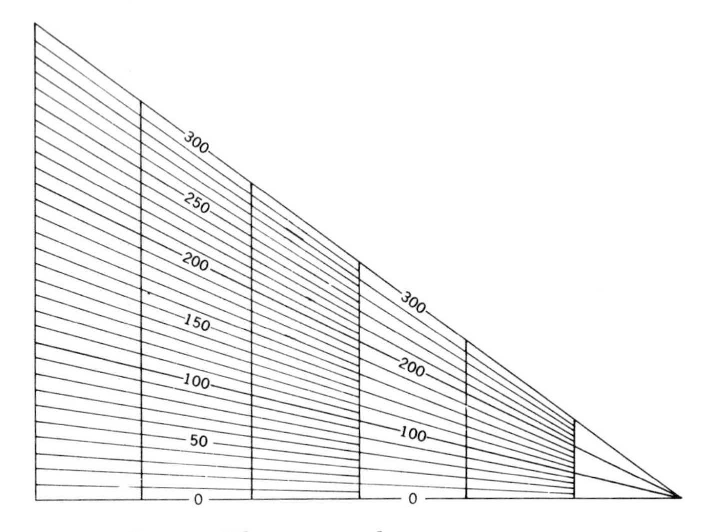

Figure 13-5B. A Loran-C linear interpolator.

on the scale marking corresponding to the exact value of the observed reading. The dividers are then shifted back to the site at which the segment is to be drawn on the chart, with care being taken not to disturb the spread determined by the interpolator, and a point is located at the spread distance from the lower-valued overprinted hyperbola. The segment is then drawn through this point, and labeled with the rate above the line and the time of observation to the nearest minute below the line, as shown in Figure 13-5C.

Two or more such Loran-C LOPs obtained from different master-secondary station pairs within a half hour can be combined to form a Loran-C fix. If the readings on which the LOPs are based were made within a reasonably short time span, such as two or three minutes, the intersection of the two plotted LOPs can be considered the Loran-C fix position. If more than a few minutes has elapsed between the readings, however, the earlier LOP(s) should be advanced to the time of the later LOP in the same manner as is done in celestial navigation. A single Loran-C LOP cannot usually be used to obtain a running fix with a later Loran-C LOP from the same secondary station in surface navigation, since the angles between successive hyperbolas are fairly small

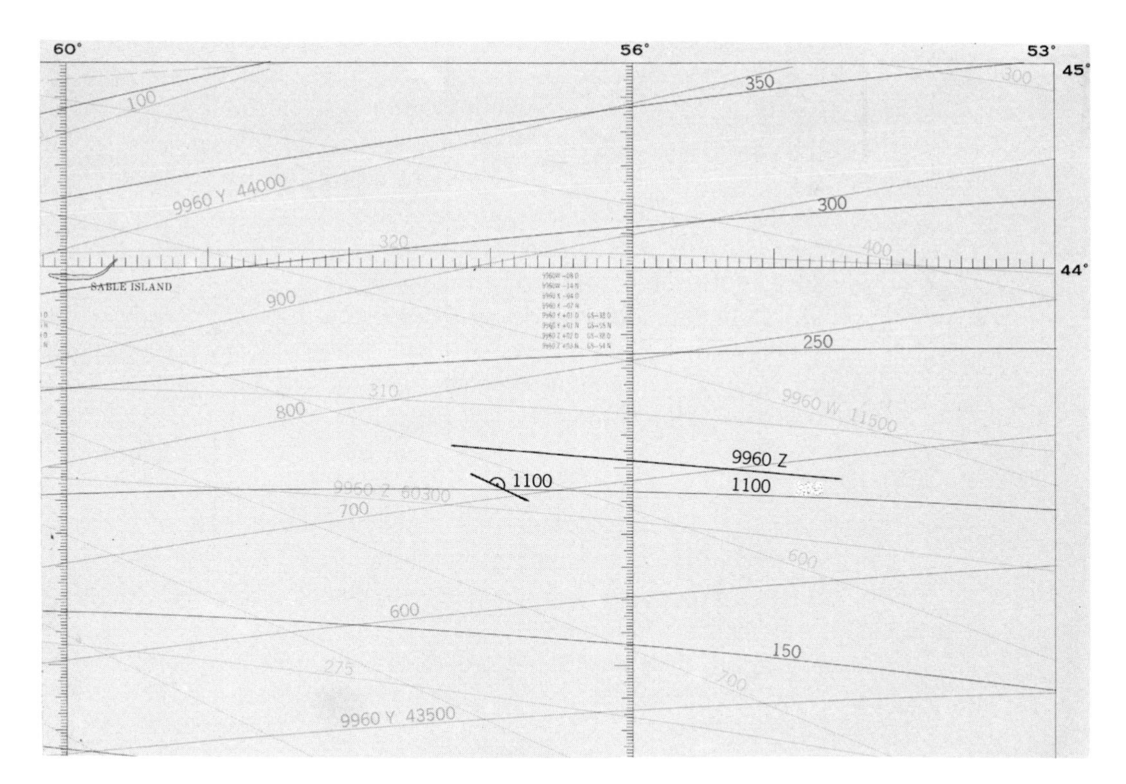

Figure 13-5C. Loran-C LOP plotted for a reading.

in most areas of coverage. It can, however, be crossed with an LOP from another source obtained either simultaneously or within three hours of the Loran-C LOP to determine a fix or running fix, respectively.

Loran-C Lattice Tables

If for some reason a suitable Loran-C chart is not available, or if the navigator desires to plot the Loran-C information on a plotting sheet, a series of Loran-C *Lattice* Tables issued by the DMAHTC as *Publication Number 221* may be used. There is a separate table issued for each secondary station of a chain. By the use of the appropriate table, a small segment of the time-difference hyperbola corresponding to the value obtained from the Loran-C receiver is plotted. If the hyperbola lies in a general east-west direction at the approximate location of the vessel, the segment is drawn by locating its latitude intercepts on two tabulated meridians bracketing the DR position of the vessel at the time of the time-difference determination. Conversely, if the hyperbola lies in a north-south direction, the segment is plotted by locating the longitude intercepts on two tabulated parallels of latitude bracketing the DR position. Depending on the amount of curvature in the hyperbolic time-difference curve, the interval between tabulated latitudes or longitudes may be a whole degree, a half degree, or one-quarter degree. Among other information, the introductory pages of each set of tables contain a diagram showing the general geographic form of the hyperbolic pattern generated by that master-secondary station pair. A quick glance at the diagram will usually indicate whether latitude or longitude intercepts may be expected at the approximate location of the navigator's vessel.

The horizontal entering argument for the *Loran-C Tables* is the nearest tabulated time difference to the observed reading, and the vertical arguments are tabulated latitudes or longitudes bracketing the DR position at the time of the observations; whether tabulated latitudes or longitudes are used depends on the trend of the hyperbolic pattern in the vicinity of the DR position. Interpolation for the observed time difference is done by multiplying the amount of increase or decrease of the observed versus the tabulated time difference by so-called "Δ values," which represent the change (to the nearest .01′) of the exact longitude or latitude intercept for a change of 1 μs of Loran reading. The resulting corrections are applied to the tabulated intercepts to obtain the interpolated intercepts corresponding to the observed readings.

The use of the tables is best illustrated by means of an example.

Suppose that at 1500 the DR position of a ship was L 25° 30.0′ N, λ 85° 30.0′ W. The ship is on course 135°T at a speed of 20 knots. The following two Loran-C time-difference readings for rates 7980-Y and 7980-W were obtained at the times indicated:

Time	Rate	Reading
1518	7980-Y	44995.0
1533	7980-W	13423.4

Since the two time-difference readings were not obtained simultaneously but rather over a 15-minute time interval, the portion of the hyperbolic LOP corresponding with the first time difference must be plotted using the *Loran-C Tables* for the master-secondary pair 7980-Y and advanced to the time of the second LOP. This has been done on the plot sheet pictured in Figure 13-6A.

Next, the second LOP must be plotted by the use of the Loran-C tables for the master-secondary pair 7980-W. The hyperbolic pattern

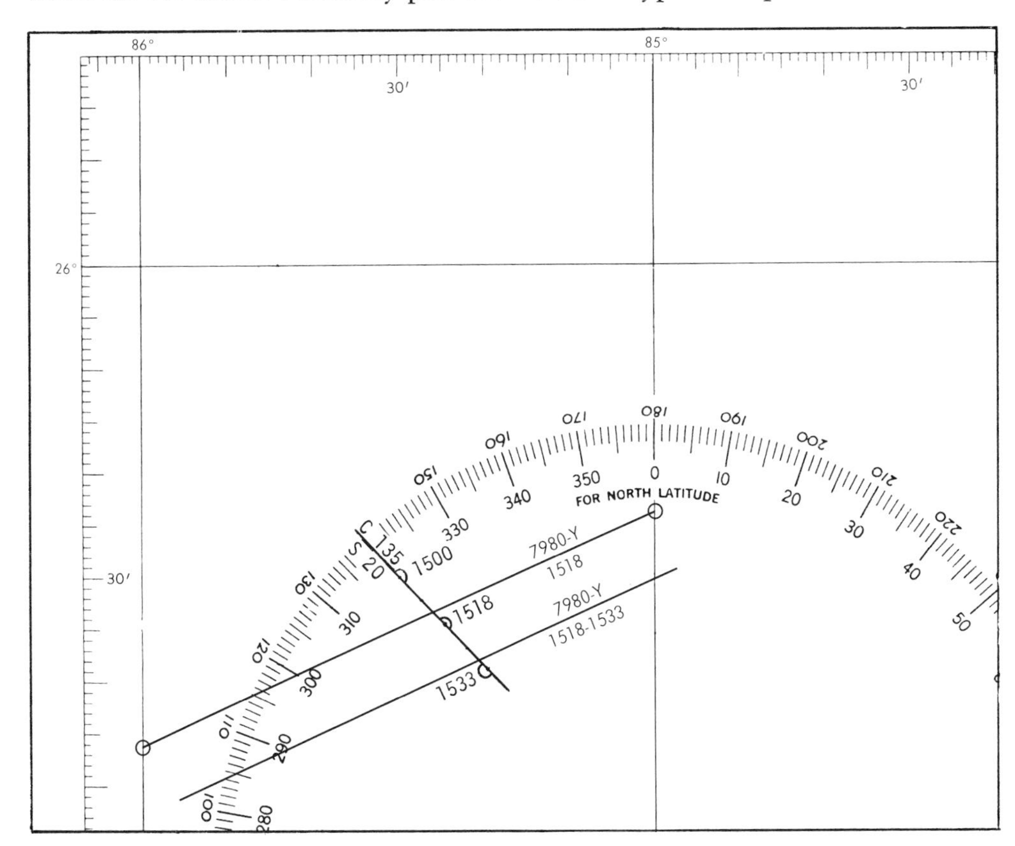

Figure 13-6A. An advanced Loran-C LOP.

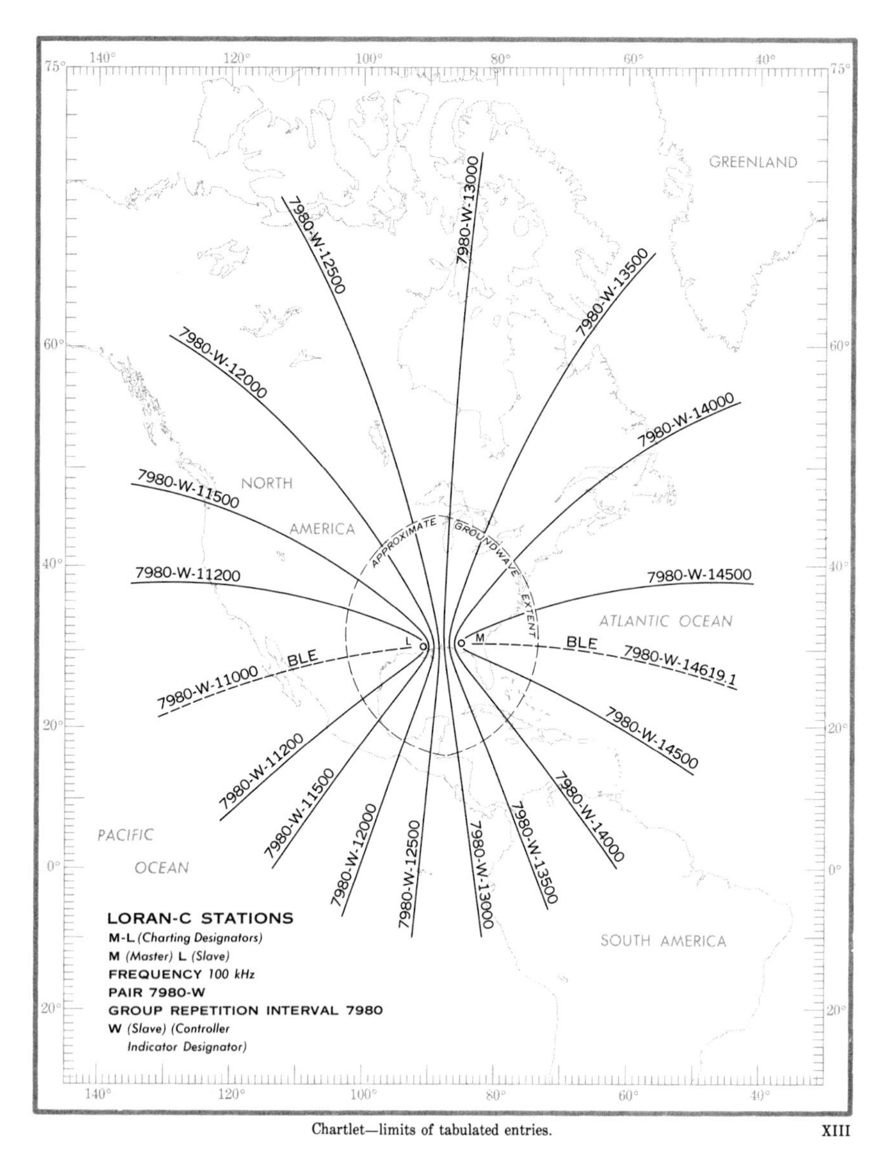

Figure 13-6B. Hyperbolic pattern generated by Loran-C master-secondary station 7980-W.

generated by this pair is shown in Figure 13-6B. From an inspection of the trend of the hyperbolic lines in this figure in the vicinity of the ship's 1533 DR position, it appears that the hyperbolic LOP corresponding to a time difference of 13423.4 μs will tend generally north and south. Thus, the tables should supply exact longitude intercepts on two tabulated latitudes above and below the 1533 DR position. Opening the tables to the page containing tabulated data for time differences near the observed value, it is found that the nearest tabulated time difference is 13420. An inspection of the tabulated intercept data in the vicinity of the DR position reveals that exact longitude intercepts are given for whole degree intervals of latitude; for the position in question, the bracketing latitudes are 25° and 26° N. The page containing these intercepts is shown in Figure 13-6C.

Inasmuch as the observed time difference is not identical with a tabulated value, interpolation must be performed to obtain the correct longitude intercepts on the 25° and 26° parallels of latitude. The arguments for this interpolation are the longitude intercepts for a time difference of 13420, the closest value to the observed time difference,

7980-W ML 195

T	13400		13410		13420		13430		13440		T
Lat ° '	° '	Δ	° '	Δ	° '	Δ	° '	Δ	° '	Δ	Long ° '
18 N	82 31.4W	- 54	82 26.0W	- 54	82 20.6W	- 54	82 15.1W	- 54	82 09.6W	- 55	
19 N	82 53.9W	- 50	82 48.9W	- 50	82 43.8W	- 50	82 38.7W	- 51	82 33.5W	- 51	
20 N	83 16.6W	- 46	83 11.9W	- 46	83 07.2W	- 47	83 02.5W	- 47	82 57.7W	- 47	
21 N	83 39.6W	- 42	83 35.2W	- 43	83 30.9W	- 43	83 26.6W	- 43	83 22.2W	- 43	
22 N	84 02.7W	- 38	83 58.8W	- 39	83 54.8W	- 39	83 50.8W	- 39	83 46.8W	- 39	
23 N	84 26.0W	- 35	84 22.5W	- 35	84 18.9W	- 35	84 15.3W	- 35	84 11.7W	- 35	
24 N	84 49.5W	- 31	84 46.4W	- 31	84 43.2W	- 31	84 40.0W	- 31	84 36.8W	- 31	
25 N	85 13.1W	- 27	85 10.3W	- 27	85 07.5W	- 27	85 04.7W	- 27	85 01.9W	- 28	
26 N	85 36.5W	- 23	85 34.2W	- 23	85 31.8W	- 24	85 29.4W	- 24	85 27.0W	- 24	
27 N	85 59.6W	- 19	85 57.6W	- 19	85 55.6W	- 19	85 53.7W	- 19	85 51.7W	- 20	
27 15N	86 05.3W	- 18	86 03.4W	- 18	86 01.6W	- 19	85 59.6W	- 19	85 57.7W	- 19	
27 30N	86 10.9W	- 17	86 09.1W	- 17	86 07.3W	- 17	86 05.5W	- 17	86 03.7W	- 18	
27 45N	86 16.4W	- 17	86 14.7W	- 17	86 13.0W	- 16	86 11.3W	- 17	86 09.5W	- 17	
28 N	86 21.8W	- 16	86 20.2W	- 16	86 18.6W	- 15	86 17.0W	- 16	86 15.3W	- 16	
28 15N	86 27.1W	- 14	86 25.6W	- 15	86 24.0W	- 15	86 22.5W	- 15	86 21.0W	- 15	
28 30N	86 32.2W	- 13	86 30.8W	- 14	86 29.3W	- 14	86 27.8W	- 14	86 26.4W	- 14	
28 45N	86 37.2W	- 13	86 35.8W	- 13	86 34.4W	- 14	86 33.0W	- 13	86 31.7W	- 13	
29 N	86 41.8W	- 12	86 40.6W	- 12	86 39.3W	- 12	86 38.0W	- 13	86 36.7W	- 12	
29 15N	86 46.3W	- 11	86 45.1W	- 12	86 43.9W	- 12	86 42.6W	- 12	86 41.4W	- 11	
29 30N	86 50.4W	- 11	86 49.3W	- 11	86 48.1W	- 11	86 46.9W	- 11	86 45.8W	- 11	
29 45N	86 54.1W	- 10	86 53.1W	- 10	86 52.0W	- 11	86 50.8W	- 10	86 49.8W	- 10	
30 N	86 57.4W	- 10	86 56.4W	- 10	86 55.3W	- 10	86 54.3W	- 9	86 53.3W	- 10	
30 15N	87 00.2W	- 9	86 59.2W	- 10	86 58.2W	- 9	86 57.2W	- 9	86 56.3W	- 9	
30 30N	87 02.4W	- 9	87 01.4W	- 9	87 00.5W	- 9	86 59.5W	- 9	86 58.5W	- 9	
30 45N	87 04.0W	- 9	87 03.0W	- 9	87 02.1W	- 9	87 01.2W	- 9	87 00.2W	- 9	
T	13400		13410		13420		13430		13440		T

Figure 13-6C. Page of 7980-W Loran-C Tables containing intercepts for 13420 μs time difference.

and the corresponding Δ values. These are customarily written as shown below.

Lat.	Tab. λ	Δ
26°N	85° 31.8′ W	−24
25°N	85° 07.5′ W	−27

The change in each tabulated longitude intercept for a 3.4 μs increase in the time difference is obtained by multiplication of the Δ values and the time-difference increase. The resultant interpolated intercepts are indicated below.

λ Change	Interpolated λ
$(+3.4)\ (-.24) = -.8$	85° 31.0′ W
$(+3.4)\ (-.27) = -.9$	85° 06.6′ W

The final step is to plot these intercepts on the plotting sheet and connect them with a straight line. The result is the segment of the hyperbolic LOP corresponding to an observed time difference of 13423.4 μs for the master-secondary pair 7980-W. The intersection of this line and the advanced 7980-Y LOP obtained and plotted earlier is the 1533 Loran-C fix. The completed plotting sheet is shown in Figure 13-6D.

For purposes of the foregoing example, it was assumed that only one master-secondary station signal was being received at a time, thus necessitating the advancement of the LOP corresponding to the earlier of the two time-difference readings. However, as stated in the section describing Loran-C receivers, most modern receivers are able to monitor the time-difference readings from two station pairs simultaneously. If the hyperbolic LOPs derived from two simultaneous readings are plotted, there is of course no need to advance or retire either; the Loran-C fix position is defined by their intersection.

If sky waves are used to obtain a Loran-C time difference, the reading obtained from the receiver must be corrected to an equivalent ground-wave reading by the application of a sky wave correction. On Loran-C charts, sky wave corrections for day and night for each rate appearing on the chart are printed at various intersections of the latitude-longitude lines over the areas where they are applicable. The correction to be applied is determined by double interpolation based on the ship's DR position. The introductory pages of each set of Loran-C tables contain, in addition to the coverage diagrams mentioned earlier, a set of tables containing sky wave corrections. Again, the proper correction is obtained by double interpolation based on the

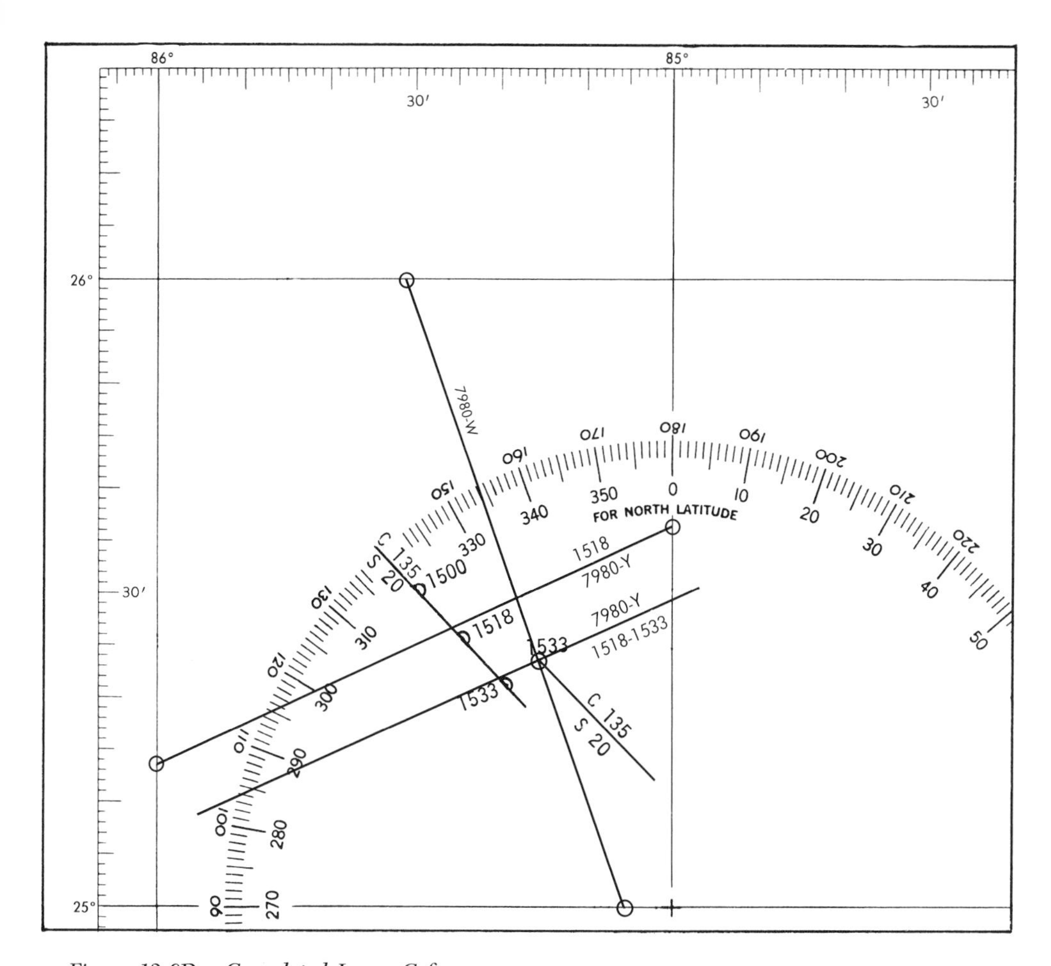

Figure 13-6D. Completed Loran-C fix.

DR latitude and longitude. The sky wave correction is always applied to the observed reading before it is plotted on a Loran-C chart or used as an argument in the Loran-C tables.

Decca

The Decca system, more properly called the Decca Navigator System, is unique in that it is owned and operated for the most part by a private enterprise, the Racal-Decca Navigator Company Limited, based in London, England. The system was originally conceived in 1937 by an American engineer, W. J. O'Brien, and developed by the British Admiralty Signals Establishment. Its first practical use was in guiding the leading minesweepers and landing craft in the Allied invasion of Normandy in 1944. The Decca Navigator Company, formed in 1945, further refined the system and established the first commercial

Decca chain in southeast England the following year. The present coverage of the system is depicted in Figure 13-7.

Decca is similar to Loran-C in that each chain uses a master station in combination with up to three slave stations, but the systems differ in that Decca employs unmodulated continuous waves (CW) rather than the pulsed waves of Loran. The characteristic hyperbolic grid pattern is formed by phase comparisons of the transmitted master and slave signals.

Figure 13-7. Areas of the world covered by Decca, 1980. (Courtesy Racal-Decca Navigator Co., Ltd.)

All stations in the Decca system transmit on frequencies in the LF band between 70 and 130 kHz. The nominal range of Decca is considered to be 240 miles from the master station both by day and by night, with sky-wave interference rendering the system unusable beyond this limit; Decca can thus be described as both a short- and a medium-range system. While the range of the system is somewhat limited, this disadvantage is compensated for by the extreme accuracy and relative simplicity of obtaining Decca LOPs within the area of coverage. The average maximum r.m.s. LOP error within the area covered by a Decca chain is given in the following table.

Nautical miles from master station	**r.m.s. error in yards**	
	Day	**Night**
100	30	100
150	60	350
200	100	700
240	150	1,200

The average Decca chain is composed of a master and three slave stations arranged in a so-called *star* pattern, with an angle of about 120° between each master-slave base line. The signals transmitted by each of the stations in a given chain are all harmonics of a single fundamental frequency (f), and chains are differentiated by assigning each a separate fundamental in the range from 14.00 to 14.33 kHz; harmonics are used to simplify the phase comparison process by which Decca LOPs are obtained. For purposes of identification and reference, each of the slave stations are designated by a color—purple, red, or green. All master stations transmit on a frequency of $6f$, all purple stations on a frequency of $5f$, all reds on $8f$, and all greens on $9f$.

The Decca receiver actually consists of four separate receivers, each of which can be set to receive one of the four signals transmitted by a given chain by simply selecting the correct fundamental frequency for that chain. Within the receiver, the signals for each master/slave pair are then electronically multiplied up to a single comparison frequency. The $6f$ master frequency is multiplied by 4 and the red $8f$ frequency by 3 to obtain a comparison frequency for the master/red-slave pair of $24f$; at the same time, in another part of the receiver the $6f$ master is multiplied by 3 and the green $9f$ slave by 2 to produce a comparison frequency for the master/green-slave pair of $18f$; and in similar fashion the master/purple-slave comparison frequency $30f$ is formed. Phase comparisons for each pair of stations within the selected chain are then made at the three comparison frequencies to yield three Decca LOPs.

Since all signals in each Decca chain are phase-locked along the base line between, say, a master and red-slave station, the comparisons in Figure 13-8 might be made at the comparison frequency $24f$. In

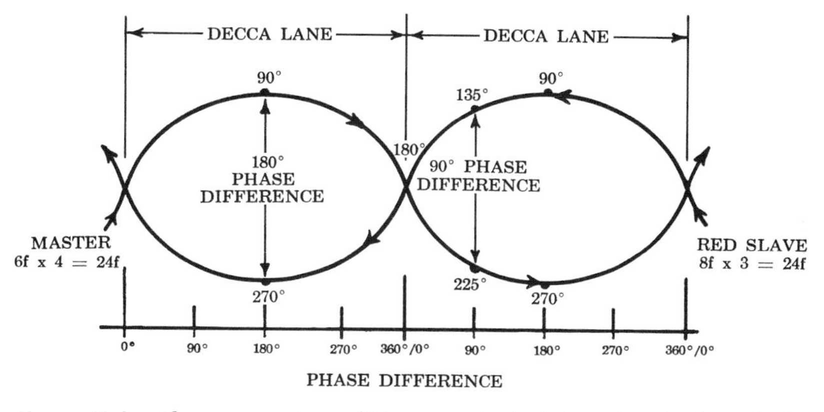

Figure 13-8. Phase comparison of Decca master/red-slave signals along the base line.

the figure, the distance between each 0° phase-difference measurement, which occurs every half wavelength, is referred to as a *Decca lane.* If the fundamental f were 14.00 kHz for this example, the wavelength would be 880 meters, and the Decca lane width measured along the base line would be 440 meters.

In the left-hand lane, a receiver is located such that a phase difference of 180° is measured between the master and red-slave comparison frequencies. Thus, the receiver must be located at a point .50 of the distance in the lane away from the master station. In the right-hand lane, a phase comparison of 90° is obtained, so the receiver is located .25 of the distance between the lanes, again measured from the master toward the slave station.

If only a phase comparison were available between two such continuous waves, it would be impossible to determine in which of the two lanes shown in Figure 13-8 the receiver was located without some additional information. In the Decca system, lane identification is accomplished by the transmission of a *lane identification signal* by each master and slave station within a given chain once each 20 seconds for a duration of .6-second. The normal signals transmitted by the chain are interrupted during these times. Briefly, during the lane identification signal period all four harmonics are radiated simultaneously, and within the receiver the harmonics are summed to derive a pulse train having the fundamental frequency f for that chain. Using this frequency, a half-wavelength *zone* is thus created encompassing 18 green lanes, 24 red lanes, and 30 purple lanes. Phase comparisons are performed on this identification signal for each of the station pairs, which in effect indicates in which lane within each of the colored zones the receiver is located. This operation is illustrated in Figure 13-9 for a

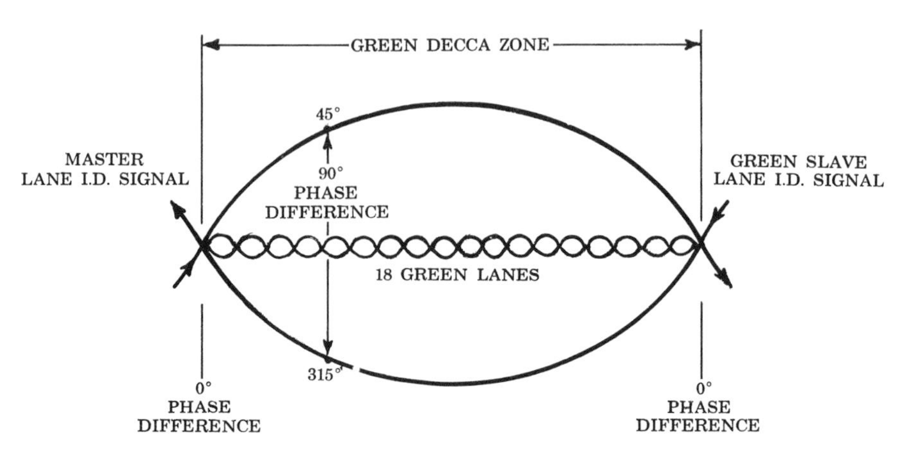

Figure 13-9. Phase comparison of Decca master and green-slave-lane identification signals.

master/green-slave station pair. In this figure, phase comparison of the master and green-slave-lane identification signals indicates that the receiver is located in the fifth green lane contained by the zone, measured from the master toward the slave station.

In addition to the lane identification signal described above, each station periodically transmits a *zone identification signal* during each 20-second cycle on a frequency of 8.2*f*. Called the orange frequency, it is combined within suitably equipped receivers with the 8*f* frequency to form a coarse hyperbolic pattern in which each 360° difference cycle embraces 5 zones. The receiver performs a phase difference measurement using this derived zone identification pattern, and indicates by a 2-character code in which of the 5 zones the receiver is located. The orange frequency is also used for system monitoring and control functions within each chain.

For identification purposes, each Decca zone is assigned a letter from *A* through *J* either clockwise or counterclockwise from the base-line extension, depending on the slave station color; the lettering is repeated every tenth zone. Each lane within a zone is identified by numbers; the numbers 0–23 are used for red lanes, 30–47 for the green lanes, and 50–79 for purple lanes.

Obtaining and Plotting the Decca Fix

In the Decca receiver, the zone letter, lane number, and fractional position within the lane to the nearest .01 are continuously indicated for each master/slave station pair on dials called *decometers.* The face of each dial indicates the fractional position within the lane, while the zone letter and lane number appear in a window just above the dial. The current model of shipboard Decca receiver, designated the Mark 21, is pictured in Figure 13-10 on the following page. It incorporates three decometers, one for each color-coded pair in the chain, plus an LED readout that displays the lane identification signal for each of the chain pairs three times each minute.

Once the receiver has been set up for a particular Decca chain, the decometers provide continuous readings of the zone, lane, and fractional lane position for each of the station pairs until the vessel passes out of range of that chain. Once obtained, the position information is used to plot segments of hyperbolic LOPs on a suitable chart overprinted with color-coded Decca hyperbolic patterns in the same fashion as is done in the case of Loran. An example Decca fix appears in Figure 13-11 on page 255.

A wide variety of receivers are available for use with the Decca system, each designed to fill the needs of a particular application. There are also several automatic graphic display devices available,

Figure 13-10. The Mark 21 Decca receiver. (Courtesy Racal-Decca Navigator Co., Ltd.)

designed to plot a continuous track over a chartlet, using information automatically supplied by the Decca receiver. This system is especially valuable in aircraft, in which it constitutes a type of automatic flight log.

Because of the limited areas of the world covered by Decca (see Figure 13-7), U.S. Navy ships are normally outfitted with Decca receivers and charts only when extended operations in an area covered by Decca are anticipated. Heavy use is routinely made of the system, of course, by merchant and fishing vessels and aircraft operating off Newfoundland, in the North Sea and the English Channel, and in the Sea of Japan. In fact, in spite of its rather limited coverage worldwide, according to the Racal-Decca Company Decca is currently used by more ships than Loran-C and Omega together.

Consol

A discussion of the long-range hyperbolic navigation systems currently available would be incomplete without a brief description of the Consol system. The system as it presently exists was developed in 1945 from an earlier German directional radiobeacon system called

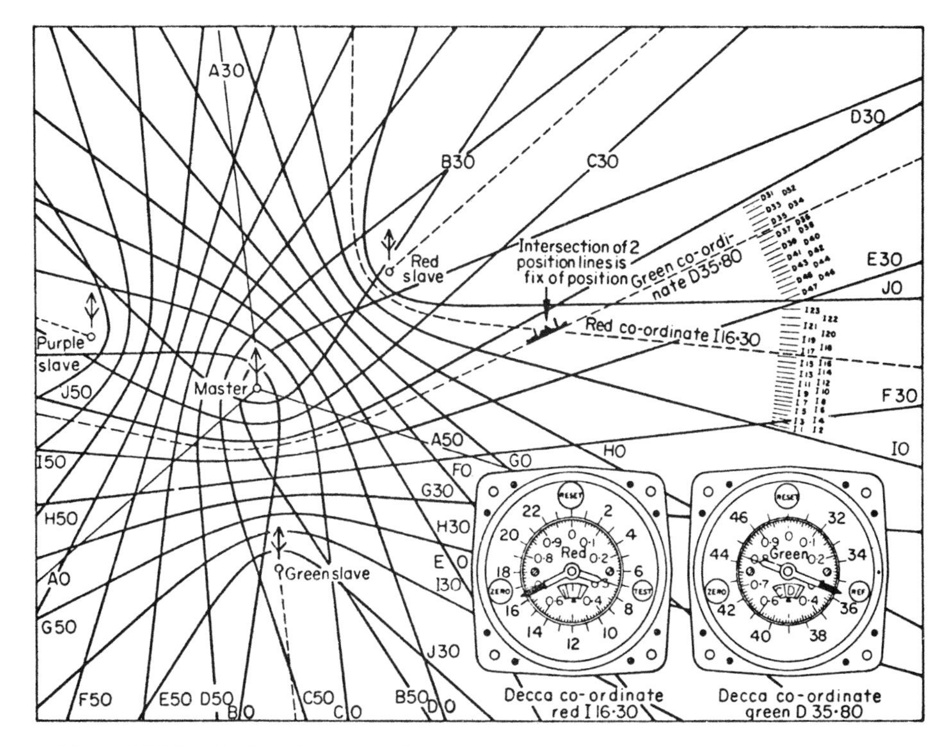

Figure 13-11. A plot of a Decca fix. (Courtesy Racal-Decca Navigator Co., Ltd.)

Sonne. Although the deployment of Consol is limited to the eastern North Atlantic, as indicated in the coverage diagram shown in Figure 13-12 on the following page, it is nevertheless worthy of mention because of the simplicity of its use.

In essence, Consol is a hyperbolic system with extremely short base-line lengths, such that a "collapsed" or "degenerated" hyperbolic pattern is formed. The curved portions of the hyperbolas close to the transmitting stations are not used; only the asymptotic portions of the hyperbolas are employed as bearings to the transmitting sites. Thus, the system is used in practice as though the Consol stations were radiobeacons of extremely long range.

The Consol system employes three towers at each transmitting site; they are located in line and spaced three wavelengths apart, and transmit on frequencies in the MF range between 250 and 370 kHz. The range of Consol is 1,000 to 1,200 miles by day, and 1,200 to 1,500 miles at night. The minimum range that the system can be used is about 25 miles. The pattern is also unusable in the vicinity of the baseline extensions.

Briefly, in operation one tower transmits a continuous wave, while the remaining towers transmit waves that undergo a 180° phase

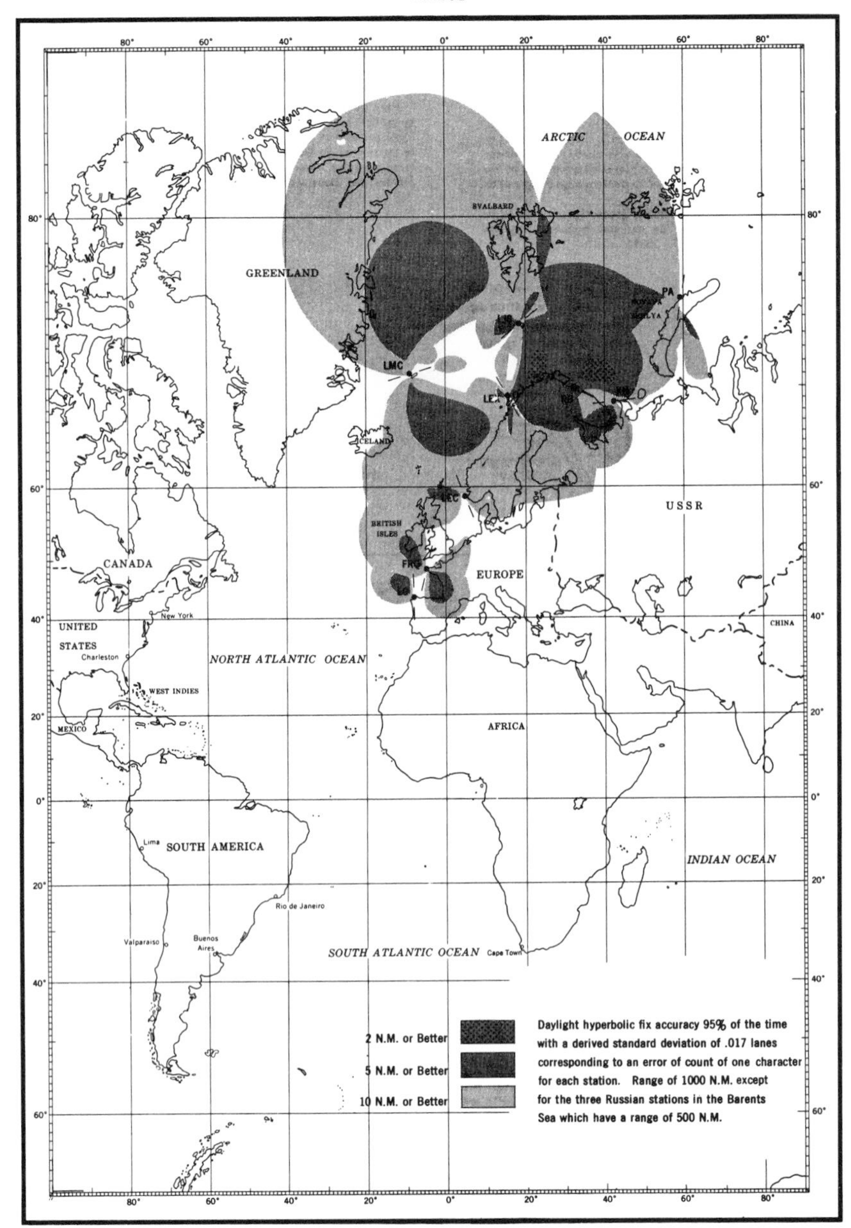

Figure 13-12. Coverage diagram of Consol, 1980.

shift with respect to the continuous wave during each so-called *keying cycle.* All signals are modulated by a system of aural dots and dashes. In effect, this produces a series of 10° to 15°-wide alternating dot-and-dash sectors radiating outward from the center of the station site, as shown in Figure 13-13. The sectors are narrowest perpendicular to the base line, and expand in width as the base-line extensions are approached.

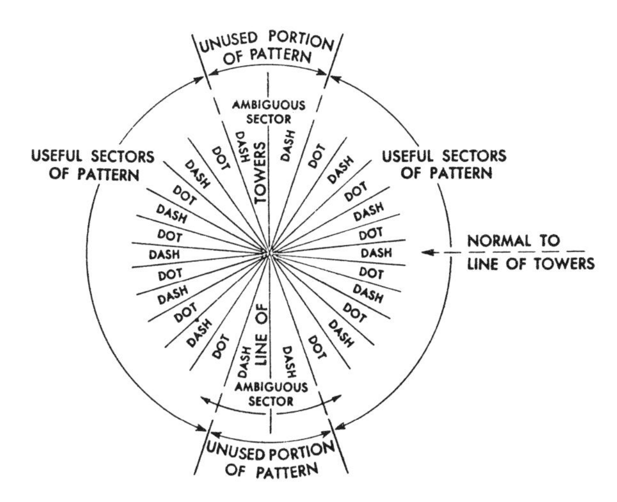

Figure 13-13. Polar diagram of Consol pattern.

The phase shift of the combined signal results in a variable number of dots or dashes being heard depending on the radial position within a sector during each keying cycle. If he is within a dot sector, the user counts the initial number of dots heard in a given cycle and determines his exact bearing to the transmitting site using this number; conversely, if he is within a dash sector, he counts the initial number of dashes heard within a cycle. There is no way to distinguish successive dot-or-dash sectors from one another, necessitating an accurate DR plot to ascertain in which sector the position lies. The determination of the proper sector grows increasingly difficult as the distance to a site decreases, because of the decreasing width of the sectors with decreasing range. The width of a sector expressed as a number of dots or dashes depends on the transmitting site; the width is either 30 or 60 dots or dashes. Any receiver able to receive signals on the carrier frequencies transmitted and demodulate the audible tones can be used, as the radial LOP depends only on the number of dots or dashes counted by the user.

After the dot-or-dash count has been obtained, either a chart overprinted with a Consol grid pattern or a set of tables giving bearings corresponding to the number of dots or dashes heard within each sector may be used. The DMAHTC *Publication 117A* contains bearing tables for all Consol stations currently in operation.

At night, Consol signals received beyond 300 miles from the transmitting stations must be treated with caution, as sky-wave interference can produce wide variation in successive counts. Should this occur in practice, the prudent navigator should not rely exclusively on the Consol bearings obtained. Accuracy of the bearings plotted varies from .2° to about 1.5° r.m.s., depending on the sector width, time of day, and range to the transmitting site.

Omega

One of the major disadvantages of all of the hyperbolic navigation systems discussed up to this point developed and deployed in the years between 1940 and 1970 is the relatively small areas of the world that they cover. Moreover, the precision of the longest-range system discussed, Loran-C, falls off rapidly beyond about 1,200 miles, although one-hop sky-wave signals are oftentimes obtainable out to about 2,300 miles. To overcome these disadvantages, in 1947 the U.S. Navy began a research and development program with the goal of developing an electronic navigation system that could achieve worldwide coverage through the use of frequencies in the LF or VLF bands. Gradually over the next twenty years the various technical problems associated with such transmissions were solved, and in the late 1960s, the last major technical obstacle was overcome with the development of atomic time standards capable of precise regulation of the signals transmitted by the eight widely separated stations that were to constitute the system. The Omega system, as it is called, will become fully operational in 1981 when the last of the eight stations begins transmitting in Australia. Transition of operational responsibilities for the system from the U.S. Navy to the Coast Guard began in the late 1970s, and was completed with the assumption of full management responsibility by the Coast Guard on 1 October 1980.

In all probability, Omega will most likely be the ultimate terrestrial navigation system that will ever be developed. Advanced satellite navigation systems now planned, most notably the NavStar Global Positioning System (GPS) slated for deployment in the late 1980s, should eliminate the need for long-range ground-based systems of any further capability.

The Omega system consists of eight stations located 5,000 to 6,000 miles apart, transmitting on frequencies in the VLF band from 10 to 14 kHz at a power of 10 kilowatts. The coverage has been continually expanded since the first several stations in the northern hemisphere were brought on line in the mid-1970s, and will be worldwide when the Australian station becomes operational. At any position on earth, the signals of at least three and usually four stations will be usable, to produce a minimum of three possible LOPs.

The 10–14 kHz frequency band was chosen specifically to take advantage of several favorable propagation characteristics, among which are the following:

At these frequencies, the earth's surface and the ionosphere form a very efficient wave guide capable of propagating the signal long distances with little loss of signal strength or distortion, enabling precise phase measurements to be made. Moreover, any small variations in the signal propagation characteristics can be predicted with reasonable accuracy.

These frequencies can penetrate into the surface layers of ocean water, enabling the signal to be received by submerged submarines.

The propagation characteristics are such that base lines of 5,000 to 6,000 miles are practical, resulting in a fairly precise grid pattern even at extremes of the coverage of any two stations.

The long range of the signals gives the navigator a choice of at least three LOPs at any position on earth.

The Omega system is similar to the Decca system discussed previously, in that for most applications the same principle of phase comparison of two CW transmissions is used to obtain an LOP, but it is different in that any two Omega stations receivable in a given geographic area may be considered as a pair for the phase comparison. It is also possible with specially designed equipment to obtain ranges to two or more Omega stations, and then determine an Omega position by plotting the intersection of the resulting range LOPs. In practice, the phase comparison method is used for the most part.

The basic frequency at which all eight stations transmit is 10.2 kHz, with atomic frequency standards being used to ensure that the transmissions of all eight stations are kept exactly in phase. Lines of zero phase differences (*isophase lines*) form a hyperbolic pattern between each pair of stations. For comparison purposes, a typical hyperbolic pattern formed by a station pair having a base-line length of 1,000 miles, such as Loran-C, is contrasted to a similar pattern formed by an Omega station pair with a base line of 6,000 miles in Figure 13-14 on the following page.

It should be evident from the figure that at locations distant from the base line the hyperbolas formed by the 6,000-mile base-line system diverge much less than do those of the 1,000-mile base-line system, contributing to greater precision of the 6,000-mile system at longer ranges. The small divergence of successive hyperbolas coupled with the accuracy with which Omega propagation corrections may be predicted is expected to result in Omega fixes being accurate to within

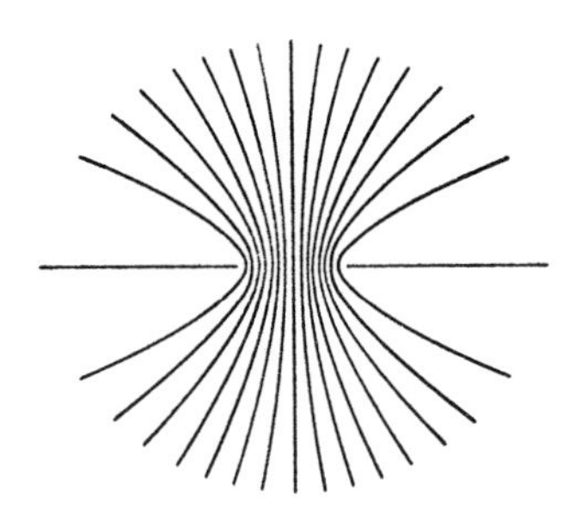

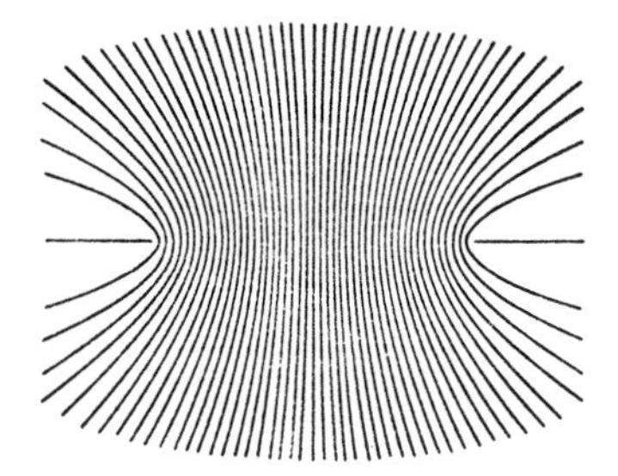

Figure 13-14. Hyperbolic patterns for 1,000-mile (left) and 6,000-mile (right) base lines.

±1 mile r.m.s. by day and ±2 miles r.m.s. by night anywhere in the world after full system validation is completed.

Two 10.2 kHz Omega continuous waves transmitted exactly in phase but traveling in opposite directions produce a series of Omega lanes; within each, a phase-difference measurement would progress from 0° to 360° as the receiver moved across the lane. Two such Omega lanes, produced by one complete cycle of two Omega signals, are shown in Figure 13-15. Position of a receiver within an Omega lane is expressed in terms of so-called *centicycle* (cec) or *centilane* (cel) units, each defined as .01 of the width of an Omega lane. A reading of 50 centicycles, for example, would indicate a position at which a phase-difference reading of 180° was obtained. If no propagation correction applied in this instance, the receiver would be located in the center of an Omega lane, midway between two adjacent isophase lines.

The wavelength of the 10.2 kHz Omega signal is 16 nautical miles; hence, along the base line, each half-wavelength Omega lane is 8 miles wide. The width expands somewhat as the distance from the base line increases. Thus, if only the basic 10.2 kHz Omega signals were com-

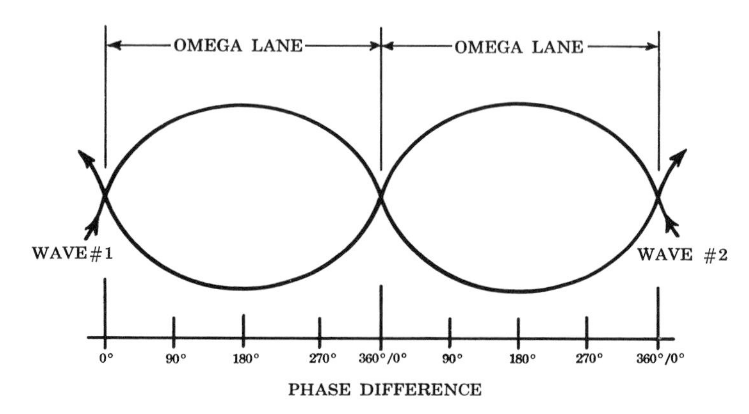

Figure 13-15. Two Omega lanes formed by phase comparison of two 10.2 kHz signals.

pared, a prior knowledge of a vessel's position within ±4 miles would be required in the vicinity of the base line to determine in which of the Omega lanes the vessel was located.

To alleviate this problem of lane identification, three other signals in addition to the basic 10.2 kHz Omega signal are transmitted by each station on a time-sharing or "multiplexed" basis. These frequencies are 11.05, 11.33, and 13.6 kHz. Within an appropriately designed Omega receiver, these three signals can be electronically combined with the basic 11.33 and 10.2 kHz signals to form so-called *difference frequencies* of .283, 1.133, and 3.4 kHz, respectively, which are $\frac{1}{36}$, $\frac{1}{9}$, and $\frac{1}{3}$ the frequency of the 10.2 kHz signal. These three difference frequencies can then be compared to establish three broader or "coarse" lane widths of 288, 72, and 24 miles, as measured along the base line. Each 3.4 kHz broad lane contains three 10.2 kHz lanes, each 1.133 kHz coarse lane contains three 3.4 broad lanes, and each .283 kHz (283 Hz) lane four 1.133 coarse lanes. Hence, a user need only be able to establish his position within ±144 miles (half the width of a 283 Hz coarse lane) to determine in which 10.2 kHz lane he is located, through four successive phase comparisons. In practice, however, since it is rare that a surface vessel would be unable to establish her position within at least ±36 miles (half the width of a 1.133 kHz lane), most shipboard Omega receivers designed to perform successive phase comparisons for lane resolution use only the 1.133 and 3.4 kHz difference frequencies, as illustrated in Figure 13-16. Moreover, after a fix has been determined and the lane number has been established, most present-day

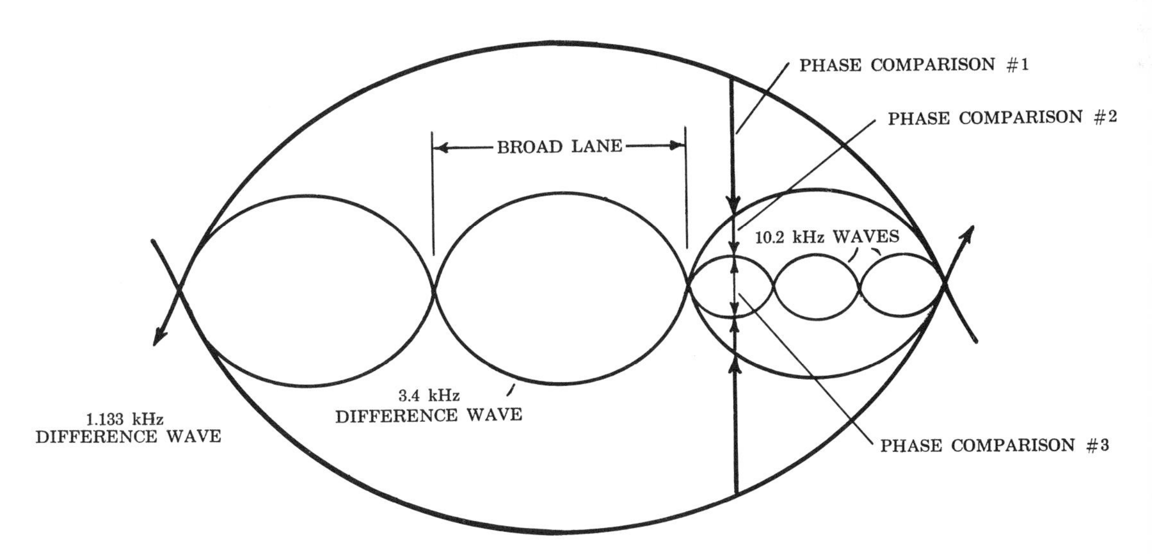

Figure 13-16. Three successive Omega phase comparisons for lane resolution on surface ships.

Omega receivers are designed to maintain an automatic count of the 10.2 kHz lanes crossed thereafter. The transmitted 11.05 signal and its associated 283 Hz difference frequency are intended for use primarily by future advanced systems for time dissemination, search-and-rescue, and balloon and free-floating oceanographic buoy tracking.

If all four Omega frequencies were transmitted simultaneously in the same order at each of the eight Omega stations, it would be difficult if not impossible to differentiate the various station signals. If an oscilloscope were available, the amplitude of the signals from a farther station might be slightly less than that of a nearer station, but two stations at the same range would be indistinguishable. Hence, the eight stations in the Omega system transmit in a 10-second format called a *commutation pattern,* shown in Figure 13-17.

Each station in the Omega system is designated by a letter from A through H; the locations of the eight stations are indicated.° Each station transmits its frequencies as shown; note that the transmission timing pattern is different for each station. These time patterns and the location and duration of the signal within the repeating 10-second commutation pattern serve to identify the individual stations. Each transmission period is interrupted by a .2-second pause, to provide the desired 10-second period for the complete commutation cycle.

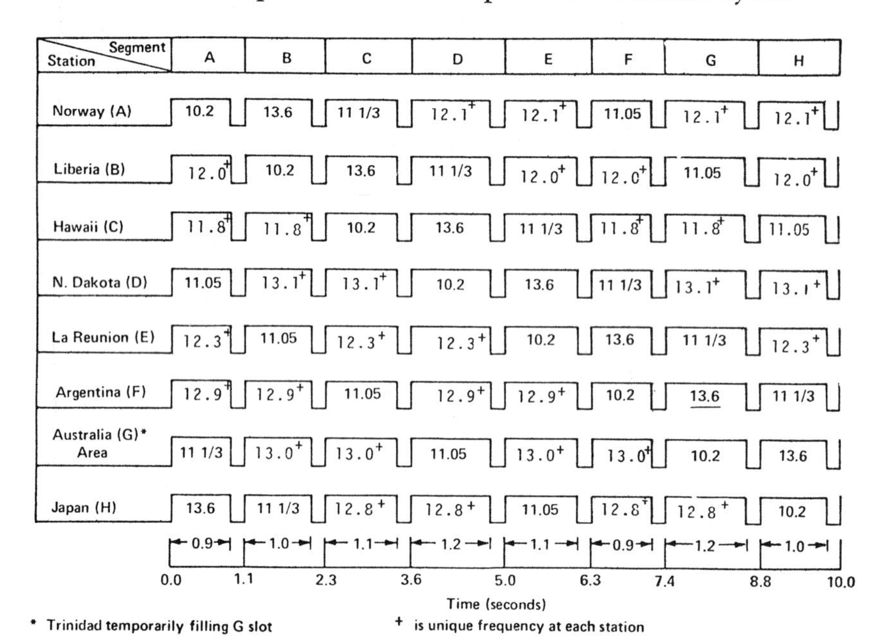

Figure 13-17. Omega station commutation pattern.

° A temporary station in Trinidad will operate in the G-segment of the commutation pattern until the Australian station is activated.

the tables in Figure 13-19B are entered to find a correction expressed in centicycles for Station B of −78 and a correction for Station C of −45. The resulting correction for this time for station pair B-C is then −78 − (−45) or −33 centicycles. After the value of the propagation correction has been determined, it is always applied to the reading obtained from the receiver before the resulting LOP is plotted by use of an Omega chart or the appropriate *Omega Tables.*

Charts for Omega navigation similar to those designed for use with Loran and Decca are issued by the DMAHTC. Families of hyperbolas for 28 different Omega pairs are currently planned for the basic 10.2 kHz Omega frequency; inasmuch as hyperbolas for the 3.4, 1.133, and .85 kHz difference frequencies will be coincident with every third, ninth, and eighteenth 10.2 kHz isophase hyperbola, respectively, a separate pattern for the transmitted 11.05, 11.33 and 13.6 kHz frequencies will not be printed. In the 10.2 kHz patterns, the isophase hyperbola bisecting the base line of each station pair is arbitrarily numbered 900. Successive lanes are consecutively numbered, increasing or decreasing from this figure as appropriate, with the lowest-numbered lane being adjacent to the station that is first in the alphabet in a given pair. Most Omega charts are printed only with every third 10.2 kHz isophase hyperbola, due to the chart scale used; only the hyperbolas for station pairs providing the best intercept angles within the geographic area covered by the chart are overprinted, with a separate color for each pair. The plot of a particular centicycle reading within the printed pattern is accomplished using a linear interpolator in the same manner as is done on Loran and Decca charts.

To obtain and plot an Omega fix on an Omega chart, the navigator has only to set his receiver for the three pairs of stations that will provide the optimum angles of intersection for three LOPs at his position, record the three readings, apply the appropriate propagation corrections, and plot the resulting LOPs. Usually the three readings can be recorded at about the same time, eliminating any necessity to advance or retire any Omega LOPs. A sample set of three Omega readings with the appropriate corrections applied is shown plotted in Figure 13-20 on the following page; note especially the labeling of the three LOPs and the fix position.

A set of *Omega Tables* similar to the *Loran Tables* discussed earlier are also available in the DMAHTC publication number 224 series. These tables are used for plotting segments of Omega hyperbolas on plotting sheets when Omega charts are not available, just as is done in the case of the *Loran Tables.* A separate *Omega Table* is planned for each of the station pairs receivable within each of the 25 Omega areas of the world. Those Omega lattice tables currently available in each

Omega Area are indicated by the station pairs listed in Figure 13-19A. An excerpt from the Area 12 *Omega Tables* for station pair D-F appears in Figure 13-21A.

The horizontal argument for the Omega Tables is the nearest tabulated integral lane count to the observed reading with the propagation correction applied; the vertical arguments found along the right or

OMEGA WORK SHEET				
Greenwich Mean Time	1200 25 JUN		ZD +1N	
Receiver	B-D 865.31	A-B 837.70	A-D 802.94	
Correction	-.01	+.20	+.21	
Line of Position	865.30	837.90	803.15	

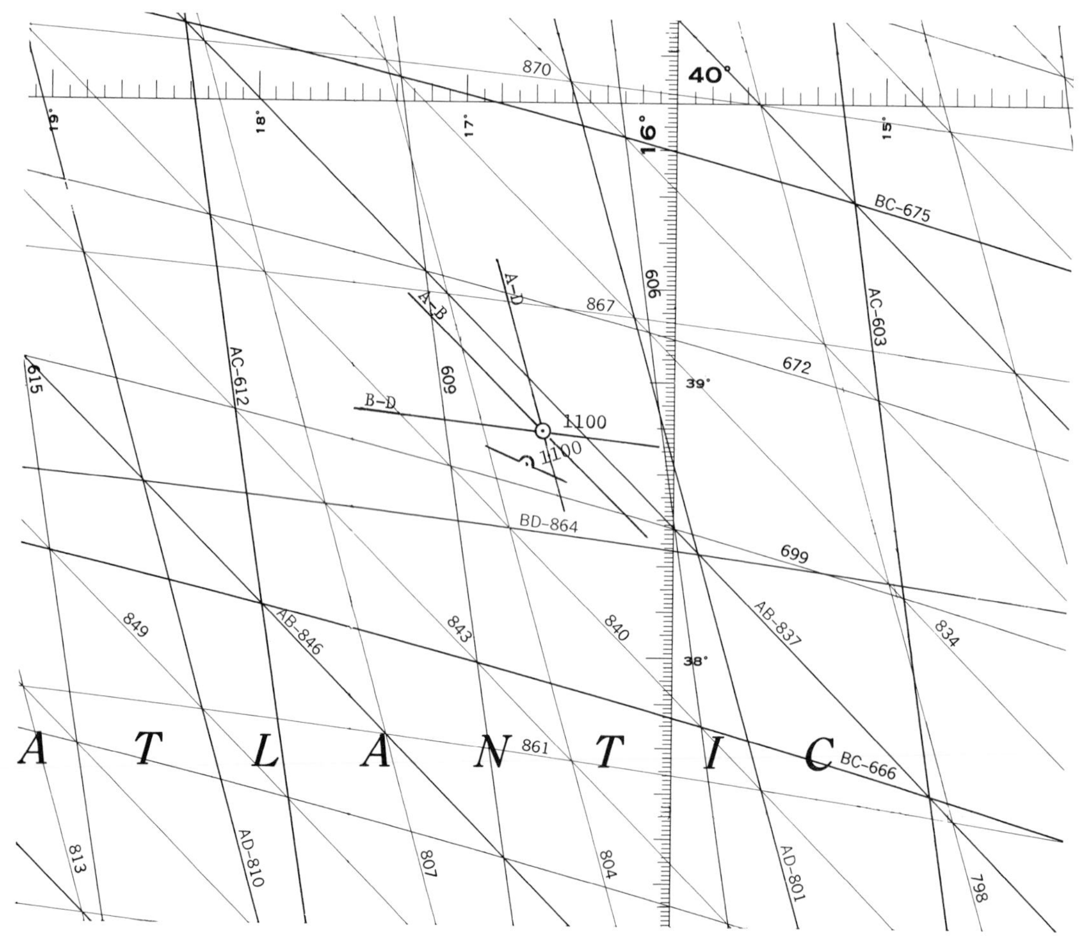

Figure 13-20. A sample Omega fix plotted on an Omega chart.

T	D-F 780	D-F 781	D-F 782	D-F 783	D-F 784	T
Lat ° ′	° ′ Δ	° ′ Δ	° ′ Δ	° ′ Δ	° ′ Δ	Long ° ′
	34 27.8N -107	34 17.0N -107	34 06.3N -107	33 55.5N -107	33 44.8N -107	30 W
	34 40.7N -107	34 30.0N -107	34 19.2N -108	34 08.4N -107	33 57.7N -107	29 W
	34 53.4N -108	34 42.5N -108	34 31.8N -107	34 21.0N -107	34 10.2N -107	28 W
	35 05.7N -108	34 54.8N -108	34 44.0N -108	34 33.1N -108	34 22.3N -107	27 W
	35 17.5N -108	35 06.7N -108	34 55.8N -108	34 45.0N -108	34 34.1N -108	26 W
	35 29.0N -108	35 18.2N -108	35 07.3N -108	34 56.4N -108	34 45.5N -108	25 W
	35 40.2N -109	35 29.3N -108	35 18.5N -106	35 07.5N -109	34 56.6N -108	24 W
	35 51.1N -109	35 40.1N -109	35 29.1N -109	35 18.3N -108	35 07.4N -108	23 W
	36 01.5N -109	35 50.5N -109	35 39.6N -109	35 28.7N -109	35 17.7N -109	22 W
	36 11.6N -109	36 00.6N -109	35 49.6N -109	35 38.6N -109	35 27.7N -109	21 W
	36 21.2N -110	36 10.2N -110	35 59.2N -109	35 48.3N -109	35 37.3N -109	20 W
	36 30.6N -110	36 19.5N -110	36 08.5N -110	35 57.5N -109	35 46.6N -109	19 W
	36 39.5N -110	36 28.5N -110	36 17.4N -110	36 06.4N -109	35 55.4N -110	18 W
	36 48.1N -110	36 37.0N -110	36 25.9N -110	36 14.9N -109	36 03.9N -110	17 W
	[illegible] 56.2N -110	36 45.2N -11[illegible]	[illegible] 34.1N -110	36 23.1N -110	36 12.1N -110	16
T	D-F 780	D-F 781	D-F 782	D-F 783	D-F 784	T

Figure 13-21A. Excerpt from Area 12 Pair D-F Omega Tables.

left margin are the tabulated latitudes or longitudes bracketing the DR position at the time of the observation. As in Loran-C, whether tabulated latitudes or longitudes are used as vertical arguments depends on the trend of the hyperbolic pattern in the vicinity of the DR position. Interpolation for the exact centicycle reading is done by multiplying the difference in centicycles between the observed versus the tabulated lane count by Δ values, which represent the change (to the nearest .001′) of the exact longitude or latitude intercept for a change of one centicycle of lane count. The resulting corrections are applied to the tabulated intercepts to obtain the interpolated intercepts on the entering latitude or longitude lines corresponding to the observed lane count corrected for propagation error.

As an example of an Omega LOP plotted by the use of the tables, consider Figure 13-21B, in which a segment of an LOP corresponding to a corrected D-F lane count of 782.65 is plotted. The DR position at the time of the reading was L 34° 06.0′ N, λ 29° 21.0′ W. A worksheet drawn up in the form shown in the figure is helpful when plotting the Omega LOP by the use of the tables. Being a single LOP, it is labeled with the station pair and the time of the reading. If two or more Omega LOPs are combined to form an Omega fix, each LOP is labeled with the station pair only, with the time of the fix being placed adjacent to the fix symbol at the intersection of the LOPs.

Station Pair: D-F $\qquad T_g - T = 782.65 - 783 = -.35$

λ	Tab Lat	Δ	Lat Change $(T_g - T) \times (\Delta)$	Interpolated Latitude
30°W	33-55.5 N	-.107	(-35)(-.107) = +3.7	33-59.2 N
29°W	34-08.4 N	-.107	(-35)(-.107) = +3.7	34-12.1 N

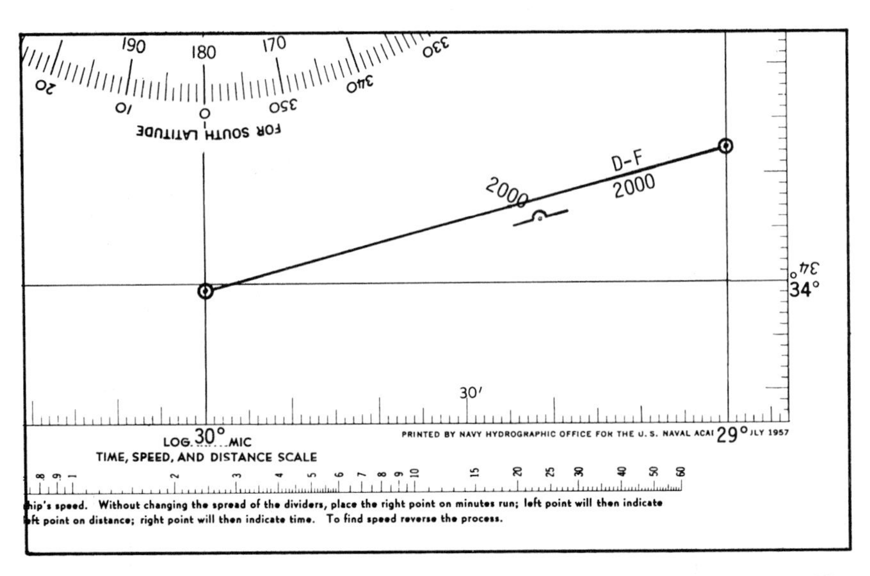

Figure 13-21B. An Omega LOP plotted on a plotting sheet using the Omega tables.

Additional examples of the use of the Omega system to obtain fixing information will be presented in Chapter 16 on a day's work in navigation at sea.

Omega System Development Efforts

Because the Omega system will not be fully implemented until 1981, much work is still in progress with various developmental aspects of the system. The propagation correction algorithms for some of the lesser traveled regions of the world are still being refined through the collection of system positioning data. In one such program, being conducted jointly by the U.S. Coast Guard and the Federal Aviation Administration, data recorders are being placed aboard selected commercial airliners and oceanographic research vessels of the United States and other cooperating nations. The data thus collected is assembled, analyzed, and disseminated to users and receiver manufacturers. As an interesting note, it has recently been determined by this effort that the Arctic and Greenland ice caps have a much more pronounced attenuation effect on the Omega signals traversing them than had originally been postulated. The Norway "A" signals, for example, are almost completely absorbed by the Greenland ice cap, causing a void in the signals from this station along most of the North Atlantic seaboard of Canada and the United States. It is anticipated that as this and other research continues, further refinements in the Omega system, and in the procedures and equipment designed to make use of it, will occur throughout the decade of the 1980s.

Summary

When the Omega navigation system is compared with the other five radio navigation systems discussed in this chapter, the fact that worldwide coverage will be possible with an accuracy of ±1 to 2 miles r.m.s. with only eight stations is a tremendous advantage. For coastal navigation, however, the greater accuracy afforded by the other systems, particularly Loran-C, close to their base lines makes their use valuable, especially when a vessel is beyond radar range to land. For this reason, it is not meaningful to attempt to compare the various systems by the use of system accuracy alone, inasmuch as the r.m.s. accuracy of all systems with the exception of Omega varies from a few hundred yards to several miles within the area of system coverage. The table below, however, may be useful in highlighting the various characteristics of the systems described herein.

System	Frequency Band kHz	Principle of Operation	Maximum Range Day/Night (Mi)
Radiobeacon	30–3000	CW transmission & RDF	20–200
Loran-A	1850–1950	Pulse time difference	800/1400
Loran-C	90–110	Pulse time difference and phase comparison	1200/2300
Decca	70–130	CW phase comparison	240
Consol	250–370	CW collapsed hyperbolic	1200
Omega	10–14	CW phase comparison	Worldwide

14

Advanced Navigation Systems

The continued rapid pace of advances in electronics technology over the last two decades has had a significant impact in the routine practice of navigation, just as it has in most other areas of modern life. The program to develop a digital sextant was mentioned in Chapter 4, and there have been several references throughout the previous chapters of this book to the impact that atomic and electronic time standards have made in the field in the recent past. The hand-held electronic calculator is gaining more capability virtually by the day (see Appendix 1); this increasing capability, plus its ever-decreasing cost, may make solution of navigational problems by tables obsolete in the not too distant future.

The impact of technology in electronic navigation has been no less profound during the last twenty years. The last chapter described several systems wholly dependent upon modern electronics technology for their effectiveness. In addition to these systems, there have appeared in the seventies and early eighties several more esoteric navigational systems. These systems have come to be called *advanced navigation systems,* because of their degree of sophistication and precision in comparison with the other electronic systems discussed thus far.

There are currently four advanced navigation systems of paramount importance either in operation or under development as of this writing. They are the Navy Transit Navigational Satellite System, the Ship's Inertial Navigational System, the doppler sonar system, and the NavStar Global Positioning System. The first three are fully operational, while the latter is scheduled to become available in the mid to late 1980s. A basic understanding of the operation of each is a necessity for the student of modern navigation. Consequently, this chapter will give a brief description of each; further information may be obtained from appropriate technical manuals.

The Navy Transit Satellite Navigation System

In the late 1950s, with the advent of the Polaris SSBN submarine, the need for a highly reliable and precise position-fixing system that would be available worldwide regardless of weather conditions or proximity of shore transmitting stations became apparent to the U.S. Navy. At that time, some efforts to extend Loran-C coverage were meeting with limited success, but Omega was still a concept in a few scientists' minds, and mid-ocean position-fixing depended largely on celestial navigation. With the launching of the first artificial satellite *Sputnik I* in 1957, however, and the rapid advances in space-age electronics and missile technology that quickly followed, it was soon realized that the means to establish the desired comprehensive navigation system were at hand.

A satellite navigation system based on the doppler shift of signals transmitted by orbiting satellites was first proposed by a team of scientists of the Applied Physics Laboratory of Johns Hopkins University in late 1957, as a result of their studies of the behavior of the signals transmitted by *Sputnik*. In the following year, a Navy contract was awarded to the Applied Physics Laboratory to develop such a system. In April 1960 a prototype system satellite was launched into an inclined orbit, and in 1962 the U.S. Navy Astronautics Group was commissioned at Point Mugu, California, to operate the system. In January of 1964, the *U.S. Navy Satellite System* (*NAVSAT*) officially became operational, following the successful launch of the first Transit satellite into polar orbit. Additional satellites were subsequently placed in orbit to complete the system configuration, and in 1967 the system was declared available for private and commercial use. Utilization of the system has progressively expanded ever since, particularly in the commercial sector, to the point where today there are considerably more private and commercial users than government and military.

Within the U.S. Navy, the system is still referred to by its original acronym NAVSAT, but users outside this community have almost universally adopted its civil designation, the *Transit Navigation Satellite System*, or simply *Transit* for short. To avoid confusion with the next generation system described later in this chapter—the NavStar system—the Transit designation will be employed throughout the remainder of this section.

The Transit system presently consists of five operational satellites, a network of ground-tracking stations, a computing center, an injection station, Naval Observatory time signals, and the shipboard receiver-computer combination. These components are all shown in Figure 14-1 on the next page.

The Transit satellites are in circular polar orbits about 600 nautical miles (1,075 km) high, as indicated in Figure 14-2, with periods of revolution of about 107 minutes. Because of the continual rotation of the earth beneath the satellites, every position on earth comes within range of each satellite at least twice a day at 12-hour intervals. As

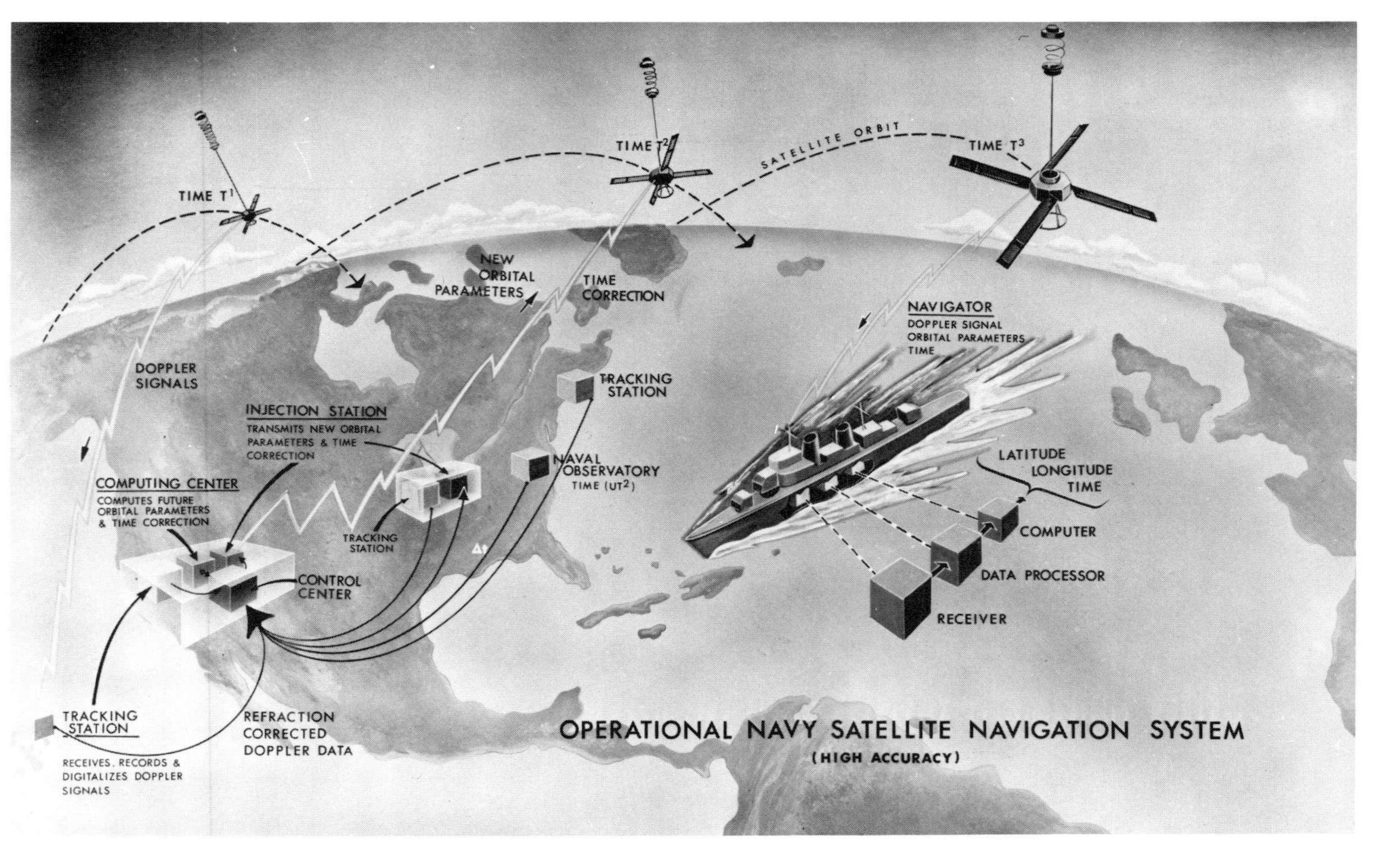

Figure 14-1. Components of the Navy Transit Satellite Navigation System.

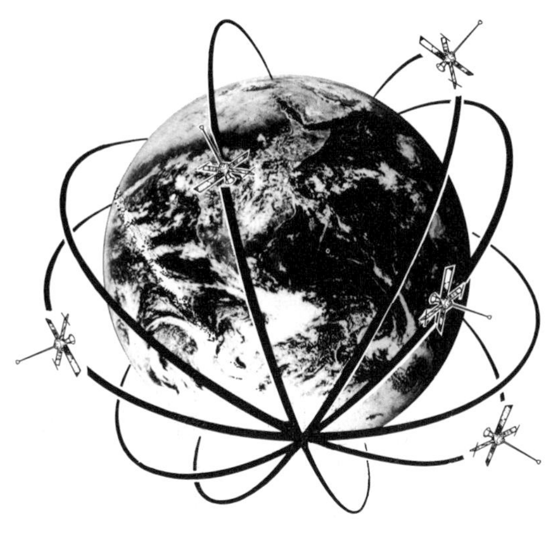

Figure 14-2. Five operational Transit system satellites in polar orbit provide an average time between fixes ranging from about 35 to 100 minutes worldwide. (Courtesy Magnavox Corporation)

originally intended, if at least five satellites were operational at any given time, the average time between fix opportunities would vary from about 95 minutes near the equator to about 35 minutes or less at latitudes above 70° north and south. Interestingly, the actual mean time between fixes in recent years has occasionally exceeded the planned 95-minute maximum at certain locations, due to the unexpected longevity of the early Transit satellites. It was anticipated that each satellite would have an average lifetime of about five years, during which time its orbit would slowly precess away from the intended optimum. As each satellite failed, a new replacement would be launched into the original orbit, to succeed the defunct bird. However, the original satellites surprisingly remained fully operational long past their expected lifetimes, with the result that their costly replacements were never launched. Hence, by 1978 the effects of orbital precession had caused gaps to occur in the coverage as indicated in Figure 14-2, with a consequent increase in the fix intervals at times to several hours or more at locations in lower latitudes. Current plans call for the launch of two second-generation satellites called *Novas* to fill this gap in 1981.

Although the follow-on NavStar GPS system may eventually render the Transit system obsolete, in all probability the steadily increasing proliferation of users and equipment should ensure the continued availability of Transit until at least the mid-1990s.

Transit System Operation

The Transit system operation is based on the doppler shift of two frequencies, 150 and 400 MHz, transmitted simultaneously by each satellite moving in its orbit at a tangential velocity of about five miles per second. Two frequencies are used so that the effects of ionospheric and atmospheric refraction on the incoming satellite transmission can be isolated and eliminated by the shipboard receiver. Each frequency is modulated by a repeating data signal lasting two minutes conveying the current satellite time and its orbital parameters and certain other information. Within the receiver, a single doppler signal is created for processing from the two signals transmitted by the satellite. By plotting the frequency of this signal on a graph versus time, a characteristic curve of the type shown in Figure 14-3 results.

Since the frequency of the transmitted signal is compressed as the satellite approaches in inverse proportion to its distance from the receiver, the curve begins at time T_1 at a frequency several cycles higher than the transmitted frequency. As the satellite draws nearer, the frequency gradually drops until it is equal to the transmitted frequency at time T_2, when the satellite is at its closest point of approach. As the satellite recedes, at T_3 the frequency of the transmitted signals is elongated, again in inverse proportion to the distance separating the satellite and the receiver, producing a gradually de-

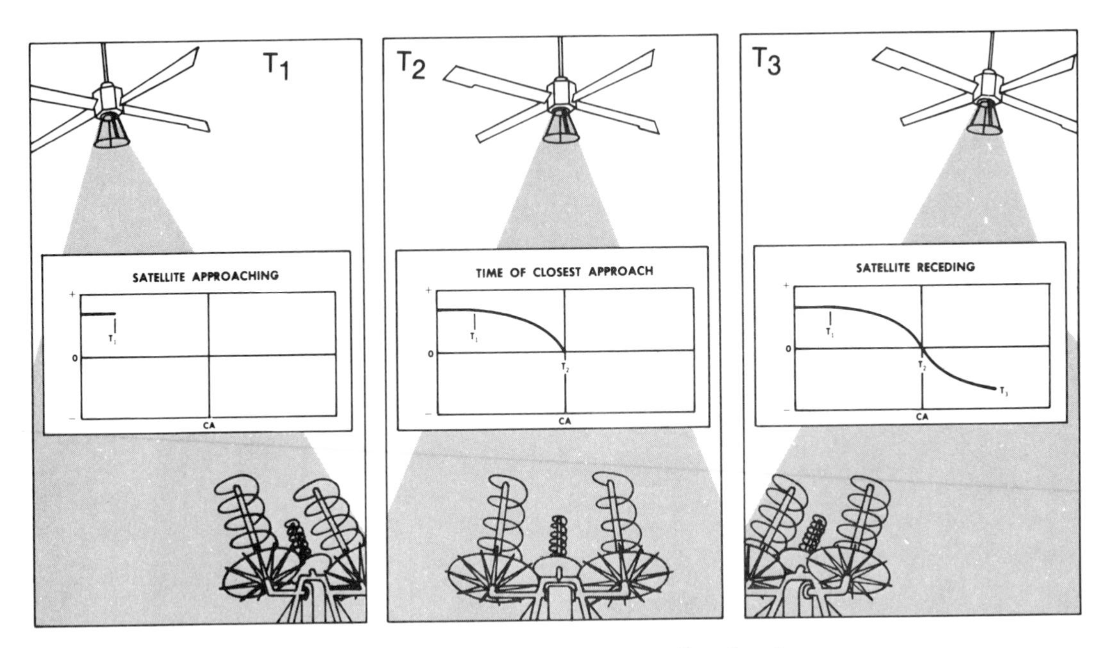

Figure 14-3. Characteristic Transit satellite doppler curve.

creasing curve. The time of zero doppler shift marks the time of closest approach of the satellite to the receiver.

Three Transit receivers currently in use aboard U.S. Navy ships are the AN/WRN-5, pictured in Figure 14-4A, the AN/SRN-9, shown in Figure 14-4B, and the AN/SRN-19, a single-channel receiver not

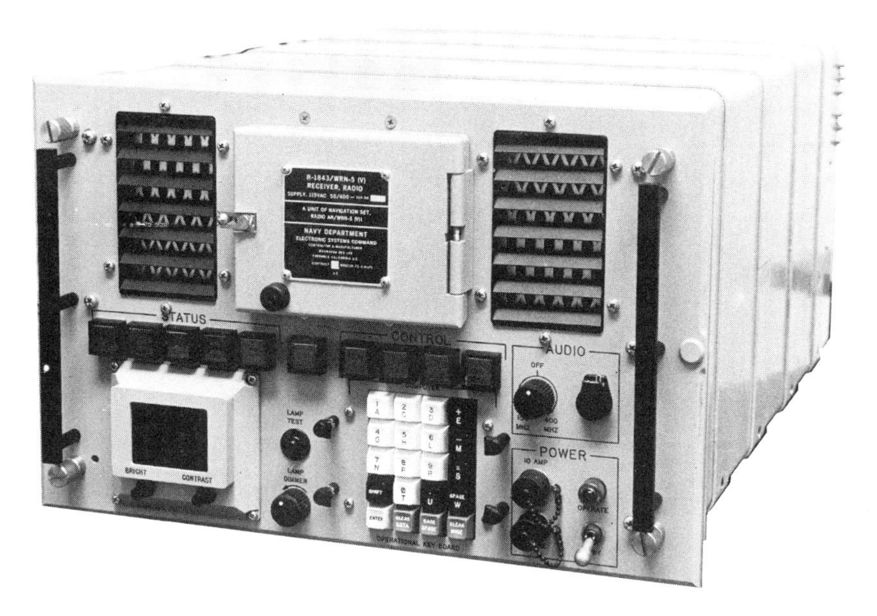

Figure 14-4A. The AN/WRN-5 Satellite Navigator. (Courtesy Magnavox Corporation)

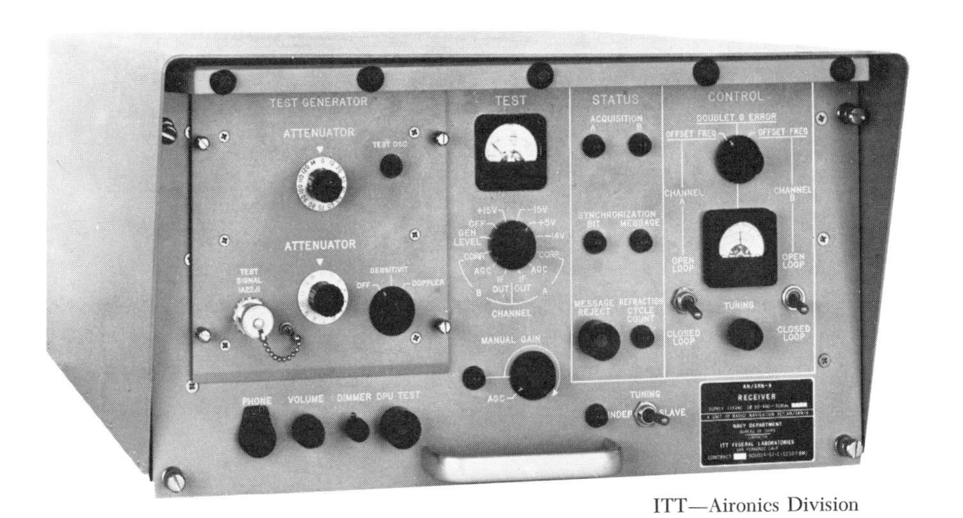

Figure 14-4B. The AN/SRN-9 receiver. (Courtesy ITT Corporation—Aironics Division)

shown. All three receive the sequential two-minute satellite broadcasts, measure the doppler shift, decode the information content of the phase modulation, and compute the fix by a so-called "integrated doppler" measurement. Briefly, this last process consists of measuring the difference in slant range from one broadcast position to the next, and then computing a resultant LOP on the earth's surface. The intersection of two such LOPs generated by at least three successive broadcasts produces a fix. In practice a minimum of four broadcast receptions is preferred, with seven broadcasts being optimum. The number of broadcasts receivable during a satellite pass depends on the maximum elevation reached by the satellite; the prescribed limits for a fix are 10° minimum and 70° maximum elevation.

There are also several models of shipboard Transit receivers available commercially, such as the one pictured in Figure 14-4C that features a CRT readout of latitude and longitude. These currently range in price from about $6,000 to $16,000 or more, depending on their degree of automatic operation and sophistication. Continued advances in micro-electronic technology may result in significant price reductions in the future, but for now this price range effectively limits the

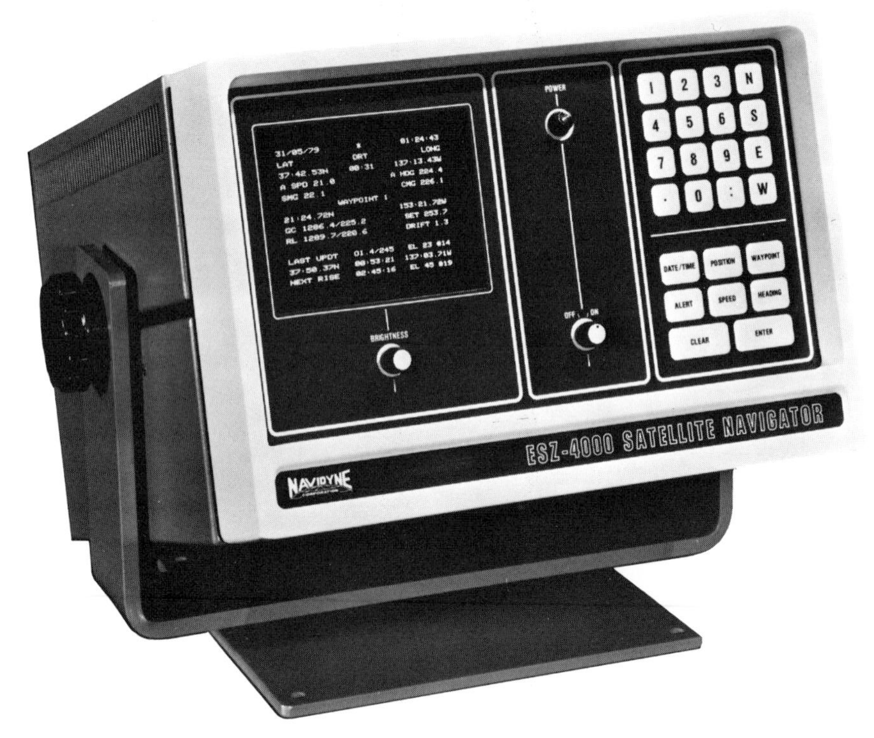

Figure 14-4C. The Navidyne ESZ-4000 Satellite Navigator receiver. (Courtesy Navidyne Corporation)

civil use of the system to wealthy private users and commercial enterprises.

The sequence of operations of the Transit system is summarized in Figure 14-5, in which a single satellite in the system is shown at different times in its orbit. Briefly, tracking stations record doppler observations and memory readouts received during each satellite pass and relay them to a central computer center via a computer-to-computer data communications link. Updated orbital position and time data communications are relayed to an injection station from the computer center for transmission to the satellite in a "burst" once each 12 hours. Enough data is supplied in this 15-second injection message to last for 16 hours of consecutive two-minute broadcasts describing the current orbital position of the satellite as it revolves around the earth.

The Transit system accuracy depends on several factors, primary among which are the accuracy of the satellite orbit computation, the effects of ionospheric refraction, and the precision of the ship's speed

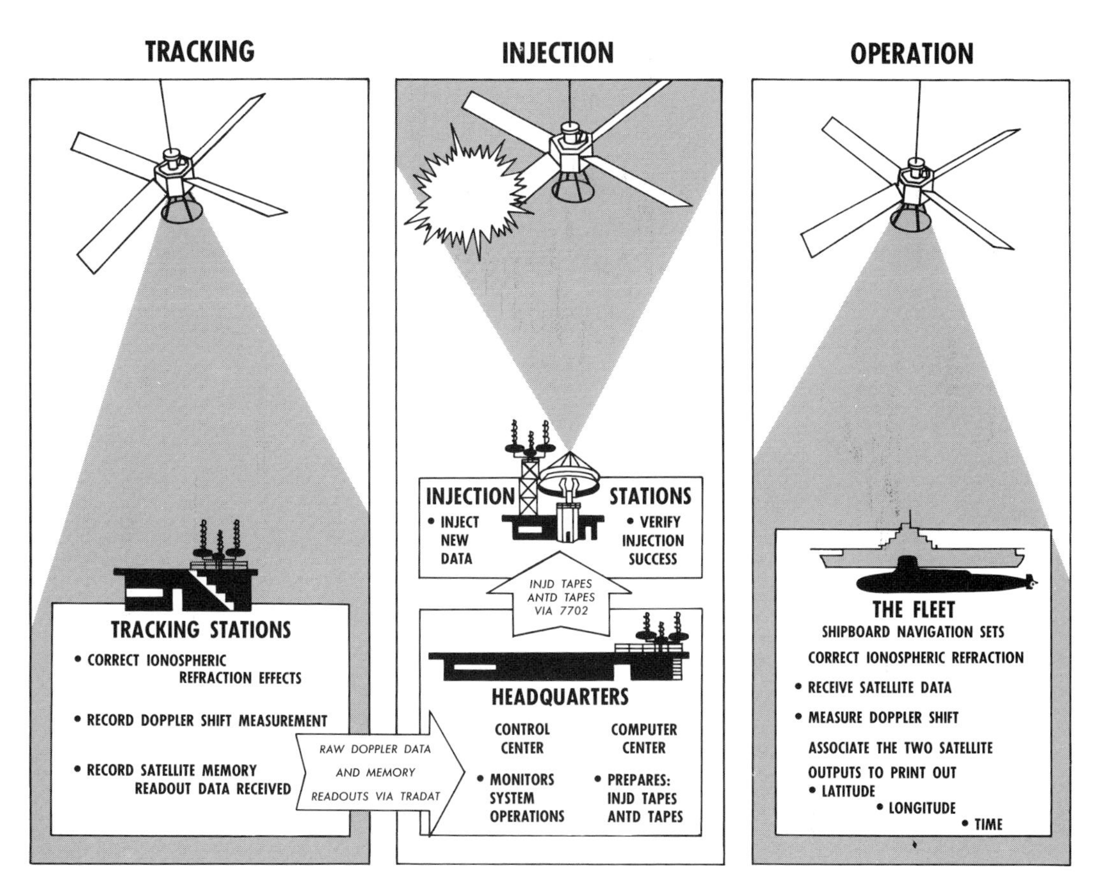

Figure 14-5. Sequential operation of the Transit navigation satellite system.

and heading determination. Under optimal conditions, the system is capable of producing fixes with a maximum r.m.s. error of about 35 meters for a stationary vessel anywhere on earth. In the case of a moving vessel, the accuracy can be comparable if the velocity and heading are precisely known. An additional .2-mile longitude error can be introduced for every knot of unknown north-south velocity relative to the earth's surface at the time of each doppler shift measurement, and a .05-mile latitude error can result from every knot of unknown east-west velocity. The overall r.m.s. error for routine fixes produced by the system is approximated by the formula

$$E = .2 + .05\, V_{E\text{-}W} + .2 V_{N\text{-}S},$$

in which $V_{E\text{-}W}$ represents the unknown east-west velocity component, and $V_{N\text{-}S}$ the unknown north-south component, expressed in knots. Typical fix errors for a sample satellite pass for a 1-knot velocity error in each of eight ordinal points of the compass are shown plotted in Figure 14-6.

As an added benefit, a time signal transmitted as a "beep" at the end of each two-minute Transit satellite transmission cycle coincides with even minutes of Coordinated Universal Time, and can be used conveniently as a chronometer check.

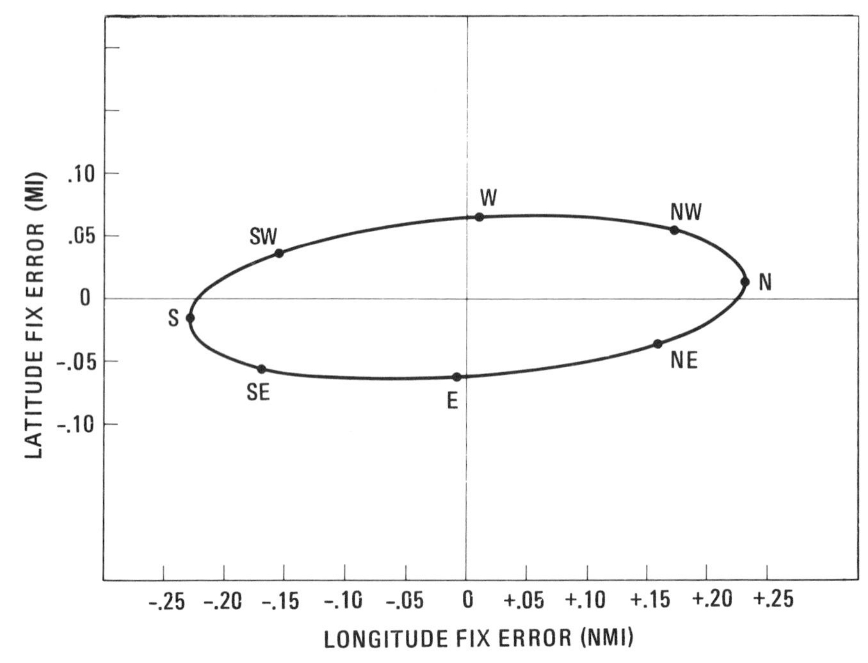

Figure 14-6. Effect of a 1-knot velocity error in eight compass directions on a typical Transit system fix gained from a 31° satellite pass. The satellite passed east of the receiver heading north. (Courtesy Magnavox Corporation)

The Ship's Inertial Navigation System

All of the electronic navigation systems discussed previously, as well as the satellite-based and sonar-based advanced systems described in this chapter, are dependent upon an electromagnetic or electromechanical (sonar) wave or series of waves transmitted either by the ship herself or by an external transmitter. In this section, a system unique in that it is independent of any such externally or internally transmitted signal will be discussed—the Ship's Inertial Navigation System (SINS).

The SINS system, like several of the other conventional and advanced navigation systems now operational, was originally developed in the late 1950s and early 1960s to meet the precision position-finding needs of the ballistic missile submarine. Since its first deployment on board the submarine USS *George Washington* in 1960, however, the SINS system has been continually refined, upgraded, and reduced in size, so that today its use has been extended to both attack submarines and to aircraft carriers, where it is used both for navigational purposes and also to set the inertial avionics equipment of the embarked aircraft. The future will probably see the system installed on increasing numbers of both Navy and commercial surface ships, if its capabilities continue to increase and its size and cost continue to decline at the same rate as they have over the last two decades.

Inertial navigation has been defined as the process of directing the movements of a vessel based on sensed accelerations in known spatial directions by means of instruments that mechanize the Newtonian laws of motion, integrating such accelerations to determine velocity and position. The basic instruments used in all inertial navigation systems are gyros, accelerometers, and the electronic computer.

Gyros were introduced in *Marine Navigation 1* in connection with the shipboard gyrocompass. Essentially, a classical gyro consists of a rapidly spinning mass that in accordance with Newton's laws maintains the orientation of its spin axis with respect to a universal reference system unless disturbed by some outside force, such as gravity or friction. An *accelerometer* is a device designed to measure acceleration (A) along a given axis by measuring the force (F) exerted along this axis upon a given mass (M), using Newton's Second Law of Motion, $F = MA$. The electronic computer used is dedicated entirely to the SINS system.

Basically, in the SINS system a platform upon which two accelerometers are mounted is stabilized by a system of three gyros in such a way that it is constantly coincident with a plane tangential to the earth's surface. The two accelerometers are continually oriented in a north-south and east-west direction. Hence, they are sensitive only to horizontal north-south and east-west accelerations. Integration of these accelerations with respect to time yields corresponding north-south and east-west velocity components, which when vectorially summed produce true ship's velocity. From this velocity, the ship's position expressed in terms of rectangular coordinates is continually converted to the spherical coordinates latitude and longitude by the computer and read out on a suitable output device.

A SINS system currently installed on several surface ships is the Mark III Mod 5 SINS, the components of which are pictured in Figure 14-7.

The accuracy of the SINS system is dependent in large measure on the precision and reliability of its instruments and supporting equipment. Some of the more significant sources of potential error in the system include the following:

- Errors caused by the daily rotational motion of the earth
- Friction in the gyro support systems
- Misalignment of the stabilized platform, causing vertical com-

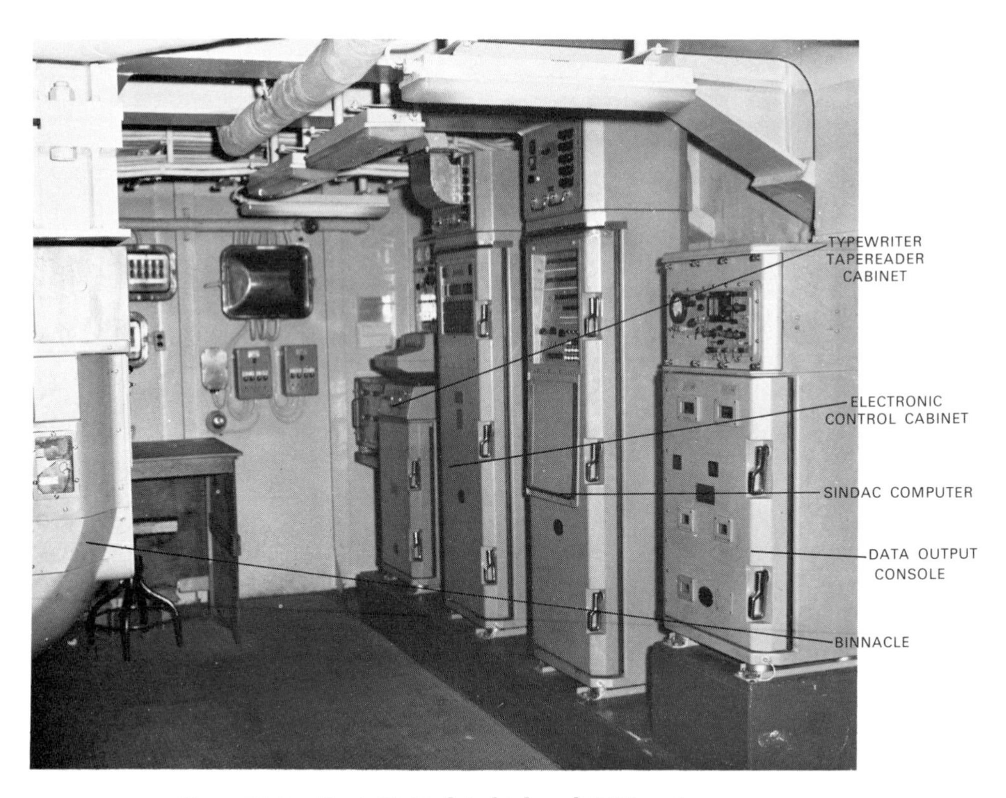

Figure 14-7. Mark III Mod 5 shipboard SINS system.

ponents of the earth's gravitational field to be falsely interpreted as horizontal components
- Imperfections in construction of the gyros and accelerometers

Because the combined error caused by these and other sources cannot be completely eliminated, all SINS systems experience some degree of cumulative error that increases with the passage of time. Thus, the position furnished by the system must be periodically compared to positions established by external means, and the system must be occasionally updated and reset. The current sixth-generation system being installed in Trident submarines, however, requires far less frequent updates than did the initial system of the early Polaris submarines. This is important since position determination by external means often necessitates cruising near the surface, where vulnerability to detection is greatly increased. When a SINS position update is needed, Loran-C, Omega, and the satellite-based systems described in this chapter can be used for this purpose, as can positions determined by celestial or bathymetric navigation.

One of the more interesting advances that has occurred during the continued efforts to refine the SINS system over the last twenty years has been the development of the *electrostatic gyro* (ESG). In this gyro, the rotor consists of a 1-centimeter diameter solid beryllium sphere spinning at 216,000 r.p.m. in a near-perfect vacuum. The rotor is supported solely by an electrostatic field, which holds the sphere suspended a few thousandths of an inch from the internal surface of the evacuated case containing it. The ESG is thus freed from the classical gyro bearing friction, as well as many of the associated random torques that a mechanical support can introduce. Hence, it represents the best approximation to a theoretically perfect gyroscope yet devised by man. In the Trident submarine SINS system, an ESG is employed to monitor continually the position being derived by the other conventional gyro subsystems, and to reset periodically the SINS position during the interim between externally determined fixes. Although in time even the ESG-monitored SINS develops a significant degree of error, the addition of the ESG lengthens the period between required external updates by a factor of about six in comparison to earlier SINS models.

Another recent development that shows great promise for incorporation in future SINS systems is the so-called *laser gyro*. In reality, this advanced device is not a gyro in the traditional sense, since there is no spinning central mass, but rather a closed geometric laser path (usually triangular) centered on an expected spin axis. Identically phased laser beams are continuously generated, which travel in opposite directions around the closed path. Any rotation about the spin axis causes an apparent phase difference in the two beams at a measurement point at one vertice of the triangle, since the path of the laser beam traveling in the direction of the rotation is effectively lengthened, while the path of the beam traveling in the opposite direction is shortened. The amount of the phase difference thus measured is directly proportional to the speed of rotation. There remain several technical problems to be solved before this system can become fully operational, but because it does not depend for its operation on a spinning mass, the laser gyro could eventually provide an order-of-magnitude increase in precision over the ESG-monitored SINS system.

The position produced by the SINS system should normally be much more accurate than a DR position derived by conventional means, but the navigator must still keep in mind that the SINS position is a sophisticated DR position, not a fix. Until such time as the state of the art has advanced to the point where all cumulative error can be eliminated, the navigator should refrain from placing complete confidence in the SINS DR position.

The Shipboard Doppler Sonar System

The *doppler sonar system* is a relatively recent development, capable of determining water depth and ship's speed over the ground with a high degree of precision and accuracy. The system is based on the doppler shift of a continuously transmitted sonar pattern.

A complication inherent in any doppler navigation system is that the propagation characteristics of a sonar signal tend to vary with the temperature, density, and salinity of the water into which the signal is transmitted. Inasmuch as the system depends on small frequency shifts of returning sonar echoes caused by the ship's horizontal motion, a method must be incorporated into any such system to compensate for random frequency shifts caused by variable water conditions. Furthermore, since only the horizontal component of a ship's velocity is of interest, any random vertical errors that might be introduced as a result of pitching or rolling must also be eliminated.

In the United States, the Raytheon Company has taken the lead in developing both military and commercial versions of the doppler sonar system. In its simplest configuration, the system consists of a fixed hull-mounted transducer that transmits pulsed sonar signals in two beams, one oriented forward and the other aft, at about a 30° angle from the vertical. This dual beam pattern is called the Janus configuration, after the two-faced Roman god of war who could look both forward into the future and back at the past at the same time. The system provides a continuous readout of ship's fore-and-aft speed precise to

.1-knot, water depth to the nearest meter, foot, or fathom, and nautical miles run to the nearest hundreth.

The dual beam pattern permits the forward and aft doppler shifts to be continuously compared in order to eliminate any vertical motion error, as well as error caused by irregular bottom features, thus allowing a very precise determination of horizontal velocity and depth to be made. In more advanced configurations, the Raytheon system employs a dual axis four-beam array, with the beams transmitted fore and aft and port and starboard 90° apart as shown in Figure 14-8A. For many military applications, the transducer is mounted on a gyro-stabilized horizontal platform that keeps the beams oriented in the four cardinal directions—north, south, east and west—to allow true ship's course and speed over the ground to be computed, for input into other shipboard navigational systems and dead reckoning plotters. For commercial applications in highly stable deep-draft ships, such as large tankers and RoRo ships, the four-beam array is rigidly affixed to the bottom, and stabilization is achieved internally by electronic means. The precise determination of athwartship's speed provided by the doppler system is of great importance to these huge ships, as this speed component is critical when maneuvering in restricted waters and while docking. The safe docking speed for a vessel exceeding 100,000 deadweight tons is about .2-feet per second; the berthing facility might collapse if contact were made with a speed greater than about 1 foot per sec-

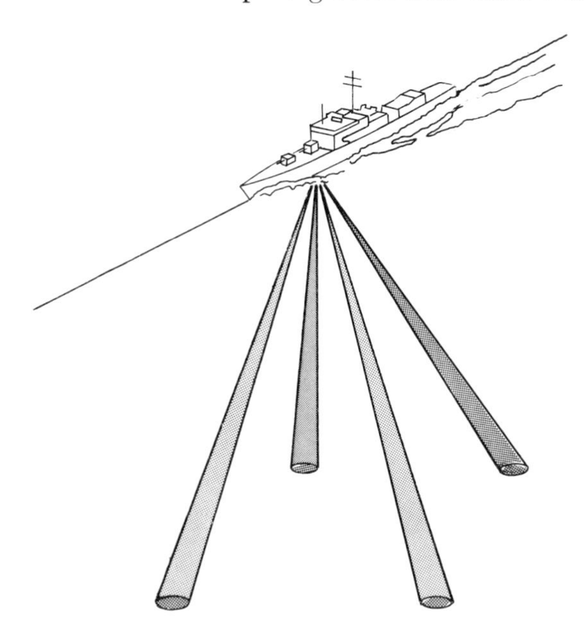

Figure 14-8A. A dual-axis four-beam Janus *sonar array of the Raytheon doppler sonar system.*

ond. A typical merchant ship doppler system output display is shown in Figure 14-8B.

Most Raytheon doppler sonar systems use a bottom-bounce mode in water depths up to about 300 meters (1,000 feet), and a volume reverberation mode wherein the signal is reflected from the water mass itself in depths exceeding this limit. The maximum error of currently installed systems has been repeatedly demonstrated to be not more than .17 percent of the cumulative distance run since the last reset, e.g., about .85 mile after a run of 500 miles. There are at present over 240 Raytheon doppler sonar systems installed in U.S. military, other government, and commercial ships, with the bulk of them aboard large tankers and freighters of the latter category. They are being used to great advantage both to maintain highly accurate DR plots in mid-ocean regions, especially in those areas not having adequate Transit, Omega, and Loran-C coverage, and during docking evolutions and piloting in restricted coastal waters and channels. System velocity data is being used as input via digital and analog interfaces to a variety of other systems, including fin stabilizers, integrated steering systems, and satellite navigation systems. Doubtlessly, continuing research and development efforts will result in ever wider usage of the system in the

Figure 14-8B. A commercial Raytheon doppler sonar system output display. (Courtesy Raytheon Company)

future in an expanding variety of surface and subsurface platforms, particularly if the rather high cost of the current equipment declines somewhat.

The NavStar Global Positioning System

By the early 1970s, the need for a precision satellite-based navigation system that would be available worldwide at all times had become acute within all U.S. armed services. Moreover, a continuous three-dimensional position-finding capability (i.e., latitude, longitude, and altitude) was stipulated as a prime system objective, in contrast to the periodic two-dimensional capability of the Transit system. Such a system would be usable not only by military ships, submarines, aircraft, and ground vehicles, but would also be of great benefit to the civil sector in a wide variety of applications, from precise topographic mapping to aircraft and ship collision-avoidance systems. In April of 1973 the Department of Defense formally initiated the development program for this second generation navigation satellite system, called the NavStar Global Positioning System, or GPS. Later that same year, the NavStar Joint Program Office was established and staffed by military and civilian representatives from all four U.S. armed services, the Coast Guard, the Defense Mapping Agency, and NATO.

Several possible final system configurations are still being considered as of early 1981. These range from a constellation of 18 satellites, with 3 in each of six orbits as depicted in Figure 14-9 on the next page, or alternatively 6 satellites in three orbits, to an originally proposed configuration of 24 satellites distributed in three orbits containing 8 satellites each. Current plans favor one of the 18-satellite arrays, which would result in considerable cost savings in comparison to the 24-satellite system, at the loss of a minor degree of system availability at certain times at various locations. Irrespective of the eventual number of orbital planes and satellites, each satellite orbit will be circular, about 10,900 nautical miles (20,180 km) high, with a period of revolution of about 12 hours. The goal of the final system configuration is to provide for at least a .95 probability of a minimum of four satellites being above the 5° minimum altitude required for good reception at any time at any position on earth. As a point of interest, all proposed configurations will provide virtually 100 percent availability of at least three usable satellites necessary to determine latitude and longitude anywhere on earth. Under each of the 18-satellite configurations, however, there would occur transitory three-dimensional fix outages lasting from 10 to 40 minutes at various locations, during which periods altitude determinations would not be possible because the required fourth satellite would not be above the minimum 5° altitude limit.

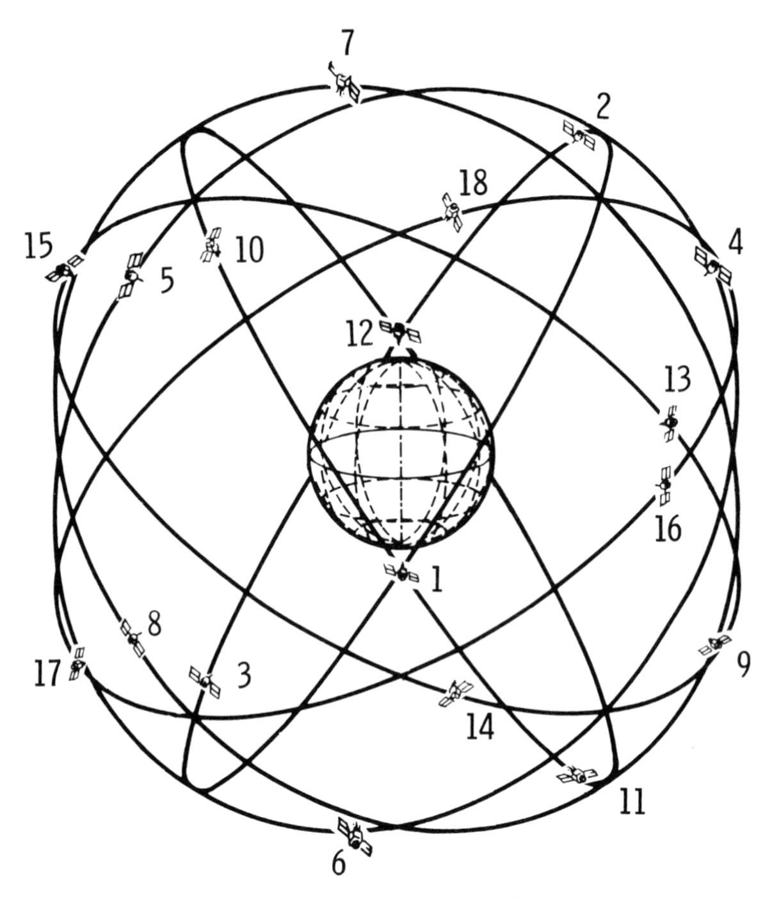

Figure 14.9. Planned GPS satellite coverage to be effective by 1987. Current plans call for 3 satellites uniformly spaced in each of six orbital planes.

NavStar System Operation

Each of the GPS satellites, like the Transit satellites, transmits simultaneously on two frequencies in order to enable the receiver to determine and eliminate the effects of refraction on the satellite transmission. However, the NavStar satellite frequencies, 1227 and 1575 kHz, are higher than those of the Transit satellites. A further difference between the satellite transmissions of the two systems is that the GPS satellites broadcast a continuous signal, in contrast to the sequential two-minute message format of Transit. The GPS receiver computes the fix by means of simultaneous electronic range determinations to each of the satellites being received. For precise latitude and longitude on the earth's surface, reception of the signals of at least three satellites is required, and for precise altitude determination, a fourth satellite signal is needed. The signals are modulated with a dual code—one for acquisition and coarse navigation (called the C/A code) and the other for precision ranging after acquisition and synchronization have been achieved (called the P code). Provisions have been made to make the second code decipherable only by specially equipped receivers, so as to deny full system accuracy to potential enemies in peacetime, and actual enemies in time of war. The coded signals contain satellite ephemeris data, atmospheric propagation correction data, and the amount of any satellite clock error. Using this information, the receiver computes the precise range to the satellite, and from this, generates a sphere of possible positions centered on the satellite. The intersection of the various satellite-centered range spheres thus determined establishes the position of the receiver with a high degree of precision.

When fully operational, the NavStar GPS system will be capable of achieving worldwide r.m.s. accuracy of less than 10 meters in latitude, longitude, and altitude using the P code, and somewhat less using C/A code signals alone. The system is scheduled to be implemented in three phases. During the first phase that ended in mid-1979, four test satellites were launched to provide periodic three-dimensional coverage in the western United States, required to facilitate development of ground-support equipment and operational procedures and prototype user equipment. In Phase II, a full-scale engineering phase scheduled to last until late 1983, several additional operational prototype satellites are being launched to extend the coverage worldwide in two dimensions. Concurrently, operational ground and control system equipment is being installed, and user equipment contractors are developing some fifty varieties of production prototype receiver equipments for evaluation on nine types of platforms within each of the four U.S. armed services. Phase III, during which the system will be expanded to full operational status, is currently scheduled to extend from late 1983 to 1987, with all system satellites to be in orbit by mid-1987. It is anticipated that the Space Shuttle will be employed beginning in 1984 to place the last several GPS satellites in orbit. Production models of receivers should begin to become available during 1985.

Because a fairly capable computer is required to perform the many calculations necessary to obtain a GPS fix, commercial GPS receivers incorporating such computers will in all probability be considerably more expensive than existing Transit receivers, at least initially. However, the continuous availability of GPS, coupled with the order-of-magnitude increase in precision and the many potential applications it offers, will undoubtedly lead to a reversal of this situation during the period from 1985 to 1990. Mass production of GPS receivers could

Figure 14-10. In the mid-1980s when Space Shuttle launch facilities become available, it is planned to place the last several GPS satellites into orbit using this vehicle.

eventually result in prices under $2,000 for basic models, if present downward price trends for computer-related electronic equipment continue.

Summary

This chapter has introduced four advanced navigation systems currently of major importance in the field of marine navigation. Three of these, the Navy Transit Satellite Navigation System, the Ship's Inertial Navigation System, and the doppler sonar system, are now fully operational, while the fourth, the NavStar Global Positioning System, is currently scheduled to achieve full operational status in mid-1987. All four systems are dependent upon the "state of the art" of the technologies upon which they depend for their effectiveness. Because the cost of their associated shipboard equipment is accordingly fairly high, at present they are normally found mainly on those military and commercial ships having a special need for the precise navigational information that the systems are capable of supplying. However, if continued operational experience, coupled with ever more rapidly advancing solid-state technology, continues to result in steadily decreasing costs of user equipment, the use of these advanced navigation systems will very probably be extended to an ever wider variety of military, commercial, and private vessels of every description over the next several decades.

15

Bathymetric Navigation

Bathymetric navigation can be defined as the art of establishing a geographic position on the open sea by use of geological features of the ocean floor. These features are located by means of an instrument called the *echo sounder;* in practice, this device is more commonly known as the *fathometer,* the name applied to an early, widely adopted model of this device by its manufacturer, the Raytheon Company.

The inexperienced navigator may question the need for bathymetric navigation in the present day when advanced techniques of celestial and electronic navigation are available. It must be remembered, however, that the professional navigator will use all means at his disposal to establish his position and safely direct the movements of his vessel. Even in the best of circumstances, bathymetric position-fixing can provide valuable back-up information. In the worst of circumstances the method may be the only means of determining the ship's position. Among the advantages of bathymetric navigation are that it uses fixed features of the ocean bottom, it is impervious to electronic jamming or weather, its application is nearly worldwide, and it is easily learned and applied.

This chapter briefly describes the nature of the ocean bottom features and the charts used to portray them, and examines the characteristics of the echo sounder and its output information. Finally, the techniques of bathymetric navigation presently in common use on many U.S. Navy surface ships are discussed, along with their advantages and limitations.

Geological Features of the Ocean Bottom

In order to make use of information received from the echo sounder in an intelligent manner, the navigator must first be aware of the basic characteristics of the geological features present on the ocean floor. In general, ocean bottom features are equivalent to topographical features seen on land, with the difference that undersea geology is usually much more subdued and gentle than land geology, owing to the more subtle erosion forces present within the oceans. Some predominant features of the ocean floor are pictured in Figure 15-1 and are described as follows:

> An *escarpment* is a long, steep face of rock, similar to a cliff on land.
>
> A *seamount* is an elevation rising steeply from its surroundings to a height of 500 fathoms (1,000 meters) or more; it is generally of relatively small horizontal extent.
>
> A *guyot* is a flat-topped seamount, rather similar to the mesas found in the southwestern United States. Guyots are especially plentiful in the Pacific Ocean region.
>
> *Submarine canyons* are similar to their dry-land counterparts and are found off most continental slopes.
>
> A *trench* is a relatively narrow canyon, distinctive because of its great depth.

The ocean bottom areas contiguous to the continents are generally devoid of any of the distinguishing features described above, but they are nonetheless useful for navigation by means of their *depth contours,* which are lines on a chart representing points of equal depth with respect to the surface datum. The zone between the emergence of a continent from the sea and the deep-sea bottom is called the *continental margin.* Within this margin three different subdivisions can generally be identified:

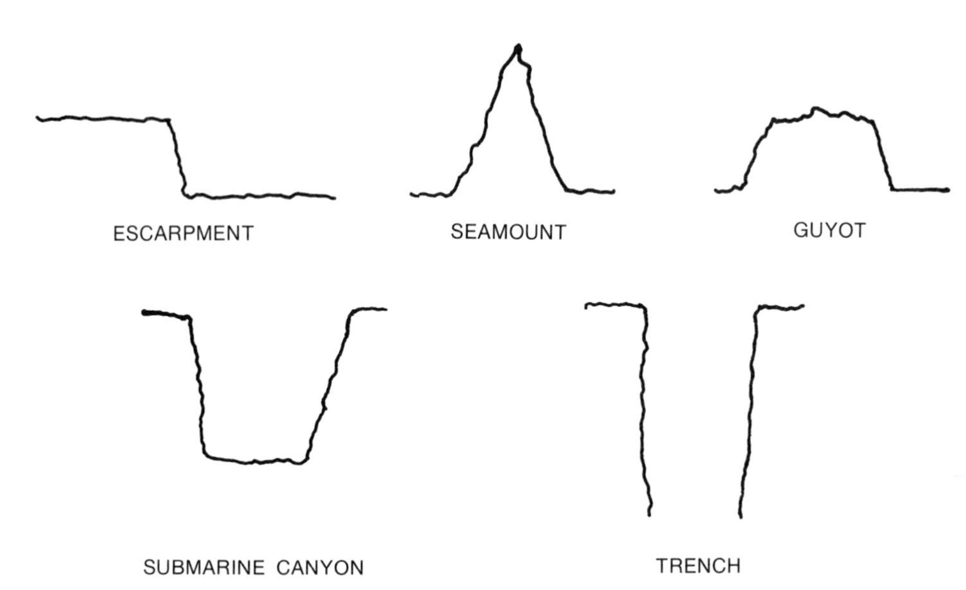

Figure 15-1. Features of the ocean bottom.

The *continental shelf* is the bottom zone immediately adjacent to a continent or island, extending from the low-water shoreline and sloping gently to an area of steeper slope. Its depth ranges from 10 or 20 fathoms down to 300 fathoms, and it may extend seaward from beyond the shoreline for a widely varying distance.

The *continental slope* is the area extending from the edge of the continental shelf into greater depth.

The *continental rise* is a gentle slope with a generally smooth surface, rising from the deep-sea floor to the foot of the continental slope.

Beyond the continental margin, the deep ocean floor generally consists of expansive plains and gently rolling hills, occasionally interrupted by distinguishing features, such as those listed in the preceding paragraph. Hence, as will be explained in more detail later, when such features can be identified by means of the echo sounder, they provide excellent positional information to the navigator, especially when used in combination with depth contours of the surrounding region.

The Bathymetric Chart

Bottom soundings and associated bottom sketches and charts have long been regularly used by the mariner; the use of the hand lead line in the Nile River well before the birth of Christ has been recorded. The first formal hydrographic office was established in Spain in 1508, and although intended primarily as a regulatory enterprise, a department was tasked with the charting of routes to and from the New World. Unfortunately, much of the information gathered by this and similar agencies of other contemporary governments was considered a secret to be closely guarded, inasmuch as a safe and rapid sea route was frequently the road to great wealth. As more and more voyages were made, however, the depths in and around the major ocean ports of the world became fairly well documented, but midocean depths were virtually unmeasurable until the invention of a mechanical sounding device known as the *Kelvin-White sounding machine* in 1878. This sounding system was in wide use until about 1930, when it was displaced by the echo sounder. The first workable echo sounder was developed about 1920, and the first transatlantic line of soundings was made in 1922. From that day to the present, continual improvements have been made in the techniques of obtaining and presenting sounding information, but it is only in the last 20 years that any systematic efforts have been made to chart the ocean bottom. For this reason it is often said that at present more is known about the details of the surface of the moon than is known about the bottoms of the lesser-explored oceans.

Bathymetric charts can be classified according to the soundings on which they are based into two broad categories—controlled and uncontrolled. Charts in the former category are produced as the result of a systematic survey of selected areas, while charts in the latter category are based on random data collected from ships that have traversed an area while transiting from one port to another. Today, controlled bathymetric charts exist for nearly all undersea areas within the mid-latitude continental margins, and for deep ocean areas beneath the more heavily traveled ocean routes between continents. Many infrequently traveled ocean areas, however, are covered only by uncontrolled charts based on random soundings first made before the turn of the century.

The DMAHTC issues a bathymetric chart series as described in Chapter 4 of *Marine Navigation 1;* they are often referred to as bottom contour charts. Bottom contours also appear on most five-digit coastal charts issued both by DMAHTC and the National Ocean Survey. Depth contours shown on charts produced by both agencies are based on an assumed sound propagation velocity of 4,800 feet per second, identical to the standard velocity to which most echo sounders are calibrated. Thus, no correction is necessary to echo sounder depths, other than the addition of the ship's keel depth in cases in which the installed equipment does not incorporate an automatic draft adjustment feature.

The contour interval of a bathymetric chart depends on the accuracy of the sounding data available and the chart scale. In general, controlled charts are produced on a scale between 1:10,000 to 1:100,000, and have contour intervals from 5 to 50 fathoms, depending on the scale. Uncontrolled charts ranging in scale from 1:1,000,000 to 1:1,500,000 are contoured at 100-fathom intervals. The importance of the proper contour interval cannot be overemphasized. The left-hand portion of Figure 15-2A depicts three 100-fathom contours, while the right-hand portion depicts 10-fathom contours in the area of the inset. If the scale of the chart were fairly large, the 100-fathom interval would be impractical for bathymetric navigation, while the 10-fathom interval provides a much more complete picture of the bottom for navigational purposes.

It should be apparent that if a very small contour interval were chosen for a small-scale chart, a morass of unintelligible contour lines would result. On the other hand, if too large an interval were selected, the chart would not be detailed enough to permit its use in bathymetric navigation.

In general, in bathymetric navigation as in piloting, the largest scale chart available depicting a given bottom area or geological feature of interest should be used. In the left-hand side of Figure 15-2B,

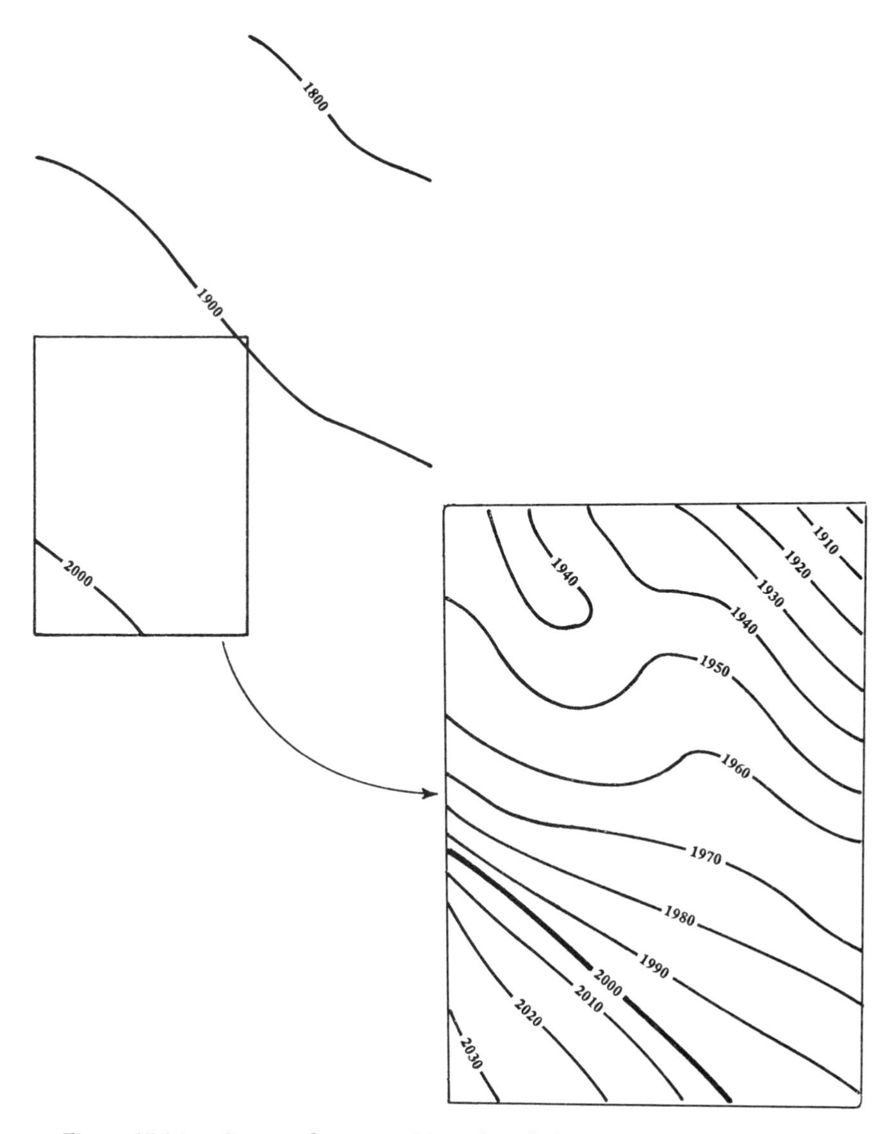

Figure 15-2A. Contrast between 100- and 10-fathom contour intervals.

a portion of an ocean area containing a seamount is contoured at a 250-fathom interval, such as might be found on a three-digit chart of the area; in the right-hand side, the same feature is depicted using 50-fathom intervals. It should be apparent that the closer interval spacing provides a much more accurate picture of the shape and size of the seamount. In fact, it is doubtful that the feature could be recognized as suitable for position-fixing from an inspection of the left-hand representation, while it is readily identifiable in the right-hand drawing.

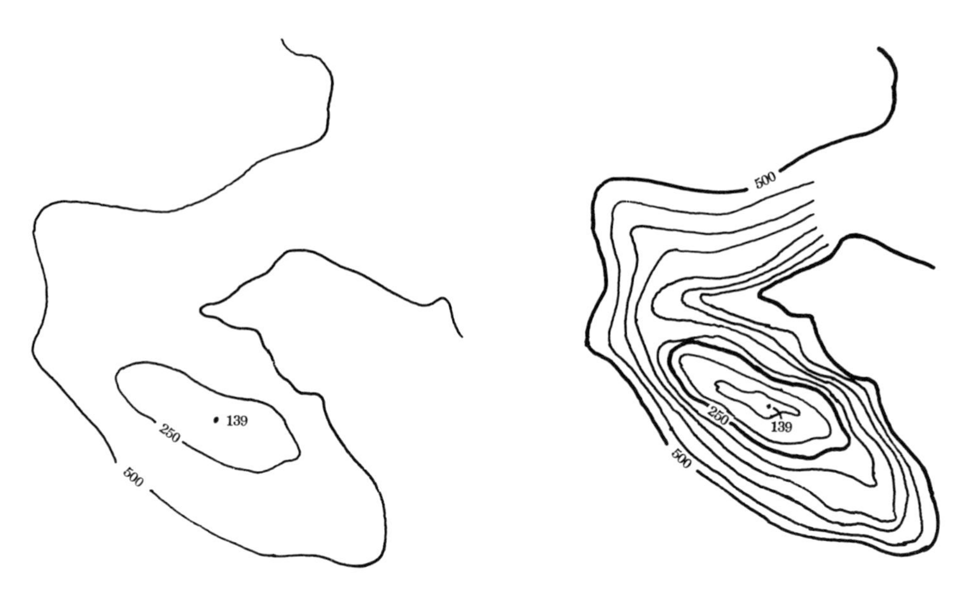

Figure 15-2B. Seamount depicted using 250- and 50-fathom contour intervals.

Characteristics of the Echo Sounder

The echo sounder or fathometer consists of two basic components—the transducer and the recorder. The *transducer* used on most ships is located at or near the keel, and contains both a *projector* for transmission of the sound signal into the water and a *hydrophone* for reception of the returning echoes. The *recorder* is located where required, usually in the chart room. The echo sounder most widely used in the U.S. Navy at the present time is the AN/UQN-4, pictured in Figure 15-3.

The AN/UQN-4 transmits on a frequency of 12 kHz. Both the pulse duration and the pulse repetition rate are variable, and several different combinations can be selected depending upon the depth of water in which the ship is operating; a digital readout indicates the depth in either feet or fathoms. The UQN-4 is also equipped with a continuous-feed graphic recorder, which produces a graphic trace of the depths encountered, and two additional very useful features: a *draft adjustment,* which automatically adds the ship's draft to the transducer depth, as well as a so-called *Lost Tracking Indicator,* which is illuminated whenever a depth varying more than 200 feet from a preset depth is recorded. The AN/UQN-4 subtends a cone 30 degrees in width, in contrast to the 60-degree cone of most other fathometers.

The signal transmitted by an echo sounder is electromechanical in nature. The transmitter, which is usually physically located in the recorder, emits a pulsed CW electromagnetic signal that is translated

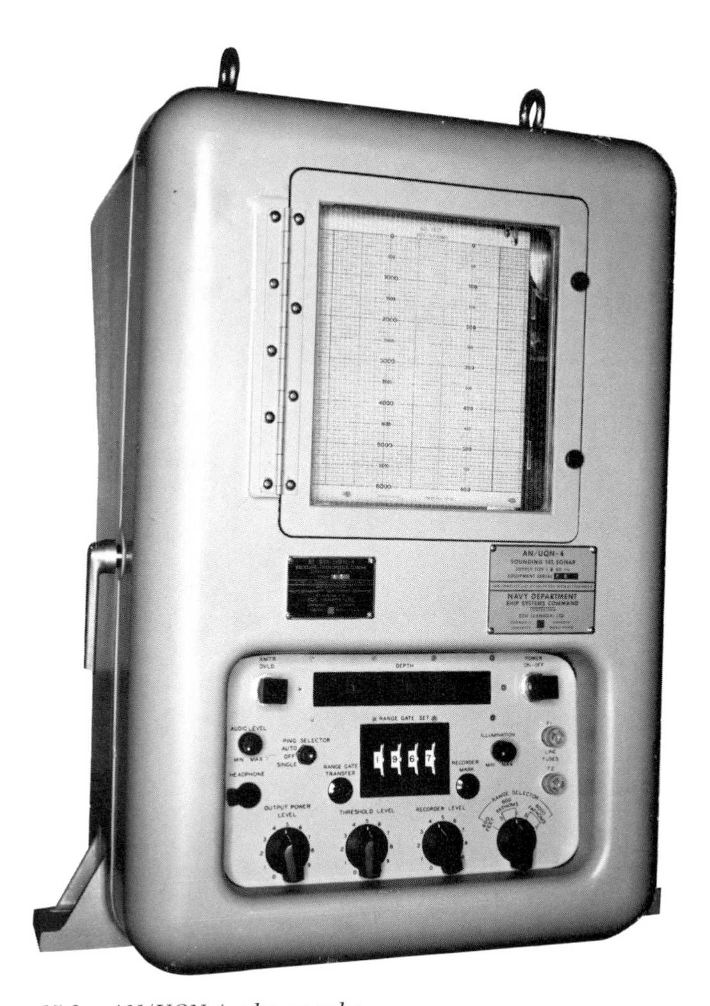

Figure 15-3. AN/UQN-4 echo sounder.

into a sound pulse by the transducer. The sound pulse train is radiated into the water in the shape of a cone, the dimensions of which are governed by the frequency of the pulsed wave and diameter of the transducer; as mentioned above, the normal cone width is 60 degrees. When a pulse strikes any surface or boundary layer of a region within which the sound propagation characteristics are different from those of the water into which the pulse was transmitted, an echo is returned. The strength of the echo depends on the quality of the reflective surface the pulse strikes. The returning echo is converted into an electromagnetic signal by the transducer, and in the recorder the depth is obtained by taking into consideration the elapsed time between the transmission of the pulse and the return of its echo and the transmission speed of sound in sea water. For purposes of the depth computation, most Navy echo sounders are calibrated to use a standard 4,800 feet per second as the sound velocity. Because the actual sound velocity is both variable and somewhat faster than this standard, varying with salinity, temperature, and pressure as shown in Figure 15-4, some difference almost always exists between the actual depth and the indicated depth. The actual depth is always greater than the indicated depth, thus providing a small variable safety factor.

There are other sources of error in the echo sounder depth as well.

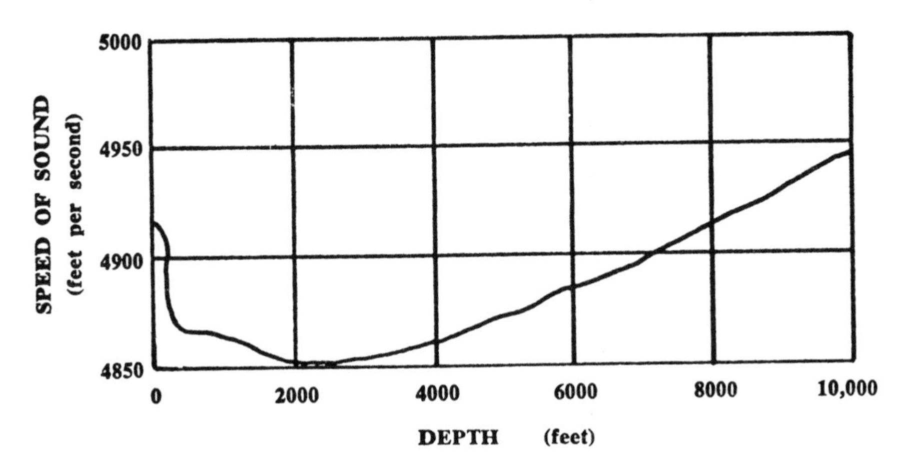

Figure 15-4. Variation in sound velocity with ocean depth.

Since the transducer pulses propagate outward in the shape of a cone, the first reflective surface that the cone encounters will produce a reflection interpreted by the recorder as the depth immediately beneath the ship. In actuality, however, the surface may lie off to one side, as shown in Figure 15-5A. Such an echo returned from an object at the edge of the sound cone is termed a *side echo*.

A third possible source of error is the rolling and to a lesser extent the pitching motion of a ship while underway. The position of the transducer on most ships is fixed, causing the transmission cone to be canted at an angle to the vertical as the ship proceeds. This effect is a factor that must be taken into account regardless of the cone width, as shown in Figure 15-5B.

Interpretation of the Echo Sounder Recording

When an echo sounder is to be used for bathymetric navigation, the continuous graphic trace is used to display and record the depth information. Most types of trace paper are marked with several scales, corresponding to the various scales on the recorder that the operator can select. As the paper feeds through the recorder at a constant rate,

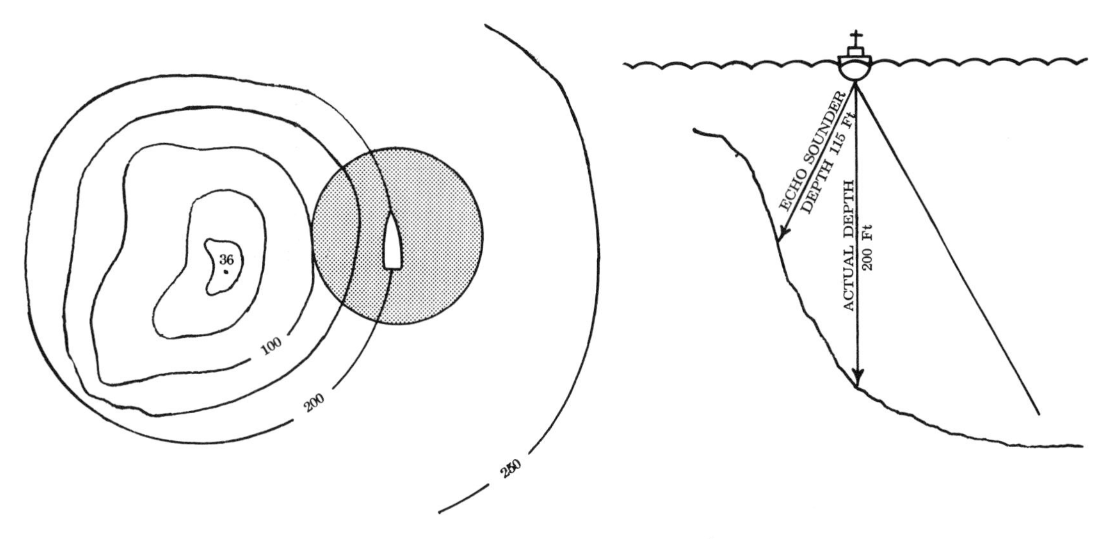

Figure 15-5A. Echo sounder depth error resulting from a side echo.

an electrostatic pen moves rapidly over the surface of the paper at a right angle to its direction of movement, shading the paper at locations corresponding to the depth measurement. Thus, a depth profile is plotted versus time.

Because of the conic sound propagation pattern of the echo sounder, a bottom trace of an object protruding above an otherwise level ocean floor appears hyperbolic in shape, as, for example, the seamount trace pictured in Figure 15-6 on the following page. The shape of the hyperbola is a function of the beam width of the transducer, the depth of the feature, and the speed of the ship.

If the ocean bottom is fairly irregular, a pattern of several hyper-

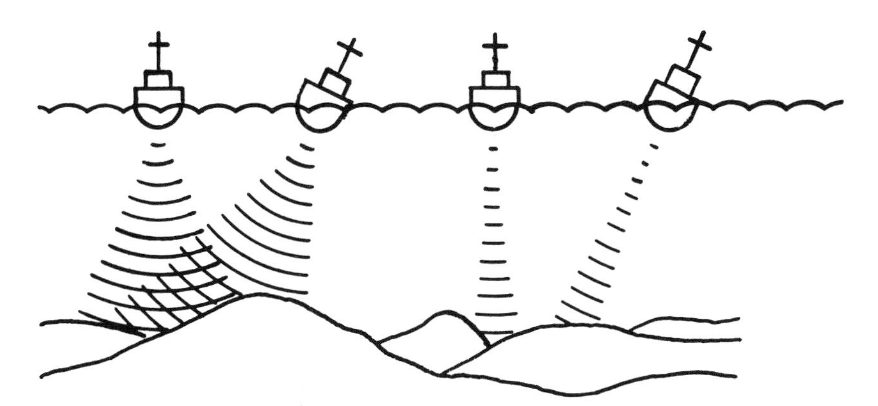

Figure 15-5B. Echo sounder depth errors resulting from ship motion.

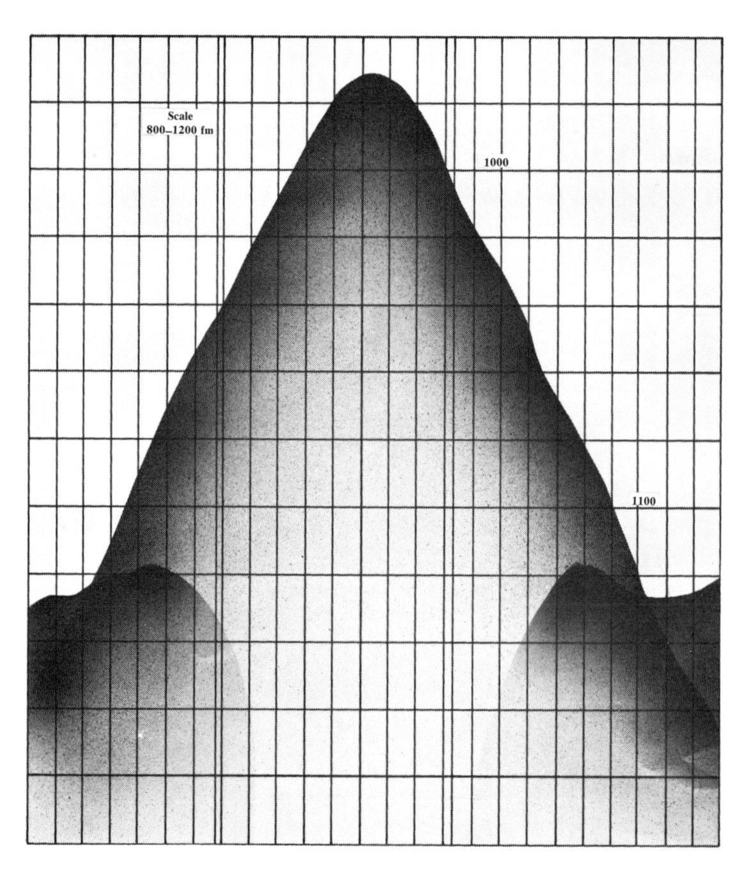

Figure 15-6. Echo sounder trace of a seamount.

bolas superimposed on one another as in Figure 15-7 may result. The multiple hyperbolas are due in large measure to the side echoes returned by the edges of the propagation cone; the effect is even more pronounced if the ship is rolling from side to side. When interpreting such a trace, the navigator must bear in mind that the minimum depths recorded to the top of each hyperbolic trace are not necessarily those directly beneath the ship. Hence, the trace cannot be regarded as a profile of the bottom along the route of the ship, but rather it is a representation of the average depths over the area swept by the sound propagation cone as the ship moves along her path.

As will be seen in the following section, the disadvantages of the conic propagation effect described above are balanced by certain other beneficial effects.

In addition to the distortion caused by the beam width, there are two other major sources of error in interpretation of the echo sounder trace. The first is the so-called *multiple bottom return,* an example of

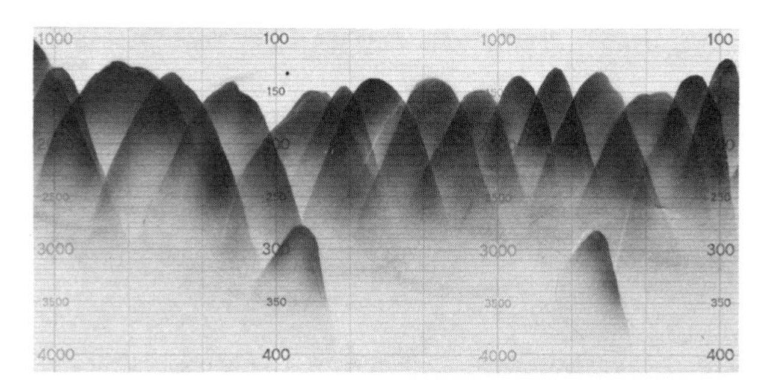

Figure 15-7. Echo sounder trace of an irregular bottom.

which is pictured in Figure 15-8A. The effect is most common in relatively shallow water with a highly reflective bottom such as sand or gravel, and is caused by reverberation of the sound pulses between the bottom and the water surface. A similar phenomenon sometimes occurs if the bottom is covered to a depth of several fathoms by a material of poor reflective qualities, such as soft mud. One return is formed by the weak reflection from the surface of the mud, while a second return results from the actual bottom reflection.

A second possible source of error in echo sounder trace interpreta-

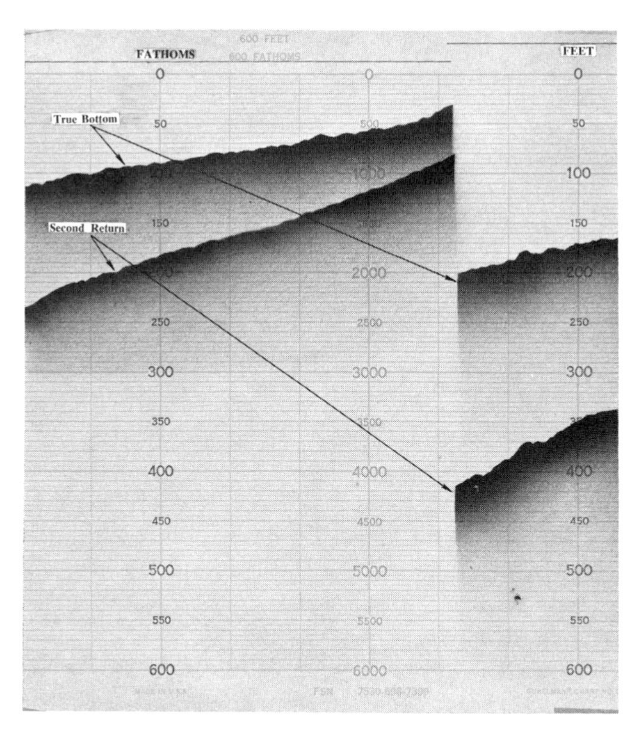

Figure 15-8A. Echo sounder trace on two scales, showing multiple bottom returns.

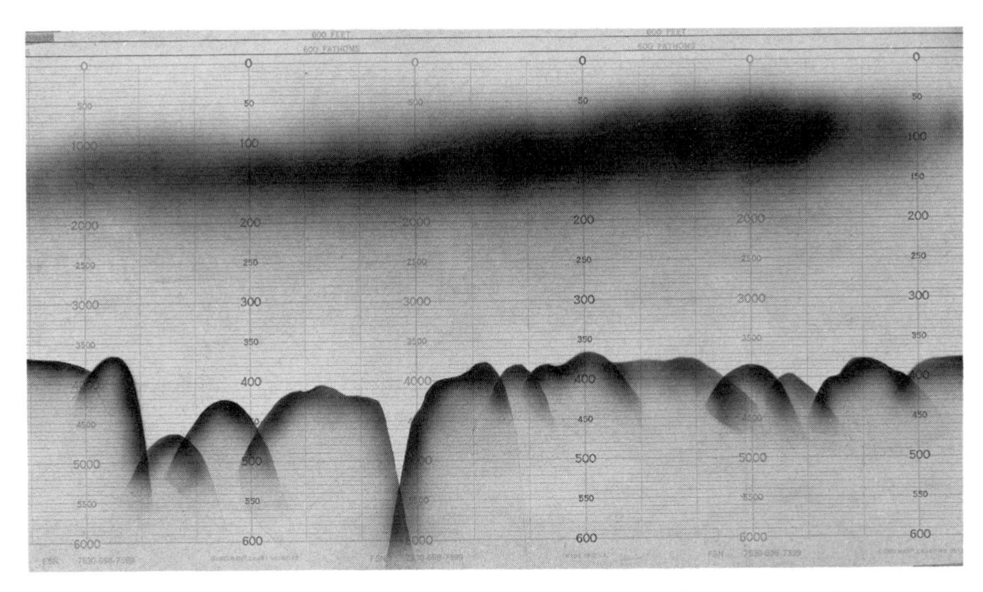

Figure 15-8B. An echo sounder trace showing the deep scattering layer.

tion is the *deep scattering layer,* a suspension of biological matter such as plankton in a layer between the surface and the bottom. The effect of this layer is illustrated on the trace pictured in Figure 15-8B.

When a layer of this type is present, it usually rises toward the surface at night and sinks somewhat by day. Often the effect is very persistent in a given area, and may lead to numerous reports of shallow water at locations where the actual depth is very great. A so-called American Scout Seamount with depths of 30 to 90 fathoms was reported east of Newfoundland so often between the years 1948 to 1964 that it was printed on several contemporary charts of the area. Controlled surveys conducted in the 1960s, however, found no depths in the area less than 2,350 fathoms. The fictitious seamount reports are believed to have been caused solely by the deep scattering layer.

Techniques of Bathymetric Position-Finding

There are several techniques available for determining the ship's position by bathymetric methods. The particular technique employed depends largely on the general geological pattern of the bottom, the type of echo sounder installed, and the skill of the navigator in interpreting the recorder output.

If the ship passes directly over an easily recognizable bottom feature such as a seamount or submarine canyon, the ship's position can be determined with a high degree of accuracy, particularly if the surrounding ocean floor is fairly uniform. Any of the features depicted in Figure 15-1 are readily identifiable on the fathometer trace, even with the distortion effects due to the propagation cone present. Seamounts are particularly well suited for position-finding, as many have distinguishable depths, sizes, and shapes. When the ship passes over such an unmistakable feature, the navigator has only to note the time of passage and commence a new DR plot from that point on the chart.

If a uniquely identifiable feature either does not exist in the area over which the ship is operating, or if the ship does not pass directly over such a feature, a more complicated technique is then necessary to determine the ship's position. There are three such procedures used for the most part—the "line-of-soundings" technique, "contour advancement," and the "side-echo" technique.

The *line-of-soundings technique* has been widely used by both commercial and navy navigators for many years. This procedure depends to a great extent on the accuracy of the depth contours depicted on the chart in use, and is essentially independent of the chart scale. In this technique, an overlay of thin transparent paper or plastic acetate is laid down over the bathymetric chart, and the DR plot is constructed on the overlay in the usual manner. As the ship proceeds, the times at which she passes over the depths corresponding to the charted contour lines are noted, and DR positions for these times are plotted on the DR course line; each is labeled with the depth recorded at the time, as shown in Figure 15-9 (next page). The length of time during which these depth recordings are made and plotted depends on the scale of the chart and contour interval.

After a number of depth recordings have been plotted, the overlay is detached and shifted about in the vicinity of the original DR track until a match between the plotted depths and the underlying charted depth contour lines is made. This matching process is represented on Figure 15-9 by lines A, B, and C; line D is the final position of the template. After the match has been made, the overlay is secured in position, and the last depth mark on the overlay DR plot is taken as either a fix or an estimated position, depending upon the degree of confidence the navigator places in the position thus determined.

It should be apparent that this method is not suitable if the ship is proceeding in a direction roughly parallel to the depth contours, or in areas where adjacent contours are parallel or nearly so, as for example where the bottom is flat, or rising or falling at a constant slope.

The *contour advancement technique* also involves the use of an overlay, and is closely related to the line-of-soundings method just discussed. First, a DR plot is laid down on the bathymetric chart and a DR position corresponding to an initial recorded depth is placed on the course line directly over the corresponding depth contour; this

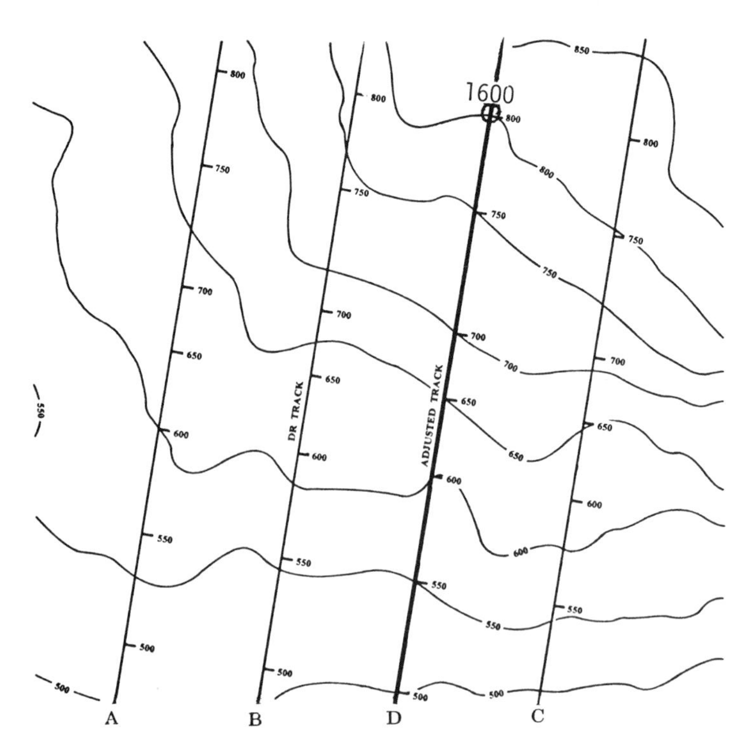

Figure 15-9. Line-of-soundings technique.

initial contour is referred to as the reference contour. Next, DR positions for times at which the ship passes depths corresponding to succeeding depth contours are plotted relative to the initial DR position, as shown in Figure 15-10A. In this figure, the 700-fathom curve is the reference contour.

After a suitable number of DR positions have been so plotted—usually four or five are sufficient—an overlay is placed over the DR plot and the reference contour is traced thereon, similar to the heavy line in Figure 15-10A.

Next, the overlay is detached and moved along the DR course line until the reference contour appears directly over the next DR position. In the case of Figure 15-10A, the overlay is advanced until the traced 700-fathom curve on the overlay crosses over the 680-fathom DR position. The contour corresponding to this DR position is now traced onto the overlay, as shown in Figure 15-10B on page 306.

At this point there are now two contours drawn on the overlay, one atop the other. Now the overlay is again moved in the direction of the DR course line until the reference contour lies over the next DR position, and the third contour is traced. After the reference contour has been so aligned with every DR position plotted and the

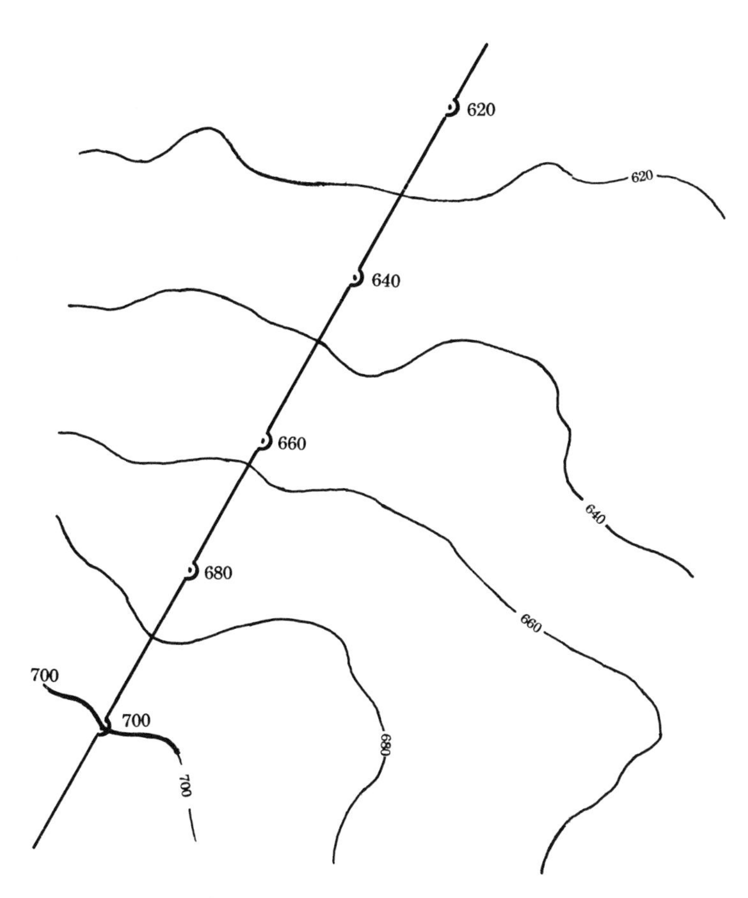

Figure 15-10A. DR plot on bathymetric chart, contour advancement technique.

corresponding depth contours traced, the overlay is fixed in position; the point on the chart over which all traced contours seem to intersect is the ship's position corresponding to the time of the last DR position plotted, as shown in Figure 15-10C on page 307. To complete the plot, the position is noted, the overlay is removed, and a new DR plot can be commenced at that position.

Again, this technique does not lend itself to use in areas where the direction of travel is parallel to the depth contours or where the contours are parallel to one another and spaced equal distances apart on the chart.

The two techiques just discussed are very useful in situations in which the bottom is fairly uneven, the depth contours are closely spaced on the chart, and the direction of travel is approximately at a right angle to the contour lines. If an isolated seamount or guyot is to be used on an otherwise level ocean floor, however, the *side-echo*

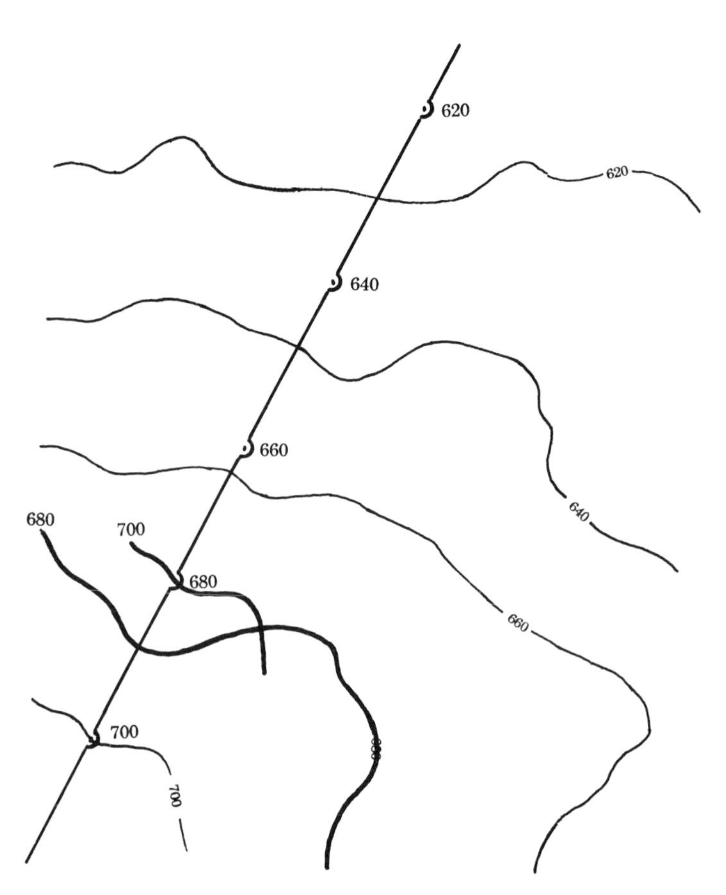

Figure 15-10B. Reference contour advanced to time of second contour recording.

technique is used if the ship's path does not take her directly over the feature.

If a seamount is to be used to determine the ship's position, a DR track is first laid down from a position within 20 or 30 miles of the feature directly across its top. As the ship approaches, continual depth readings are recorded. If the minimum depth agrees with the charted minimum depth over the top of the feature, it can be assumed that the ship has passed directly over it; a fix symbol is plotted, and the DR plot is continued from this point. If the minimum depth is more than the charted depth, however, it can be assumed that the ship passed near but not directly over the top of the seamount. In this case, the side-echo technique is applied as described below.

First, a construction line is drawn perpendicular to the DR course line through a DR position corresponding to the time of the minimum depth reading obtained. After a downward trend in the depths being recorded is observed, the ship's course is changed in a wide 270° turn

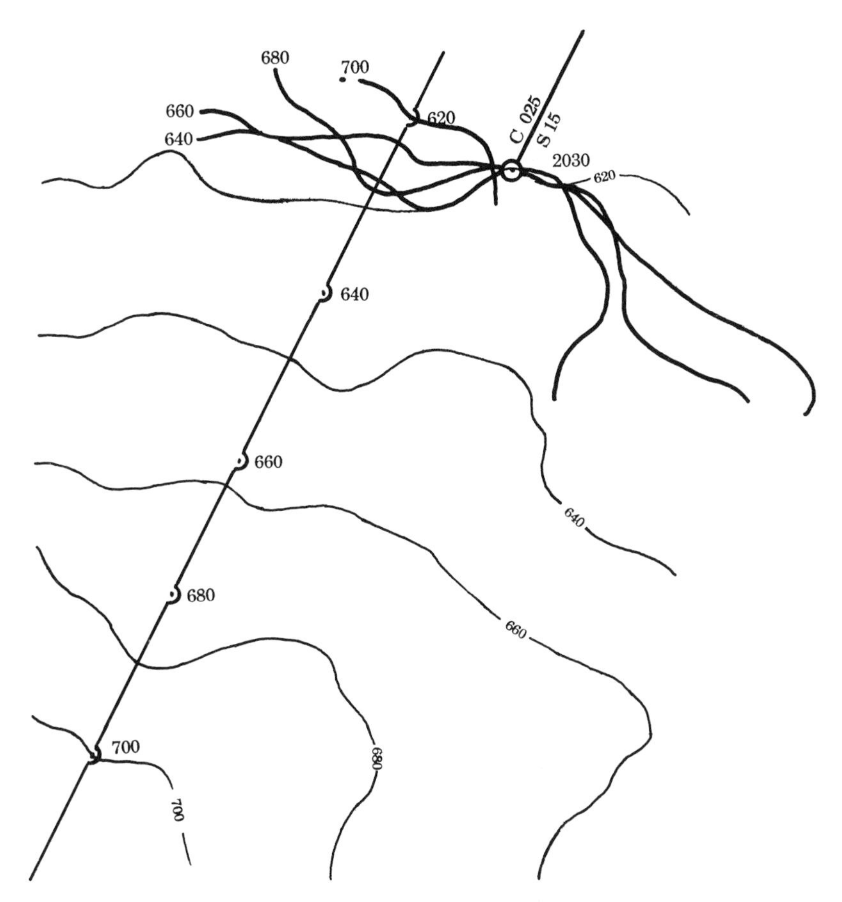

Figure 15-10C. Completed contour advancement overlay.

to either side, such that the projected ship's track will cross the original DR course line at a right angle near the minimum depth reading. Echo sounder readings are again carefully recorded, and a DR position for the time at which a minimum depth is again recorded is plotted. A perpendicular construction line is then drawn through this point, so that the plot now resembles that depicted in Figure 15-11 on the following page.

The point at which the two construction lines intersect locates the seamount top relative to the DR plot. An estimated position can then be plotted by shifting the DR position used for the second construction line through the distance separating the intersection of the two construction lines and the charted position of the top of the seamount. The ship's course is then adjusted as desired, based on the estimated position.

Although the use of the side-echo technique is disadvantageous because of the course change required, it has the compensating benefit that use of the technique takes full advantage of the conic propagation zone of the echo sounder. If the cone were fairly narrow like a radar beam, the ship would have to pass directly over a raised feature to be able to find and use it for navigational purposes. The expanding cone allows features to be readily located, and the side-echo technique makes possible the use of the feature for position-fixing without the necessity of passing directly over its top.

New techniques of position-finding by means of undersea topographical features are now being developed that incorporate the electronic computer to resolve difficulties arising in situations more ambiguous than the rather straight-forward examples given in this section. Computers will provide the ability to perform statistical analysis upon a large amount of bathymetric data that the navigator now has the ability to gather, but cannot at the present time conveniently analyze for positional information over complicated bottom topography.

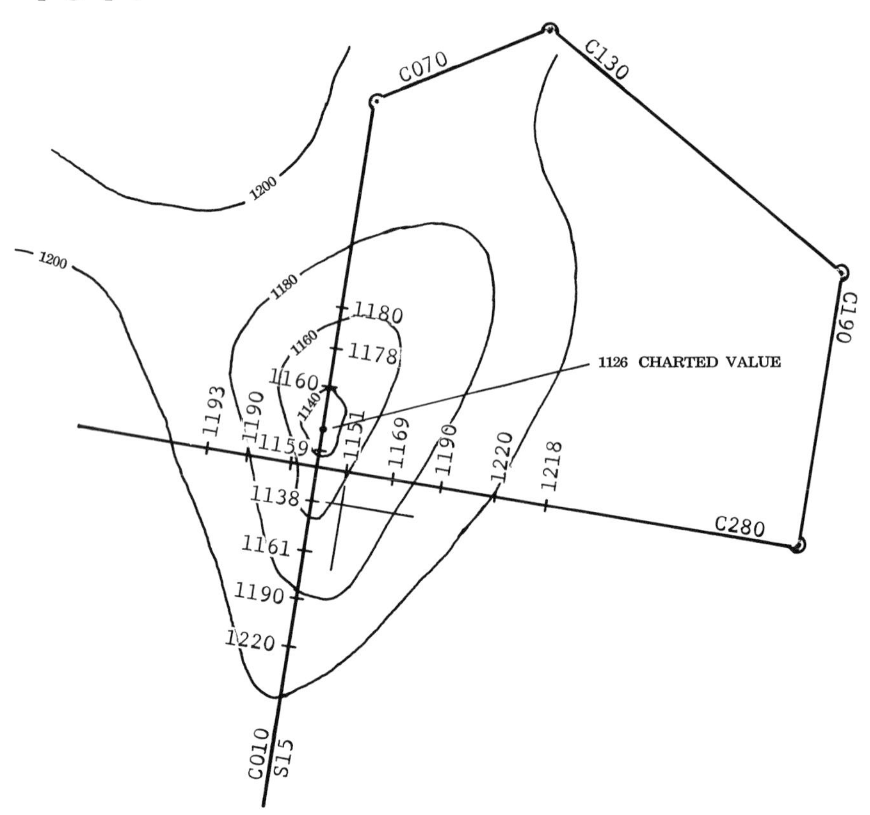

Figure 15-11. Side-echo technique.

Summary

This chapter has discussed the present state of bathymetric navigation on surface ships of the U.S. Navy. Although the techniques described here are not widely used when celestial observations and electronic aids to navigation are available, they nonetheless represent another potential source of positional information available to the navigator of any type of vessel. The techniques of bathymetric navigation should not be overlooked, as positional information supplied from this source can be a valuable back-up to information obtained by other means at sea. Bathymetric information is one of the few kinds of navigational data obtainable by the navigator in any weather, independent of outside facilities.

16

A Day's Work in Navigation at Sea

In this final chapter of *Marine Navigation 2,* the routine of a typical U.S. Navy navigator's day at sea will be briefly outlined, with special emphasis on the interplay of celestial and electronic means of position-finding. While nonmilitary readers may find that some of the auxiliary duties and activities described herein are not applicable to their past experience and probable future circumstances at sea, nevertheless most of the actual work of navigating will be common to almost all vessels operating on the high seas. Many practicing navigators of commercial and private vessels, in fact, have often been heard to remark that the only real difference between Navy and civilian or Merchant Marine navigators at sea is that the former usually have many more assistants. Be this as it may, there is undeniably a great deal of similarity in the routines of all professional navigators of ocean-going vessels.

The transit from Norfolk, Virginia, to Naples, Italy, preplanned as an example in the last chapter of *Marine Navigation 1,* will form the framework of the discussion in this chapter. Before examining the events of a typical day en route on this voyage, it is first necessary to outline the normal routine of a Navy navigator's day's work at sea.

The Routine of a Day's Work

For purposes of this discussion, the typical day's work in navigation while on an extended voyage can be considered to commence with the preparation of the rough draft of the Captain's Night Orders, which were briefly described in Chapter 15, *Marine Navigation 1.* In actual practice, the navigator usually begins this task after the evening celestial fix has been plotted and the ship's DR track for the night has been laid down from it.

The following is a fairly comprehensive listing of the tasks generally accomplished by the navigator of a U.S. Navy ship, with the help of his staff of quartermasters, in the course of a typical day's work in navigation at sea:

1. Prepare the rough draft of the Captain's Night Orders.
2. Compute the commencement of morning star-time (civil twilight), the time of sunrise, and compile a list of the most favorable stars and planets suitable for observation.

3. If no unusual events are anticipated during the night, turn in until morning star-time.
4. Arise prior to morning star-time, and make the sextant ready for the day's observations.
5. Observe, plot, and record the morning celestial fix.
6. Observe sunrise, obtain and record a sun amplitude observation, and compute and record the resulting gyro error.
7. Construct the day's DR track, incorporating any corrections necessary to regain the preplanned intended track based on the morning fix.
8. Prepare the 0800 ship's position report slip (see Chapter 15, *Marine Navigation 1*) based on the morning celestial fix.
9. Wind the ship's chronometers and determine and record chronometer error by radio time signals.
10. While the sun is at low altitude, observe its azimuth to determine gyro error if necessary.
11. Obtain a mid-morning sun observation for a morning sun line.
12. Compute the time of LAN.
13. Observe the sun at LAN and plot a running fix by advancing the mid-morning sun line.
14. Prepare the noon position report slip.
15. Obtain a mid-afternoon sun observation for an afternoon sun line and an azimuth for determination of the gyro error if necessary.
16. Compute the time of sunset and the ending of evening star-time (ending of civil twilight) and compile a list of the most favorable stars and planets for observation.
17. Observe, plot, and record the evening celestial fix.
18. Prepare the 2000 position report slip based on the evening celestial fix.
19. Lay down the ship's DR track for the night from the evening fix.
20. Prepare the rough draft of the Captain's Night Orders.

Additional responsibilities that are interspersed throughout the routine tasks listed above are as follows:

1. Obtain, plot, and record additional celestial LOPs and fixes of

opportunity, such as additional sun lines and daytime observations of the moon and Venus when available.

2. Obtain, plot, and record electronic fixes as available when required.
3. Obtain, plot, and record any bathymetric fixes of opportunity.
4. Perform routine administrative duties.

While it is not possible to specify the optimum intervals between fixes and running fixes at sea under all conditions, a set of rough guidelines has been developed over the years by many practicing navigators of this author's acquaintance. They appear below:

Range to nearest hazard	Recommended Minimum fix interval
More than 300 miles	Every 4 to 8 hours
From 300 to 200 miles	Every 2 to 4 hours
From 200 to 50 miles	Every hour to 2 hours
Less than 50 miles	Same as in piloting

If long-range radio navigation aids or advanced navigation systems are available to the navigator, it may be possible to obtain fixing information much more frequently than the minimum intervals listed above. Under these conditions the navigator or his staff may obtain fixes every hour or half hour for comparison with the projected DR track. The frequency with which these fixes are plotted and used to adjust the DR track depends on the circumstances and the desires of the individual navigator.

As mentioned above, the navigator does not personally have to accomplish each of the tasks listed, as his quartermasters are skilled enough to perform many if not all of them. Exactly which tasks will be carried out and by whom is in the last analysis the prerogative of the individual navigator, based on his experience and the ability of his subordinates. Because he is responsible to the commanding officer for the safe navigation of his ship, however, the professional navigator will, as a minimum, usually desire to personally observe morning and evening stars, lay down the ship's DR track, and approve and sign all ship's position reports prior to their submission to the commanding officer.

The Captain's Night Orders

The Captain's Night Orders have already been introduced in "Voyage Planning," Chapter 15 of *Marine Navigation 1.* As was explained at that time, the night orders are written primarily for the benefit of the night OODs, to ensure that they will be informed as to the important details of an operational and navigational nature expected to occur during their watches. Although there is no established format for the night orders, most include the following information:

1. The authority under which the ship is operating.
2. The status of the ship, including:
   a. Material condition (X-ray, Yoke, Zebra).
   b. Status of the engineering plant.
   c. Condition of readiness set.
   d. Major equipment casualties.
   e. Times of anticipated changes to ship's status.
3. Formation status and command structure, including:
   a. Range and bearing to formation guide.
   b. OTC, SOPA, and screen commander.
   c. Vessels joining or departing formation during the night.
   d. Base course and speed.
   e. Anticipated changes to formation or command structure.
4. Navigational information, including:
   a. Times of planned base course and speed changes.
   b. Times and approximate bearings and ranges at which aids to navigation should be observed either visually or on radar.
   c. Frequency at which fixes are to be obtained and plotted.
5. Pertinent tactical information.
6. Times at which the navigator and commanding officer are to be called the following morning.

The navigator's staff can normally fill in the more routine items included as part of the night orders, but the navigator usually adds the pertinent navigational and operational data. After the rough draft of the night orders has been completed, they are submitted to the commanding officer for annotation and final approval. On most ships, after the night orders have been signed by the captain, the OOD is authorized to make the listed course and speed changes at the times indicated, using the night orders as his authority to do so. The OOD will then notify the CO of the action completed.

Most well-run commercial ships and private vessels will use some written equivalent of the Captain's Night Orders, so that the night watch officers will have a clear idea of any action intended to be accomplished during their watches. On sailing craft, the ship's log will often be used as the medium for recording and relaying this information.

A typical example of the Captain's Night Orders that might have been written for the night of 24–25 June, while en route on the transit preplanned in Chapter 15 of *Marine Navigation 1,* appears in Figure 16-1.

Laying Down the Ship's DR Track

After the navigator or his staff has obtained and plotted a celestial or electronic fix, a DR track should be laid down from this fix to take the ship back to the preplanned track, if in fact the fix shows the ship to be displaced from it. As was mentioned in the voyage planning chapter, if the ship is en route on an extended transit, the navigator will normally desire to maintain a position about three to four hours ahead of the position of intended movement (PIM) along the track until the ship is near the end of the voyage. The slack time thus built up allows for delays en route resulting from such causes as maneuvering exercises, engineering drills, equipment casualties, and rough weather.

The Interplay of Celestial, Electronic, and Bathymetric Fixing Information

The modern-day navigator has available a wide variety of sources from which to obtain position-fixing information at sea. As recently as the 1950s, the navigator had to rely almost exclusively on celestial observations to direct the movements of his vessel on the open sea. With the advent of worldwide navigation systems such as Omega and the Transit and NavStar navigation satellite systems, celestial navigation may soon become for the first time in history a back-up technique rather than the primary method of position-finding at sea. Even when these advanced systems become fully operational, however, the navigator should continue to observe morning and evening stars and the sun at local apparent noon, if for no other reason than to remain proficient in celestial navigation in the event of malfunction or destruction in wartime of the electronic aids. Furthermore, observation of a celestial body either as an amplitude or azimuth sight remains the primary method of determining gyrocompass error at sea, an essential daily task of the navigator. The professional navigator should always take full advantage of all means at his disposal to determine his ship's position at sea.

An Example of a Day's Work in Navigation at Sea

As an example of a day's work in navigation at sea, let us return to the mock voyage discussed in Chapter 15 of *Marine Navigation 1.*

CAPTAINS NIGHT ORDERS

SHIP	TIME ZONE	DATE
USS EXAMPLE	+1 N	24-25 JUNE

ENROUTE FROM NORFOLK, VA TO NAPLES, ITALY

OPERATING WITH	AREA
N/A	NORTH ATLANTIC

OTC	FLAG SHIP
C.O. THIS SHIP	N/A

STANDARD TACTICAL DATA

FORMATION N/A

BASE COURSE		SPEED
102 °T	118 °PSTGC	16 KTS 118 RPM

FORMATION AXIS	GUIDE	BEARING	DISTANCE
N/A °T	N/A	N/A °T	N/A YARDS

SCREEN DATA

TYPE SCREEN	SCREEN	CIRCLE	NO	SHIPS UNASSIGNED

OWN SHIP DATA

ENGINES ON THE LINE	GENERATORS ON THE LINE	PLANT	SHIP DARKENED
2	2	2+4	YES ( ) NO (X)

EQUIPMENT CASUALTIES	ETR		ETR
SPN 40 AIR-SEARCH RADAR	UKN		

WEATHER DATA

SUNRISE	SUNSET	MOONRISE	MOONSET
250447	241920	242334	251115

NAVIGATION AND WEATHER REMARKS

1. SHIP IS 3 HRS. 17 MINUTES AHEAD OF PIM AT 242000N JUN.
2. NO UNUSUAL WEATHER CONDITIONS EXPECTED 24-25 JUN.
3. CALL NAVIGATOR FOR MORNING STARS AT 0400.

NIGHT INTENTIONS

1. CHANGE COURSE TO 109° T AT PT "G" AT MIDNIGHT (250000 N).

1. CARRY OUT STANDING NIGHT ORDERS. CHECK THEM OVER TO REFRESH YOUR MEMORY.
2. ETERNAL VIGILANCE IS THE PRICE OF SAFETY
3. CALL ME WHEN IN DOUBT AND IN ANY EVENT AT 0630

SIGNATURE (COMMANDING OFFICER) T. F. Epley

SIGNATURE (NAVIGATOR) R R Hobbs SIGNATURE (EXECUTIVE OFFICER) D. C. Cuberti

WATCH	OOD	JOOD
20-24	CS	C.T.
00-04		
04-08		

WATCH REMARKS

Figure 16-1. Captain's Night Orders, Norfolk-Naples transit, 24-25 June.

At that time, the voyage planning process was illustrated by means of an example based on an independent transit from Norfolk, Virginia, to Naples, Italy. It was determined that the ship should depart Norfolk at 1000R 17 June and arrive at Naples at 0800A 30 June. The preplanned track is shown plotted in Figure 15-5A of *Marine Navigation 1.*

For purposes of this discussion, it will be assumed that all applicable bathymetric, Omega plotting sheets, and large-scale charts are on board, and that the ship is equipped with the AN/UQN-4 echo sounder and AN/SRN-12 Omega receiver. Furthermore, it will be assumed that all Omega stations receivable in the North Atlantic area are operational on the 10.2 kHz transmission frequency. As several different examples of the plotting of positions determined by celestial, electronic, and bathymetric methods have already been presented in earlier chapters, detailed solutions of the various fixes obtained in the course of this example will not be presented. Rather, the LOPs and fix positions are plotted as they might appear on the chart at the end of the day's work. For purposes of clarity in this example, all fixes are labeled to show the source from which they were derived, and Omega LOPs are labeled with the corrected readings received. In practice, celestial and bathymetric fixes would be labeled only with the time in the usual manner, and Omega LOPs would be labeled only with the station pair letters.

Let us begin the example of the day's work at midnight (0000N) 25 June. From the Captain's Night Orders shown in Figure 16-1, the ship is steaming approximately three hours ahead of PIM, to allow for any unexpected delays that might occur during the remainder of the transit.

The navigator has been using as his primary working charts the bathymetric chart series covering the route of his transit across the North Atlantic. In addition to these, the 7600-series Omega charts of the area have been used to plot his Omega fixes; once plotted, the Omega fix positions have then been transferred to the bathymetric charts. The navigator has also been recording the ship's progress every four hours on the applicable intermediate-scale, three-digit general purpose charts of the North Atlantic, for use by bridge and CIC watchstanders. A portion of Chart No. 126 indicating the ship's position at 250000N June appears in Figure 16-2. Note that at this time the ship has just arrived at Point G after completing the F-G segment of her track. She has turned to course 109°T at a speed of 16 knots, the track and SOA of the G-H segment, as per the instructions in the night orders.

The previous evening the navigator computed the time of the

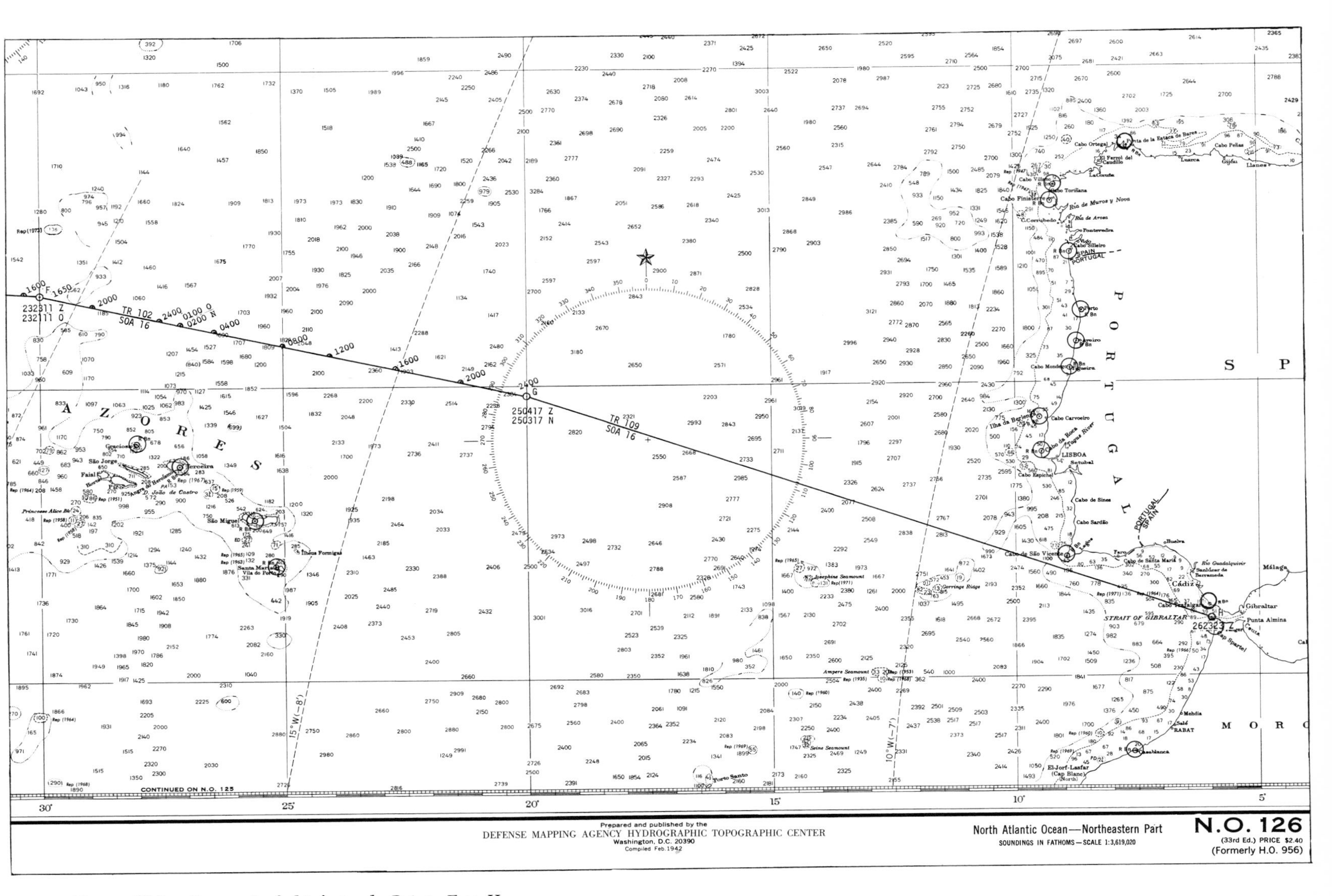

Figure 16-2. Segment of ship's track, Points F to H.

commencement of civil twilight to be 0413, and sunrise to be 0447 (unless otherwise specified, all times given henceforth will be time zone N times). Consequently, he instructed the QMOW to call him at 0400 to observe the morning stars, which were selected earlier by use of the Rude Starfinder.

After observation of the morning stars, three of those listed in Volume I *Tables No. 249* were plotted and used to determine the morning celestial fix, recorded at 0413. This fix is shown in Figure 16-3, which depicts the portion of the bathymetric chart used for the navigation plot from 0000 to 1200N. The ship's position is about 13 miles northeast of the 0413 DR position.

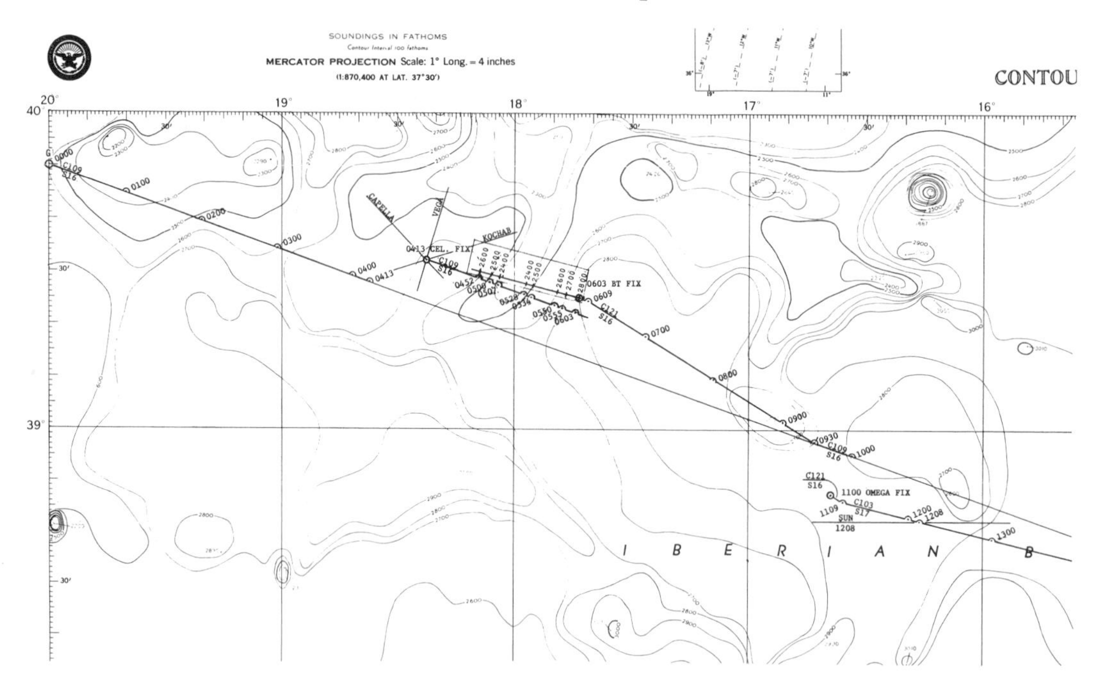

Figure 16-3. Ship's navigation plot, 0000-1200N 25 June.

At sunrise, the azimuth of the sun was observed on the gyrocompass repeater and compared with the computed amplitude to determine the gyro error. The gyro azimuth was 059° pgc, and the amplitude angle computed using Table 28 of *Bowditch* was E 31.1° N or 058.9°T. Thus, the gyro error was considered negligible.

Shortly after plotting the 0413 celestial fix, the navigator noticed that the ship's course 109°T would take her over a series of charted depth contours ideal for obtaining a bathymetric (BT) fix. Therefore, he decided to delay recommending a course change to take the ship back to the intended track until after the BT fix was obtained. A DR position was plotted for the times at which the ship crossed over each charted contour, and the BT fix was plotted at 0603 using the line-of-soundings overlay technique. Shortly thereafter at 0609 the navigator recommended a course change to 121°T to take the ship back onto track by 0930. The 0800 position report slip was then made up based on the DR plot extended from the 0603 BT fix.

At 1000 the ship went to general quarters and commenced a series of maneuvering and engineering casualty drills. During this time the navigator shifted the DR plot to a small-area plotting sheet, and he obtained several fixes using the Omega system. At 1100 the ship secured from general quarters and resumed course 121°T at a speed of 16 knots. At 1109 the navigator recommended a course change to 103°T based on an 1100 Omega fix and a speed increase to 17 knots to make up some of the distance lost during the general quarters drills. The 1100 Omega fix is illustrated as it was plotted on the Omega chart No. 7624 in Figure 16-4.

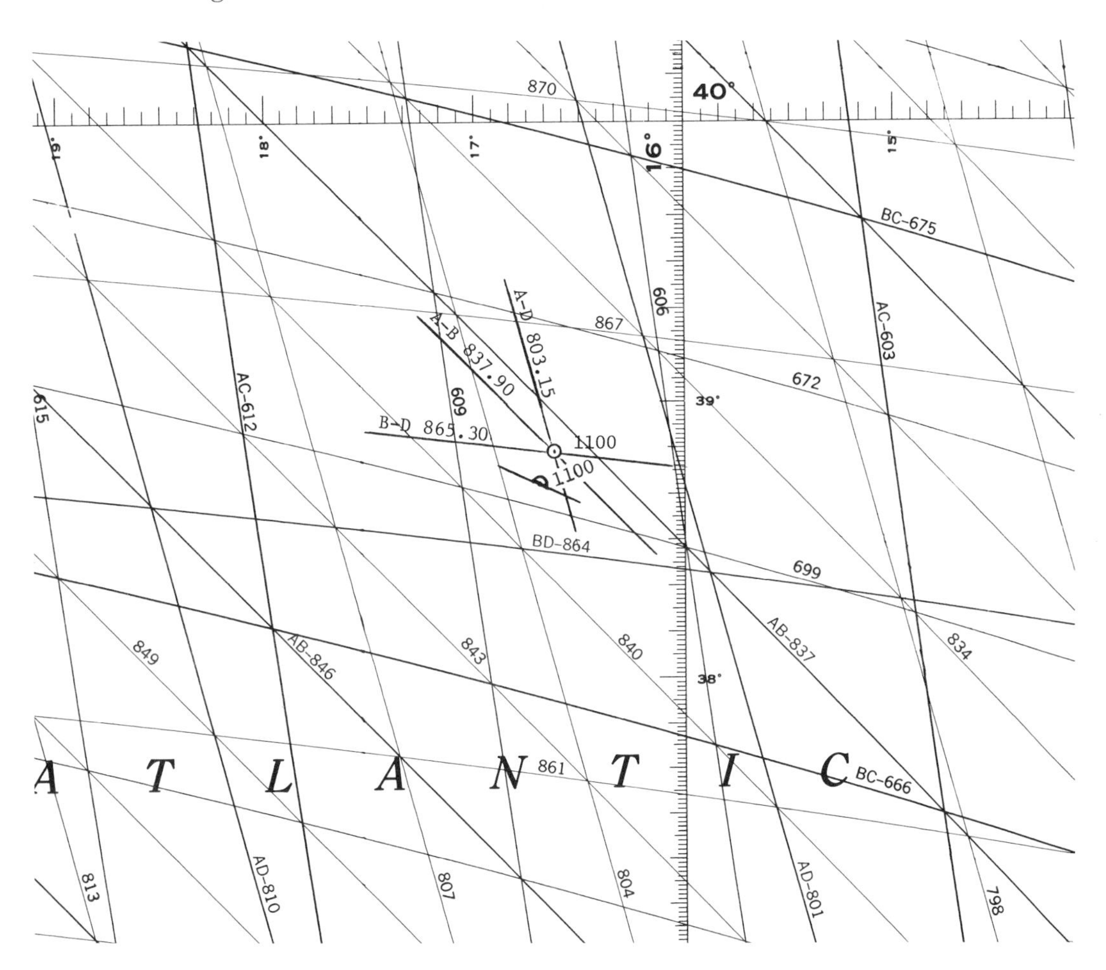

Figure 16-4. The 1100 Omega fix plotted on Omega chart No. 7624.

Between 1100 and 1200 the ship's radio communication facility patched a radio time signal into the chartroom for a chronometer time check, and the results were entered into the chronometer record book. The 1200 position report shown in Figure 16-5 was made up based on the 1100 Omega fix and projected 1200 DR position. At 1208 the sun was observed at LAN; the resulting sun line passed through the 1208 DR position.

At 1500 an afternoon sun line was observed, and the 1208 LAN line was advanced to form the 1500 running fix. This is shown in Figure 16-6, which depicts the navigation plot from 1200 to 2400N. An Omega fix obtained at 1530 was coincident with the 1530 DR position; it is shown plotted on the Omega chart in Figure 16-7 on page 322.

At 1600 the navigator noted that the intended track would again take the ship through an area of bottom contours suitable for a BT fix. From 1649 to 1903′ (19-03-30) DR positions were plotted for the times at which the contoured depths were indicated on the echo

SHIP'S POSITION
NAVSHIPS-1111 (REV. 5-62)

TO:
COMMANDING OFFICER, USS EXAMPLE

AT (Time of day)	DATE
1200 N	25 JUN

LATITUDE	LONGITUDE	DETERMINED AT
38° 43.0′ N	16° 19.2′ W	1100

BY (Indicate by check in box)
☐ CELESTIAL ☒ D. R. ☒ OMEGA ~~LORAN~~ ☐ RADAR ☐ VISUAL

SET	DRIFT	DISTANCE MADE GOOD SINCE (time) (mi.)
070°	1.5 kn	0800 N - 46.8

DISTANCE TO	MILES	ETA
PT H	524	262148Z

TRUE HDG.	ERROR	VARIATION
103°	FOR. GYRO 0° AFT GYRO 0°	12° W

MAGNETIC COMPASS HEADING (Check one)
☐ STD ☒ STEERING ☐ REMOTE IND ☐ OTHER 117°

DEVIATION	1104 TABLE DEVIATION	DG: (Indicate by check in box)
2° W	2° W	☐ ON ☒ OFF

REMARKS
1 HR. 35 MIN AHEAD OF PIM

RESPECTFULLY SUBMITTED
R R Hobbs

CC:

Figure 16-5. The 1200 position report, 25 June.

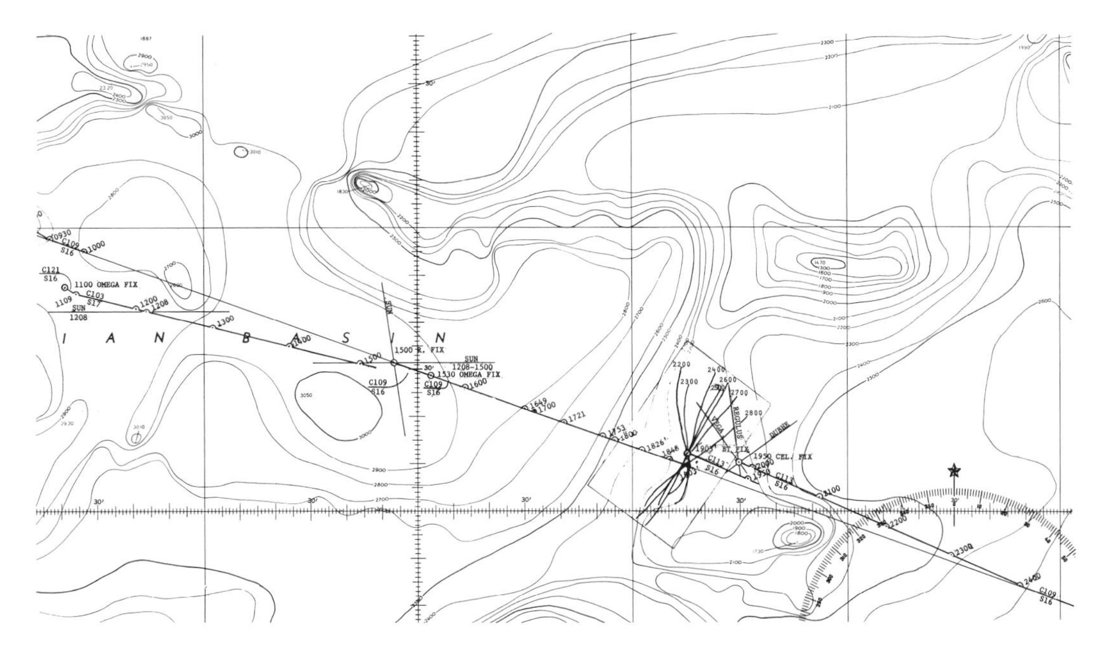

Figure 16-6. Ship's navigation plot, 1200-2400N 25 June.

sounder trace, and the 1903′ BT fix was plotted using the contour advancement overlay technique. While these depth marks were being taken, the navigator computed the times of sunset and ending of civil twilight as 1922 and 1951 respectively, and he made up a list of selected stars for the evening celestial sight using the Rude Starfinder. Since the 1903′ BT fix showed the ship's position to be north of the intended track, the navigator recommended a course adjustment to 113°T.

Just before sunset a second sun amplitude observation was made to check the gyro error once again. The evening stars were then observed and the complete solutions of three of them were worked using the *Nautical Almanac* and *Tables No. 249,* Volume I. The resulting 1950 celestial fix was plotted and used as the basis for the 2000 position report. Because this position also placed the ship north of the intended track, the navigator decided to remain on course 113°T at 16 knots until midnight.

As his final action in the day's work prior to retiring to his sea cabin for the night, the navigator made up the rough draft of the Captain's Night Orders for the night of 25–26 June. The planned course change at midnight was noted, as well as the fact that the ship would be approaching the coast of Spain and the Straits of Gibraltar the following day.

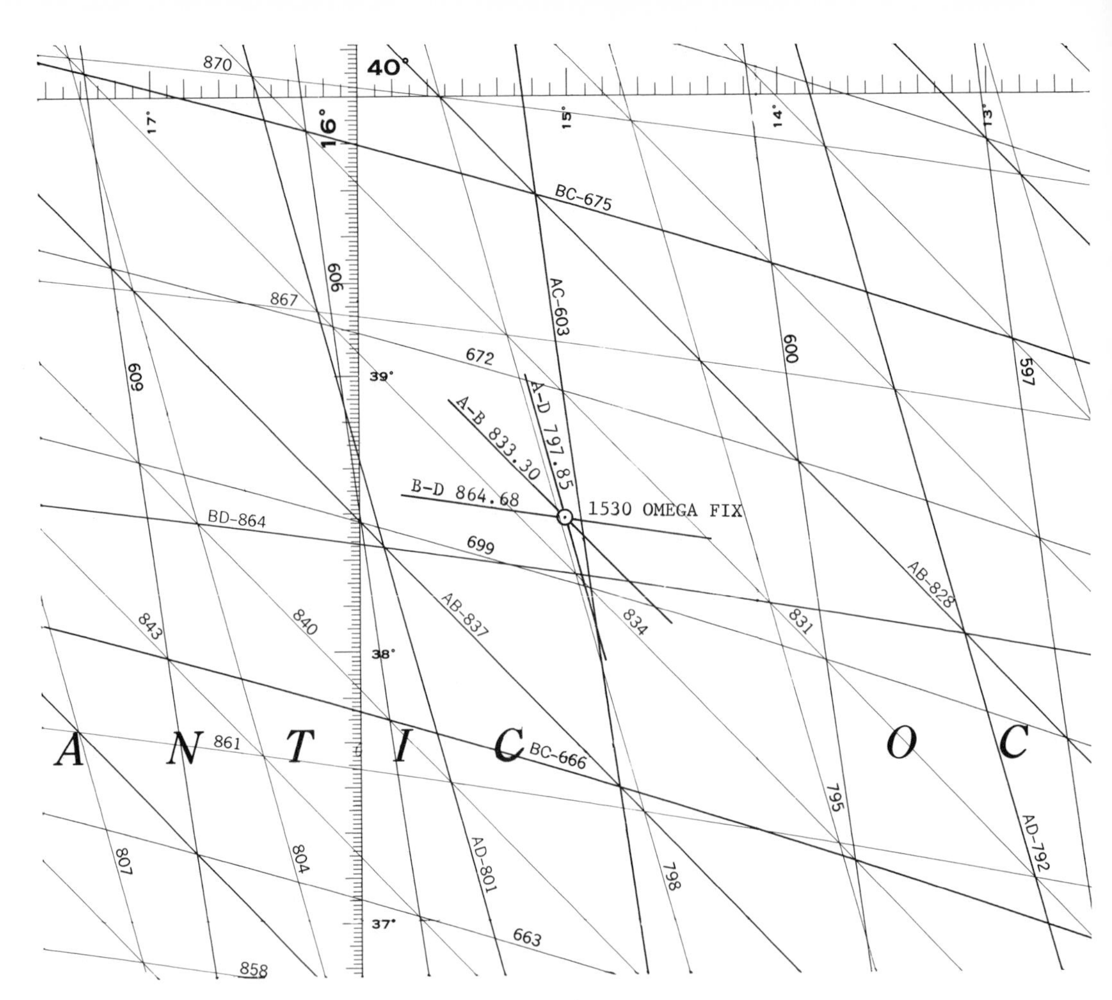

Figure 16-7. The 1530 Omega fix.

Notes on the Day's Work

In the example of a day's work in navigation just given, only the more noteworthy events were described in the interest of brevity. A great many other routine items accomplished daily by the navigator and his staff of quartermasters were omitted, among which were winding the chronometers, preparing the sextant for use, and obtaining and reporting weather observations. Although only two Omega fixes were plotted, in practice such a widely available and accurate means of position-fixing would be used much more often. Other electronic aids to navigation also available in the area were disregarded, the most notable of these being Loran-C. The example given, however, should provide the inexperienced navigator with a good feel for the pace of the navigator's activity at sea.

Summary

This chapter has concluded *Marine Navigation 2: Celestial and Electronic* with a brief outline of a typical day's work in navigation at sea. While the emphasis as of this writing remains on the practice of celestial navigation as the primary method of determining the ship's position at sea, it is very probable that within the decade of the 1980s a historic shift in emphasis toward electronic and satellite navigation as the prime method of position-fixing at sea will take place. Even in these circumstances, however, celestial observations will probably still be the principal method of determining gyrocompass error at sea, and celestial navigation will doubtless remain of foremost importance to navigators of smaller Navy as well as commercial and private vessels in which installation of sophisticated and rather costly electronic receivers is not feasible. Celestial navigation, with the possible exception of the ship's inertial navigation system, is the only reliable method of determining the ship's position at sea using information independent of externally transmitted electromagnetic or electromechanical radiations. And celestial navigation remains the sole reliable method of navigation at sea that does not require an electrical power supply—a point of prime importance to all sailboat navigators, most of whose craft can barely generate and store enough power to keep their running lights shining for the duration of a voyage at sea.

If he has mastered the material presented in *Marine Navigation 1* and 2, the student of navigation can feel confident that he is familiar with the basic knowledge he needs to safely direct the movements of a surface vessel anywhere in the world. However, in navigation, be it visual piloting, celestial, or electronic, there is no substitute for experience in developing the expertise and fine judgment required and expected of the professional navigator.

Appendix A

Electronic Calculators in Marine Navigation

The rapid proliferation of the low-cost hand-held electronic calculator during the decade of the 1970s has certainly been one of the most remarkable technological developments of this century. The incorporation of trigonometric functions in many calculators, and the programmable features of several, has initiated a great deal of renewed interest in sight reduction of celestial observations by formulas derived from the old Cosine-Haversine trigonometric formulas, as an alternative to the solution by use of sight reduction tables. Any calculator capable of handling trigonometric functions can be used in this endeavor. Programmable models, however, especially those with the capability of storing user-oriented programs on magnetic storage devices, greatly facilitate the calculator solution of the sight reduction problem, as well as a great many other common navigational computations.

Many of the more sophisticated calculators now come equipped with various "applications programs" in the field of navigation. These programs, written by the manufacturers, can be fed into the calculator memory from one of several kinds of external magnetic storage media. After having read the program into the calculator, the operator then keys in the values of the appropriate variables called for by the particular program, pushes the "execute" key, and receives the answer. The two largest calculator manufacturers in the United States, Hewlett-Packard and Texas Instruments, both offer several models of programmable calculators that come with a supply of navigational applications programs. Two of these are shown in Figure A-1A; both are currently priced under $400.

In addition to the general purpose programmable calculators that can be set up to solve various specific navigational problems, there are also available several makes of so-called *navigation calculators* that feature permanently built-in navigation programs. The more sophisticated of these are actually microcomputers, with perpetual almanacs, sight-reduction algorithms, and other navigational routines permanently resident in their internal memories. They are capable of calculating a single LOP, or a fix from two LOPs, from sights of the sun, moon, or navigational planets and stars, simply by entering into

Figure A-1A. Two models of programmable trigonometric calculators.[*]

the calculator the sextant IC, the height of eye, the sextant angle, the GMT and the DR position for that time, and an identifying number assigned to the body. The less sophisticated models will perform the same calculations but require, in addition to the above variables, some appropriate almanac data such as the SHA of the body observed. These navigational calculators currently range in price from about $1,500 for the most capable models, such as that shown in Figure A-1B on the next page, to under $400 for one of the less automatic calculators such as the one shown in Figure A-1C on page 327.

Some Basic Formulas for Calculator Navigation

Probably the most basic of all formulas in the field of celestial navigation are those used to solve the navigational triangle in order to compute a celestial LOP. There are several variations of these; almost all are based upon the classic Cosine-Haversine formulas developed in

[*] Mention of particular calculator models or manufacturers herein is not to be construed as an endorsement of them either by the author or by the publisher.

Figure A-1B. The NAVICOMP Navigation Computer®. *It is capable of computing a fix from two celestial observations with no manual almanac or data table look-ups required.* (Courtesy Weems & Plath)

the late nineteenth century. The formulas appearing below were derived from them:

$$Hc = \arcsin(\sin L \sin D + \cos L \cos D \cos LHA)$$

$$Z = \arccos[(\sin D - \sin L \sin Hc)/(\cos Hc \cos L)]$$

Hc = computed alt
Z = azimuth angl
L = assumed latit
D = exact declina
LHA = local hour angle

Rule: If latitude is contrary to declination, enter declination as negative.

When the formulas are used to solve the navigational triangle, the DR latitude and longitude can be used as the assumed latitude and longitude; with some models of calculator, minutes of both these quan-

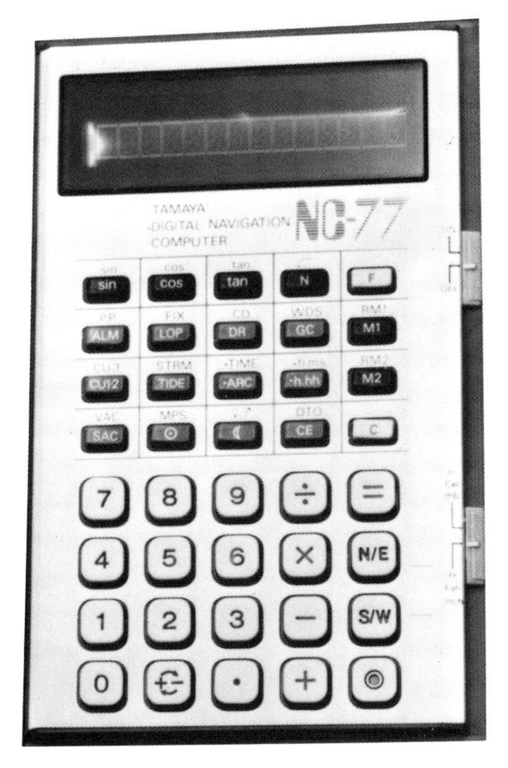

Figure A-1C. The Tamaya NC-77 Digital Navigation Computer. *It incorporates a "prompting" feature consisting of sequentially displayed symbols that identify the various data to be entered.* (Courtesy Weems & Plath)

tities must first be converted to tenths of a degree by division by 60. The declination and GHA of the body observed are then obtained from either the *Nautical* or *Air Almanac,* and recorded to the nearest tenth of a degree. Comparison of the assumed longitude with the exact GHA yields the local hour angle (LHA) of the body to the nearest tenth.

With a little practice, the use of these formulas can effectively substitute for the contemporary method of sight reduction by either *Tables No. 229* or *249.* For those desiring to get away from traditional methods altogether, the U.S. Naval Observatory has recently begun publication of an annual *Almanac for Computers* that can serve essentially as a complete alternative to the *Air* and *Nautical Almanacs* for the purpose of obtaining the necessary time-dependent coordinates of all celestial bodies of navigational interest. The almanac contains data tables and procedures for the precise calculation of all the quantities chronologically arranged in the daily pages of the *Air* and *Nautical Almanacs,* including the times of the various rising and setting phenomena. There are also formulas and tables given for the precise calculation of all the corrections required to be applied to the sextant altitude of the various bodies to obtain the observed altitude Ho, including dip, refraction, semidiameter, and parallax.

In addition to sight reduction, there are a large number of other navigational computations that can be performed using the calculator, some of which are extremely difficult if not impossible to do using manual methods. A partial list of these includes the following:

- Linear regression of time sights
- Speed-time-distance conversions
- Distance by sextant angle
- Mathematical calculation of the sailings
- Identification of an unknown body by computation of its SHA and declination
- Determination of GMT and longitude by observation of lunar distance
- Determination of optimum sailing tacks given the true wind

It is not possible to set forth in this small appendix these and all the other useful navigational formulas that can be solved to great advantage by use of the electronic calculator. There are several excellent books available to which the navigator interested in pursuing calculator navigation can refer, to obtain many of these useful formulas and algorithms. Two of the best of these are *Calculator Afloat: A Mariners Guide to the Electronic Calculator* by H. H. Shufeldt and K. E. Newcomer (Annapolis: Naval Institute Press, 1980) and *Calculator Navigation* by M. Rogoff (New York: W. W. Norton & Co., 1979). Each contains a wealth of both elementary and more sophisticated quantitative navigational techniques that before the advent of the calculator were beyond the capabilities of any but the most astute of mathematically minded navigators. Many of these navigational formulas and algorithms are also available in the applications packages mentioned earlier that are supplied by the various manufacturers of programmable calculators.

A Final Cautionary Note

Lest the student and the inexperienced navigator come away from a reading of this appendix with the idea that the electronic calculator is the ultimate panacea for the modern practitioner of celestial navigation, one note of caution must be related. That is simply that the calculator, although capable of delivering extremely fine *accuracy* of computation of navigational formulas, can only deliver *reliability* consistent with the reliability of the data keyed into it by the operator. It is very easy to press a wrong key, or to insert a variable out of the proper sequence on calculators not having a prompting feature, especially under conditions of poor lighting and an unstable chart table that often prevail on board a vessel at sea. Moreover, many experienced navigators have reported that the process of entering the many quantities required to solve the navigational triangle using even a programmable calculator often takes as much if not more time than does the traditional solution by means of the almanacs and sight reduction tables. And, unlike a pencil, the calculator does demand an electrical power source, however small. Thus, it is well to keep in mind that while the calculator can be a powerful tool in the navigator's inventory, it too has its limitations as well as its capabilities.

Appendix B

Abbreviations and Symbols Used in Celestial and Electronic Navigation

A	amplitude; away (altitude intercept)
a	altitude intercept
a_0, a_1, a_2	Polaris sight corrections
ADR	average daily rate (of chronometer)
aL	assumed latitude
AM	amplitude modulation; ante meridian (before noon)
AP	assumed position
aλ	assumed longitude
Bn	beacon
BT	bathymetric
C	Centigrade (Celsius); chronometer time; compass; course
CE	chronometer error; compass error (magnetic)
cec	centicycle
cel	centilane
CIC	Combat Information Center
cm	centimeter
CO	commanding officer
COG	course over the ground
$corr^n$	correction
CPA	closest point of approach
cps	cycles per second
CRT	cathode ray tube
CW	continuous wave
D	deviation; drift (of current)
d	altitude difference
d $corr^n$	correction for change in declination
dec	declination
DG	degaussing
DMA	Defense Mapping Agency
DMAHTC	Defense Mapping Agency Hydrographic/Topographic Center
DR	dead reckoning
DS $corr^n$	double-second difference correction
DST	daylight savings time
DUT1	UTC/UT1(≅GMT) difference correction
dλ	longitude difference
E	east
EHF	extremely high frequency

EP	estimated position
ETA	estimated time of arrival
ETD	estimated time of departure
F	Fahrenheit; fast
f	frequency; multiplication factor (*Air Almanac*)
FM	frequency modulation
fm	fathom
ft	foot
G	Greenwich; Greenwich meridian (upper branch)
g	Greenwich meridian (lower branch)
GE	gyro error
GHA	Greenwich hour angle
GMT	Greenwich mean time
GP	geographic position
GPS	Global Positioning System
GRI	group repetition interval
h	hour
HA	hour angle
ha	apparent altitude
Hc	computed altitude
HF	high frequency
Ho	observed altitude
HP	horizontal parallax
hs	sextant altitude
ht	height
Hz	Hertz
IC	index correction
in	inch
kHz	kilohertz
kn	knot
L	latitude
λ	longitude; wave length
LAN	local apparent noon
LAPS	large area plotting sheet
LF	low frequency
LL	lower limb
LMT	local mean time
LOP	line of position
M	magnetic; observer's meridian (upper branch); nautical mile
m	observer's meridian (lower branch); meters; minutes
MF	medium frequency
μs	microsecond
MHz	megahertz
N	north
NAVSAT	Navy Navigation Satellite System
NOS	National Ocean Survey
NWP	Naval Warfare Publication
OOD	Officer of the Deck
OTC	Officer in Tactical Command
P	pole; parallax
pgc	per gyrocompass
P in A	parallax in altitude
PM	post meridian (after noon); pulse modulation
P_n	north pole; north celestial pole
PRI	pulse repetition interval

PRR	pulse repetition rate
P_s	south pole; south celestial pole
psc	per standard compass
p stg c	per steering compass
Pub	publication
Q	celestial equator
QMOW	Quartermaster of the Watch
R	relative
RA	right ascension
RDF	radio direction finding
RF	radiofrequency
R Fix	running fix
rms	root mean square error
RPM	revolution per minute
S	slow; south; speed
s	second
SAPS	small area plotting sheet
SD	semidiameter
SHA	sidereal hour angle
SHF	super high frequency
SINS	ships inertial navigation system
SOA	speed of advance
SOG	speed over the ground
SOPA	Senior Officer Present Afloat
T	toward (altitude intercept); true
t	meridian angle
TB	temperature-barometric correction
Tg	ground wave
Ts	sky wave
TR	track
UHF	ultra high frequency
UL	upper limb
UTC	Coordinated Universal Time
V	variation
v corr^n	correction for irregular orbital motion
VHF	very high frequency
VLF	very low frequency
W	watch time; west
WE	watch error
yd	yard
Z	azimuth angle
z	zenith distance
ZD	zone description
Zn	true azimuth
ZT	zone time

MISCELLANEOUS SYMBOLS

Δ	delta (unit change)
°	degrees
′	minutes of arc
″	seconds of arc
▬	remains below horizon
▭	remains above horizon
////	twilight lasts all night
♀	Venus
☉	sun
☾	moon
☆	star
♈	First Point of Aries
⊡	estimated position
⊙	fix, running fix
—◠—	DR position
λ	longitude; wavelength
μs	microsecond

Various Sight Forms Used in Celestial Navigation

Sight Reduction using Tables 229 C: S:		M W E m	M W E m	M W E m	M W E m
Body					
IC		+ –	+ –	+ –	+ –
Dip (Ht	')				
Sum					
hs					
ha					
Alt. Corr					
Add'l.					
H.P. (	)				
Corr. to ha					
Ho (Obs Alt)					
Date					
DR Lat					
DR Long					
Obs. Time					
WE (S+, F–)					
ZT					
ZD (W+, E–)					
GMT					
Date (GMT)					
Tab GHA	v				
GHA incr'mt.					
SHA or v Corr.					
GHA					
±360 if needed					
aλ (–W, +E)					
LHA					
Tab Dec	d				
d Corr (+ or –)					
True Dec					
a Lat (N or S)		Same Cont.	Same Cont.	Same Cont.	Same Cont.
Dec Inc	(±)d				
Hc (Tab. Alt.)					
tens	DS Diff.				
units	DS Corr.	+	+	+	+
Tot. Corr. (+ or –)					
Hc (Comp. Alt.)					
Ho (Obs. Alt.)					
a (Intercept)		A T	A T	A T	A T
Z					
Zn (°T)					

Figure C-1. Sight Reduction by Tables 229

Sight Reduction using Table 249 Vol. I C: S:	M W E m	M W E m	M W E m	M W E m
Body				
IC				
Dip (Ht ′)				
R_o				
Sum				
hs				
Ho (Obs Alt)				
Date				
DR Lat				
DR Long				
Obs. Time				
WE (S+, F−)				
ZT				
ZD (W+, E−)				
GMT				
Date (GMT)				
Tab GHA γ				
GHA incr'mt.				
GHA γ				
aλ (−W, +E)				
LHA γ				
a Lat (N or S)				
Hc (Tab Alt.)				
Ho (Obs. Alt.)				
a (Intercept)	T A	T A	T A	T A
ZN (°T)				
P&N Corr.				

Figure C-2. Sight Reduction by Tables 249, Vol. I

Sight Reduction
using Tables 249
Vols. II and III

C:

S:

	M W E m	M W E m	M W E m	M W E m
Body				
IC				
Dip (Ht ')				
R_o				
S.D.				
Sum				
hs				
P in A (Moon)				
Ho (Obs Alt)				
Date (GMT)				
GMT (Obs Time)				
DR Lat				
DR Long				
Tab GHA				
GHA incr'mt				
SHA (Star)				
GHA				
±360 if needed				
a λ (-W, +E)				
LHA				
Tab Dec				
a Lat (N or S)	Same Cont	Same Cont	Same Cont	Same Cont
Dec Inc (±)d				
Hc (Tab Alt)				
Dec Corr'n				
Hc (Comp Alt)				
Ho (Obs Alt)				
a (Intercept)	A T	A T		
Z				
Zn (°T)				

Figure C-3. Sight Reduction by Tables 249, Vols. II and III

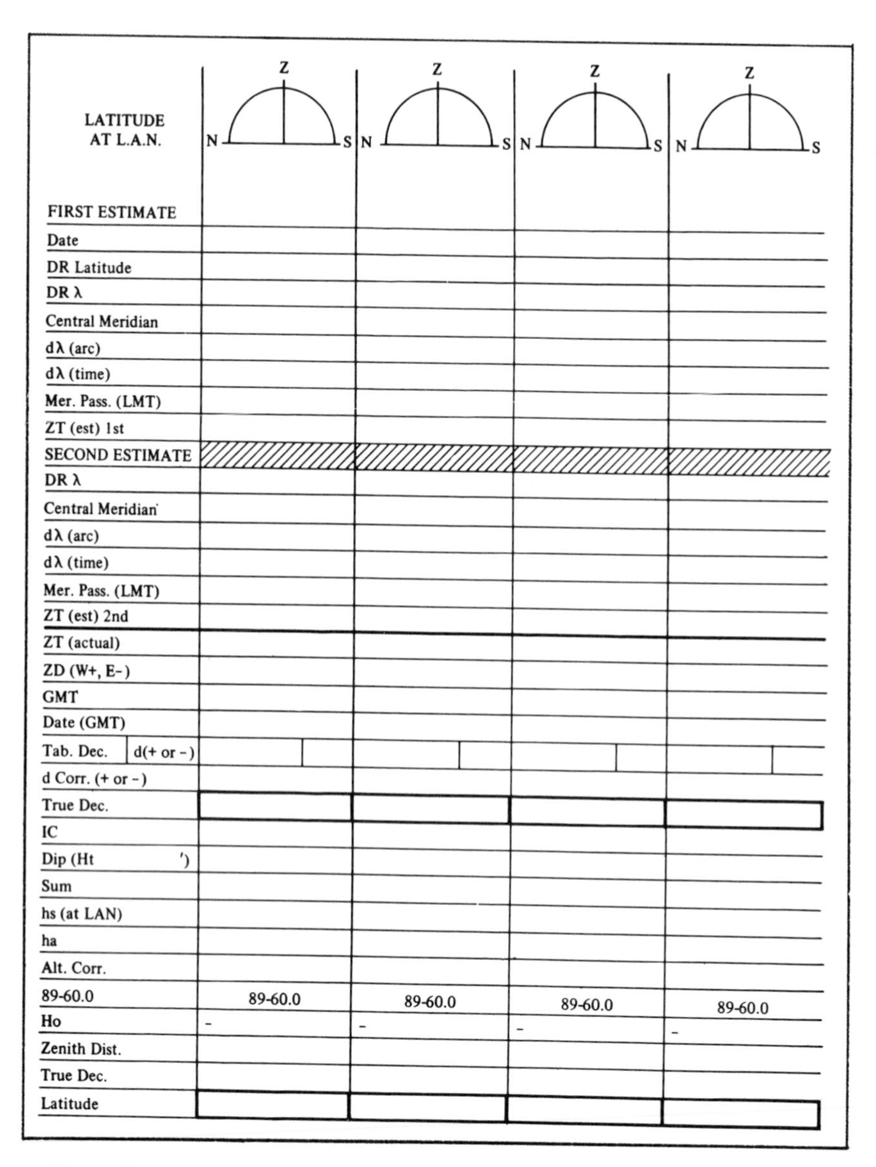

LATITUDE AT L.A.N.	Z N S	Z N S	Z N S	Z N S
FIRST ESTIMATE				
Date				
DR Latitude				
DR λ				
Central Meridian				
dλ (arc)				
dλ (time)				
Mer. Pass. (LMT)				
ZT (est) 1st				
SECOND ESTIMATE				
DR λ				
Central Meridian				
dλ (arc)				
dλ (time)				
Mer. Pass. (LMT)				
ZT (est) 2nd				
ZT (actual)				
ZD (W+, E−)				
GMT				
Date (GMT)				
Tab. Dec. / d(+ or −)				
d Corr. (+ or −)				
True Dec.				
IC				
Dip (Ht ')				
Sum				
hs (at LAN)				
ha				
Alt. Corr.				
89-60.0	89-60.0	89-60.0	89-60.0	89-60.0
Ho	−	−	−	−
Zenith Dist.				
True Dec.				
Latitude				

Figure C-4. Latitude at L.A.N.

LATITUDE BY POLARIS	M W E m		M W E m		M W E m		M W E m	
DR Lat.								
DR Long.								
Date								
ZT								
ZD (W+, E-)								
GMT								
Date (GMT)								
Tab. GHA ♈								
Incr'm't ♈								
Total GHA ♈								
±360 if needed								
DR λ (-W, +E)								
LHA ♈ (Exact)								
IC								
Dip (Ht. ')								
Sum								
hs								
ha								
Refr. Corr.		-		-		-		-
TB (ha < 10°)	+	-	+	-	+	-	+	-
A_0	+		+		+		+	
A_1	+		+		+		+	
A_2	+		+		+		+	
Add'n'l		-60.0		-60.0		-60.0		-60.0
Sub Total	+	-	+	-	+	-	+	-
Total Corr. to ha (±)								
ha								
Latitude		N		N		N		N
True Azimuth		°T		°T		°T		°T
Gyro Brg.		°pgc		°pgc		°pgc		°pgc
Gyro Error		°(E or W)		°(E or W)		°(E or W)		°(E or W)
NOTES:								

Figure C-5. Latitude by Polaris

EXACT AZIMUTH USING TABLES 229

Body __________

DR L __________

DR λ __________

Date (L) __________

ZT __________

ZD (+ or −) __________

GMT __________

Date (G) __________

Tab GHA __________

Inc'mt __________

GHA __________

DR λ __________

LHA __________

	d(+/−)
Tab Dec	

d corr __________

Dec __________

	EXACT Deg	EXACT Min	Z DIFF. (+ or −)	CORR. (+ or −)
LAT				
LHA				
DEC				
			Total (±)	

Tab Z __________

Exact Z __________

Exact Zn __________

Gyro/Compass Brg __________

Gyro/Compass Error __________

NORTH LAT

LHA greater than 180° . . . . Zn = Z
LHA less than 180° . . . . Zn = 360° − Z

SOUTH LAT

LHA greater than 180° . . . . Zn = 180° − Z
LHA less than 180° . . . . Zn = 180° + Z

Figure C-6. Exact Azimuth by Tables 229

Index